Nanobiotechnology for Green Environment

Nanobiotechnology for Green Environment

Edited by

Amit Kumar and Chhotu Ram

CRC Press
Taylor & Francis Group
Boca Raton London New York

CRC Press is an imprint of the
Taylor & Francis Group, an **informa** business

First edition published 2021
by CRC Press
6000 Broken Sound Parkway NW, Suite 300, Boca Raton, FL 33487-2742
and by CRC Press
2 Park Square, Milton Park, Abingdon, Oxon, OX14 4RN

© 2021 Taylor & Francis Group, LLC
CRC Press is an imprint of Taylor & Francis Group, LLC

Library of Congress Cataloging-in-Publication Data

Names: Kumar, Amit (Assistant professor in biotechnology), editor. | Ram, Chhotu, editor.
Title: Nanobiotechnology for green environment / edited by Amit Kumar and Chhotu Ram.
Description: First edition. | Boca Raton : CRC Press, 2021. | Includes bibliographical references and index. | Summary: "The book examines environmental issues and their solutions with advancements in biotechnology and nanotechnology. This book will focus on environmental friendly waste management, wastewater treatment and utilization of wastes for energy"-- Provided by publisher.
Identifiers: LCCN 2020041193 (print) | LCCN 2020041194 (ebook) | ISBN 9780367460686 (hardback) | ISBN 9780367461362 (ebook)
Subjects: MESH: Waste Management--methods | Nanotechnology | Green Chemistry Technology | Biodegradation, Environmental | Nanostructures
Classification: LCC TD793 (print) | LCC TD793 (ebook) | NLM WA 778 | DDC 363.72/8--dc23
LC record available at https://lccn.loc.gov/2020041193
LC ebook record available at https://lccn.loc.gov/2020041194

ISBN: 978-0-367-46068-6 (hbk)
ISBN: 978-0-367-69538-5 (pbk)
ISBN: 978-0-367-46136-2 (ebk)

Typeset in Times LT Std
by KnowledgeWorks Global Ltd.

Contents

Editors

Dr. Amit Kumar currently works as an Assistant Professor at the Department of Biotechnology, College of Natural and Computational Sciences, Debre Markos University, Ethiopia. He completed his Doctorate in Biotechnology from the Indian Institute of Technology Roorkee, India. He is extensively involved in research on industrial enzymes, pulp and paper biotechnology, bio-fuels production, and environmental biotechnology. He has published several research and review articles in various reputed international journals. He has also co-edited three books and published several book chapters. He has guided several graduate and post-graduate projects.

Dr. Chhotu Ram is an Assistant Professor at the Department of Chemical Engineering at Adigrat University, Ethiopia. Dr. Ram did his M. Tech. in Environmental Technology from Thapar University Patiala, India. Further, he earned his Doctorate in Environmental Engineering from Indian Institute of Technology Roorkee, India. He has published several research papers in reputed journals worldwide and research contribution to many international conferences. He has research interests in the area of water quality and wastewater engineering and materials performance. Presently, he deals with various R&D projects in the fields such as wastewater treatment, solid waste management, and water quality for a sustainable environment.

Contributors

Nalluri Abhishek
Center for Fuel Cell Innovation,
    School of Materials Science
    and Engineering
Huazhong University of Science
    and Technology
Hubei, China

Diwakar Aggarwal
Department of Biotechnology
Maharishi Markandeshwar (Deemed
    to be University)
Mullana, Ambala, India

Ayantika Banerjee
Korea Atomic Energy Research
    Institute
University of Science and Technology
Daejeon, South Korea

Yilkal Bezie
Department of Biology, College
    of Sciences
Bahir Dar University
Bahir Dar, Ethiopia

Amit Kumar Bharti
Department of Paper Technology,
    Indian Institute of Technology
    Roorkee, Saharanpur Campus
Saharanpur, India

Bhabatush Biswas
Department of Bio Engineering,
    National Institute of Technology
Agartala, India

Arghya Chakravorty
School of Bio Sciences and Technology,
    Vellore Institute of Technology
Vellore, India

Moharana Choudhury
Voice of Environment (VoE)
Guwahati, India

Rajesh Kumar Jena
Department of Science and Humanities,
    Darbhanga College of Engineering
Aryabhatta Knowledge University
Madhpur, India

Navneet Kaur
School of Energy and Environment,
    Thapar Institute of Engineering
    and Technology
Patiala, India

Amit Kumar
Department of Biotechnology, College
    of Natural and Computational Sciences
Debre Markos University
Debre Markos, Ethiopia

Amrish Kumar
Indian Institute of Technology Roorkee,
    Saharanpur Campus
Saharanpur, India

Dushyant Kumar
Center for Rural Development and
    Technology, Indian Institute of
    Technology, Delhi
New Delhi, India

Jagdeesh Kumar
Department of Hydrology, Indian
    Institute of Technology Roorkee
Roorkee, India

Neeta L. Lala
Centre for Nanofibers and
    Nanotechnology, Department
    of Mechanical Engineering
National University of Singapore
Singapore

Sandeep K. Malyan
National Institute of Hydrology
Roorkee, India

Vimala Raghavan
Centre for Nanotechnology Research,
    Vellore Institute of Technology
Vellore, India

Chhotu Ram
Department of Chemical Engineering,
    College of Engineering and
    Technology
Adigrat University
Adigrat, Ethiopia

Seeram Ramakrishna
Centre for Nanofibers and
    Nanotechnology, Department
    of Mechanical Engineering
National University of Singapore
Singapore

Rehab A. Rayan
Department of Epidemiology, High
    Institute of Public Health
Alexandria University
Alexandria, Egypt

Siva Sankar Sana
School of Chemical Engineering
    and Technology
North University of China
Taiyuan, China

Vikas Kumar Sangal
Department of Chemical Engineering,
    Malaviya National Institute
    of Technology
Jaipur, India

Kulbir Singh
Department of Environmental Science
    and Engineering
Guru Jambheshwar University of
    Science and Technology
Hisar, India

Steffi Talwar
Department of Chemical Engineering,
    Thapar Institute of Engineering
    and Technology
Patiala, India

Mengistie Taye
College of Agriculture and
    Environmental Sciences
Bahir Dar University
Bahir Dar, Ethiopia

Anoop Kumar Verma
School of Energy and Environment,
    Thapar Institute of Engineering
    and Technology
Patiala, India

Bushra Zaman
College of Engineering
Utah State University
Logan, Utah

1 Global Environmental Issues and the Role of Nanobiotechnology for the Sustainable Environment

Chhotu Ram
Adigrat University

Amit Kumar
Debre Markos University

Yilkal Bezie
Debre Markos University
Bahir Dar University

CONTENTS

1.1 INTRODUCTION

Everything that surrounds or affects an organism during its lifetime is known as its environment. It consists of both living (biotic) and nonliving (abiotic) components. Human civilization and globalization are the dominant culprits of continuous change in the present global environment (Singh and Singh, 2017). As resources have become limited due to rapid growth in human population, the availability of pollution-free technologies and clean energy for sustainable development of human society is the need of the hour. Biotechnology and nanotechnology have shown potential for the development of 'cleaner' and 'greener' industrial processes with substantial health and environmental benefits (Khan, 2020). Nanotechnology uses novel nanomaterials and has shown potential for groundwater, surface water, and wastewater remediation by heavy metal ions, organic-inorganic solutes, and a wide range of microorganisms (Khan, 2020). Nanomaterials are very small in size and they have a very high surface area to volume ratio. Therefore, nanomaterials can be used for the detection of very sensitive contaminants. Environmental applications of nanotechnology are classified as (1) remediation and purification of contaminated material, (2) pollution detection (sensing), and (3) pollution prevention (Indarto et al., 2008; Yunus et al., 2012).

Biotechnology offers new possibilities for the development of a safe, eco-friendly, and sustainable ecosystem. Some of the modern biotechnological techniques are used to attain the goals of sustainability in different fields such as food production, industrial practices, agricultural processes, waste management, bioremediation, and utilization of renewable raw materials (Kaur, 2017). A large number of microbial and plant enzymes have been proven to be involved in biodegradation of toxic organic pollutants. Several extracellular hydrolytic enzymes like amylases, proteases, lipases, DNases, pullulanases, and xylanases are applicable for diverse usages in different areas such as the food industry, animal feed additives, biomedical science, and chemical industries in a sustainable manner. Cellulases and hemicellulases are essential for the degradation of lignocellulosic biomass. Microbial oxidoreductases oxidize toxic compounds to harmless compounds through an oxidation-reduction reaction. Microbial enzymes have also been exploited for decolorization and degradation of dyes (Husain, 2006; Karigar and Rao, 2011; Sánchez-Porro et al., 2003). This chapter deals with the challenges ahead and various strategies related to nanotechnology and biotechnology that have the potential to achieve a sustainable environment.

1.2 GLOBAL ENVIRONMENTAL ISSUES

The global environmental issues present a systematic and stimulating introduction to the major environmental challenges that are of utmost importance to preserving our natural environment. It was observed that in the past two decades environmental challenges have become a global issue, not only due to the civil society movement but all around the planet. In the recent decades and present scenario, an interesting institutionalization process of environmental problems has been observed in the world, and this has further increased the interest of the scientific community globally. This section discusses briefly the major global environmental issues in the present aspects and relevant to the scope of this book. The evolution of the Earth's ecosystem (geosphere, hydrosphere, biosphere, and ecosphere) provides the required understanding of scientific concepts, processes, and background to environmental issues. The anthropogenic activities are at the center of the debate on the environmental changes, and are constituted as unprecedented challenges to contemporary societies (Biermann, 2014). Man-made activities, their impacts, and their management of the natural environment are a concern for maintaining the natural ecosystem balance. Rapidly increasing population and industrial development are leading to several environment-associated problems worldwide.

According to the United Nations' latest report (2019), the world's population will continue to grow from an estimated 7.7 billion people in 2019 to around 8.5 billion in 2030, 9.7 billion in 2050, and 10.9 billion in 2100. Another interesting fact from the projected population indicated that India will be the most populous country in the world and China will be in second position by the year 2050. Thus, the increasing population will exert more pressure on natural resources. The environmental problem is a global problem, rather than local, and thus its effects will manifest in varied ways such as water shortage, soil exhaustion and erosion, deforestation, water pollution, climate change, and waste management. Water and air pollution do not have boundaries and poor soil quality in one country may affect the food quality and production in another country (Bisgrove and Hadley, 2002). The major environmental challenges include forest and agriculture land, agro-waste generation, municipal solid waste (MSW) generation, xenobiotic compounds from industries, effluent and sewage generation, impacts of environmental degradation, associated public health issues, etc. (Anand, 2013). A study from Solomon Islands showed that people are facing tremendous impacts of sea level rise which is leading to community relocation. Thus, it was an observance of the consequences of climate change affecting the historical developments in unsustainable manner (Albert et al., 2016). The contaminants can affect the environments at other local, regional, or even international levels (Akimoto, 2003), the biggest example being the ozone pollutants traveling from China to the Western United States (Verstraeten et al., 2015). It is also predicted that freshwater will decline, and by 2050 an urban population ranging between 1.3 and 3.1 billion is expected to live in the seasonal water shortage areas (McDonald et al., 2011). Several countries such as Uganda could face environmental issues such as drought that will affect agricultural productivity and livelihood in these countries (Epule et al., 2017).

Literature also cites an important link between environmental problems and important concerns to public health impacts which are seen over decades. From 1960 to 2000, public health was highest concern for local environment preservation, as exemplified by water and air pollution (Mackenbach, 2007). However, it is a tough task to measure the direct health impacts and correlate it with environmental issues. This has benefitted in terms of resolving the 20th century localized environmental

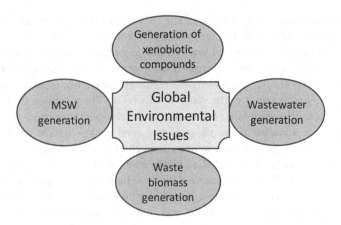

FIGURE 1.1   Global environmental issues.

problems in developed parts of the world (Diez-roux et al., 1999). One article investigated Brazil and China on environmental policies that have been noted for their international importance, and the environmental issues have been at the core of their actual political processes. Finally, an interdisciplinary analysis is capable of promoting dialogue between the social theory and environmental sociology, and sustainability is proposed by considering it at international levels (Ferreira and Barbi, 2016). A previous study also showed the relation of social workers and their role toward the mitigation of the human consequences of global environmental change, and clarifies further needs as an evidence-based approach for the appropriate response strategies (Mason et al., 2017). The scope of the present chapter is to identify the most important environmental issues such as solid waste (SW), agro-industrial waste, wastewater, and xenobiotic compounds; their generation and impacts are briefly discussed in the following sections (Figure 1.1).

1.2.1   Municipal Solid Waste Generation and Its Impact

The fast-growing world population and progressive increase in public living standards has led to a further increase in consumption of goods, energy, and SW generation. SW generation and its improper management is a big environmental issue and contaminates environment, social inclusion, and economic sustainability (Gupta et al., 2015; Vitorino de Souza Melaré et al., 2017), which need to assess, work, and use the holistic approaches to solve these real issues. MSW is non-hazardous, biodegradable or non-biodegradable, reusable or unusable SW that is generated every day from domestic activities, commerce, office buildings, trade, yards, gardens, and streets (Abbasi, 2018; Ngusale et al., 2017). Mismanagement of generated MSW leads to ever-increasing concentrations of greenhouse gases (GHGs) in the atmosphere and greater risk to public health (Zhang et al., 2019). Literature has shown that developing countries generate significant waste quantity, and accordingly face increased waste management and disposal problems. Other works also draw a similar conclusion that MSW management problems are more acute in developing countries than in developed countries (Albert et al., 2016; Ferronato, 2019). As per the economic scenario, middle- and low-income countries with a rapid population growth, urbanization, and industrialization are facing more acute waste management problems (Alam and Ahmade 2013; Damtew and Desta 2015). Thus, we can conclude that the MSW generation rate is increasing with an increase in population, technological advancements, and change in the public lifestyle (Ali 2009; Monavari et al. 2012). As stated previously, the present world population of 7.7 billion could reach around 9.7 billion by 2050 (United Nations, 2019). The current world population around 7.1 billion generates around 1300 million tons of SW per year (1.2kg/capita/day) which is mostly generated in the urban centers of the world (Vaish et al., 2016). Another world bank report predicts that the SW quantity is expected to increase and reach up to 2200 million tons per year by 2025 (Hoornweg and Bhada-Tata, 2012). Hence, tackling this enormous amount of SW generation and its proper management are the need of the hour. The negative impacts of MSW include land occupation, environmental pollution, and the spread of disease. Improper and inefficient MSW management leads to the emission of GHGs and odors which degrade the environment and pose serious public health risks.

One study predicted that emission of GHGs from waste management in developing countries is increasing exponentially (Friedrich and Trois, 2011). Approximately 5% of the global emission of greenhouse gases (GHGs) occurs from waste disposal landfills, and hence it is a significant contributor to man-made climate change (Stocker et al., 2013). GHGs such as methane, nitrous oxide, and carbon dioxide are generated from the aerobic and anaerobic biodegradation of SW. Methane production from landfill process is the largest source of GHGs, and accounts for 1–2% of total GHG emissions (Bogner et al., 2011). Net annual CH_4 emission from landfills in India is continuously increasing since 1999. It was also shown a high correlation of methane emission between gross state domestic product (GSDP) of states and gross domestic product (GDP) of the country, which is an indicator of human wellness. Overall emphasis should be given to improve the MSW management policy, utilization of CH_4 as an energy source (Singh et al., 2018). A leachate treatment facility is the second-largest source of methane and nitrous oxide production (Bernstein, 2008). Extensive efforts have been conducted to minimize the GHG emissions from landfill sites and a major aim to propose the methane utilization and reduction of other GHGs generation from landfills. Some of the landfill leachate constituents are toxic and harmful chemicals such as heavy metals, bio-recalcitrant organic compounds, and microorganisms, which are the major source of groundwater pollution (Dan et al., 2017). Another research highlights the strong correlation between groundwater quality and leachate pollution from landfills. The finding shows the increased value of certain pollutants such as Cd, E_c, and total organic carbon in groundwater determine the negative impacts of leachate (Przydatek and Kanownik, 2019; Vasanthi et al., 2013). An assessment was carried out to see the influence of the discharge of SW on the soil quality in Algeria. The results obtained from the characterization testing of soil indicated the observance of higher amounts of organic matter in the soil and increasing heavy metals concentrations (Zn, Cu, Pb, Cd, Cr, and Ni). This is further correlated to the unscientific MSW disposal and pollute the surrounding environment (Mouhoun-Chouaki et al., 2019).

Leachate is known to cause the ecological problems along with the water blooms and soil salinization. Water borne diseases are also caused by the exposing to the human body through drinking water or bathing. The human body exposed to heavy metals contaminated drinking water for a long period will increase the risk of cancer, infant death, and cognitive dysfunction in children (Parvez et al., 2011; Rahman et al., 2010; Wasserman et al., 2004). Nitrate ions are present in the MSW landfills leachate, and previous research has shown to correlate to the blue baby syndrome, higher risk of abortion, and non-Hodgkin lymphoma (Gurdak and Qi, 2012; Martínez et al., 2017). Further, previous studies investigated the emerging pollutants impact with reproductive toxicity, genotoxicity, and embryo toxicity, including hormones, antibiotics, nanoparticles, and personal care products in the landfill leachate and nearby groundwater source (Egen and Ouyang, 2010; Toufexi et al., 2013). Thus, MSW management is becoming a big environmental challenge that attracts scientific communities in the world to manage SW sustainably. Background studies also suggest the various techniques and options such as thermal treatment, biological treatment, landfilling with gas recovery, and recycling for effective and sustainable waste management (Shekdar, 2009). Recently, energy generation methods in the form of electricity, heat, or transport fuels received special attention and were considered as waste to energy (WtE) options. The WtE technique includes mainly combustion, gasification, pyrolization, anaerobic digestion, and landfill gas recovery, of which their use is increasing to manage waste without affecting the surrounding environment (Kumar and Samadder, 2017).

1.2.2 WASTEWATER GENERATION AND ITS IMPACT ON THE ENVIRONMENT

Water (H_2O) formation takes place due to the catalytic combination of two molecules of hydrogen and one molecule of oxygen; still, the theory on the origin of the water is debated. Water is an elixir of life and the human body constitutes about 72% of water (Singh, 2015). Water pollution is the biggest challenge in both developed and developing countries globally alongside economic growth and public health issues for billions of people (Mateo-Sagasta et al., 2017). The major sources of water pollution are human settlements, industries, and agriculture. Worldwide, around 80% of domestic wastewater is discharged untreated in the water bodies, and industrial effluents constituents such as heavy metals, toxic sludge, solvents, and other wastes are dumped into the water sources annually (WWAP, 2017). Agricultural activities account for 70% of water abstractions globally and play a significant role in water pollution. Thus, the resultant water pollution poses a greater risk to the aquatic ecosystem, public health, soil, and other productive activities (UNEP, 2016). Wastewater generated from industries causes an imbalance in the natural ecosystem by discharging without proper regulation. Globally, water resource needs are more critical due to their physical scarcity and also progressive degradation of water quality in several countries (Mateo-Sagasta et al., 2017). The Sustainable Development Agenda 2030 will strongly influence future environmental policies and strategies and to ensure water pollution control as the national and international priorities (SDG Target 6.3, 2017). It is also predicted that around 70% of the world's population is expected to live in urban areas by 2050, and most of the cities do not have adequate treatment systems to manage the wastewater efficiently and sustainably (United Nations, 2014). WHO and UNICEF both report that a large segment of the population is still consuming contaminated drinking water with feces and facing challenges associated with related diseases such as typhoid, dysentery, cholera, and polio (WHO/UNICEF, 2015). Many developing countries have water-related diseases, such as cholera and schistosomias, that are more prominent where only a small fraction of domestic and urban wastewater is treated before its releases into the environment (UNESCO, 2017).

Industrial development is part of economic development and has an adequate role in the GDP and prosperity of any nation, yet it could degrade the environment. Rapid industrial growth is believed to be one of the main contributors to the pollution of environmental resources globally (Meyer et al., 2019). Freshwater availability is a shortage in various countries including South Africa where the country is currently classified as water-stressed (Moodley et al., 2011). In South Africa, just over 1200 m^3/person/year of freshwater is available for a population of around 50 million. Further, industrial effluent contamination to the environment is also a big issue in these countries. One study from the Middle East and North Africa (MENA) has shown that total water consumption is approximately 3400 million m^3 and MENA releases around 793 million m^3 of wastewater per annum from processes used in the production of domestic goods and exports. It is also reported that around 409 million m^3 of virtual water flows from MENA to the European Union (EU) 27 by crossing boundaries with Libya being the largest water exporter (Sakhel et al., 2013).

Water demand for the industrial sector in India is considered to be about 8–10%. However, it also shows a wide variation in data due to scarcity in national and international organizations and inconsistency in the adopted methods for data generation (Joseph et al., 2019). A recent report from China on water consumption indicated that 604.02 billion cubic meters of the total water was used, whereas 130.8 billion cubic meters for specific industrial water was used (Wang et al., 2015). Other interesting data showed that China discharged 73.53 billion tons of wastewater in 2015 that included industrial effluent discharges of around 19.95 billion tons (Liao et al., 2016). The industrial wastewater and sewage discharge reaches to the Yellow River of China at different places with different quantities. The Center for Sustainable Systems, University of Michigan (USA), estimates that industrial water consumption is highest at around 45.9%, including thermoelectric water consumption, in 2015 in USA. Surface water resources as freshwater are considered under high pressure due to its demand among all other water resources (CSS, 2019). Water-use intensity for a specific economic sector is the water used in volume per unit of gross value that is used in Europe and measures the pressure of the economy on water resources concerning its economic influence. One study (Forster, 2014) from European countries presents that water consumption in manufacturing, mining, and quarrying are much higher than the construction industry. Thus, it correlates with the manufacturing industry in Germany, which reported the highest values, i.e. 202.8 m^{3}per thousand EUR (2010 data), and Finland 90.7 m^3 per thousand EUR (Water supply data, 2011) indicates heavy water requirements. The mining and quarrying areas observed the highest values, i.e. 159.1 m^3 per thousand EUR (2010 data) in Germany and 139.6 m^3 per thousand EUR (2009 data) in Belgium. Thus, the industrial sector is the major source of pollution in the environment. The various types of processes in different industries discharged various pollutants into the environment through the public sewerage system. The industrial effluents mainly constitute waste from manufacturing or processing units, washing water, and less polluted water from boilers and cooling tower operations. Wastewater composition contains a maximum 99.94% of water by weight and 0.06% materials dissolved or suspended in the water (Water Pollution Control Federation: year book, 1980). This dissolved or suspended matter contains organic and inorganic materials. Further, organic matter may constitute the fats, grease, oils, proteins, surfactants, pesticides and agricultural chemicals, volatile organic compounds, and other toxic chemicals. Inorganic materials may contain the heavy metals, nutrients (nitrogen and phosphorus), chlorides, sulfur, alkalinity, and other inorganic pollutants (Lee and Lin, 2000). These chemicals are considered as non-biodegradable compounds and are persistent in nature. Various industries such as paper mills, textile, and pharmaceuticals generate different amount of wastewater, and treatment is important before discharge into environment. Table 1.1 represents the water consumption and wastewater generation from several industries.

1.2.3 GENERATION OF XENOBIOTIC COMPOUNDS AND THEIR EFFECT ON THE ENVIRONMENT

A xenobiotic is a chemical substance found within an organism. These are synthetically prepared compounds produced from large industrial, agriculture, and domestic use (Atashgahi et al., 2018). Another interesting fact is that the natural compounds can become xenobiotic if they are taken up by another organism. For example, if natural hormones found in downstream of wastewater treatment plants outfalls are taken up by fish, or the chemical defenses produced by some organisms as protection against predators (Mansuy, 2013). Further, the term xenobiotic is derived from the Greek words (*xenos*) means foreigner, and (*bios*) means life. Xenobiotic is understood as a substance foreign to an entire biological system, or artificial substance, which did not exist in nature before their man-made synthesis. Xenobiotics may be grouped as environmental pollutants, drugs, carcinogens, food additives, hydrocarbons, and pesticides. Dioxin, furan, and polychlorinated biphenyls (PCB) are the pollutants often used as xenobiotics, and their effects on the biota are also reported in various studies (White and Birnbaum, 2009). Many xenobiotics are potentially hazardous chemicals to the organisms that are exposed due to the interaction with the environment. The author reported (Maenpaa, 2007) that any xenobiotic compound toxicity is related to the bioaccumulated chemical residue present in the organism. Xenobiotics may persist in nature and stay in the environment for months to years. One example, lignin, has a polymeric structure or cell wall of the spores of new fungi that are resistant to the natural environment (Fetzner, 2002). In an aquatic ecosystem, hydrophobic pollutants stored in the sediments become hazardous on exposure to the benthic organisms. The lower trophic levels have greater chances to the exposure of xenobiotics that could result in the biomagnification or more serious toxic impacts at the higher trophic levels (Newman, 1998; Streit, 1992).

The release of xenobiotic compounds into the environment comes from industries such as fossil fuels, pulp and paper mills, pharmaceuticals, agricultural chemicals manufacturing, mining, solvents production, electroplating, textile dyeing, etc., and

TABLE 1.1

Water Consumption and Wastewater Generation in Various Industries

S. No.	Industries	Freshwater Consumption (per ton/liter of production)	Wastewater Generation (per ton/liter of production)
1.	Pulp and paper mills	200–250 m³ per ton (Saadia and AshfAq, 2010)	220–380 m³ per ton (Badar and Farooqi, 2012)
2.	Textile industry	36,000 liters of water consumed for 20,000 pounds per day of fabric (Shaikh, 2009)	93.16 m³ per ton (in case of processing industry) (Sarayu and Sandhya, 2012)
3.	Distillery industry	–	10–15 liter per liter of ethanol (Bhardwaj et al., 2019)
4.	Sugarcane processing	1.5–2.0 m³ per ton (Sahu and Chaudhari, 2015)	1 m³ per ton (Sahu and Chaudhari, 2015)
5.	Tractor industry	–	0.58 m³ effluent generation per tractor manufactured (Ansari et al., 2013)
6.	Automobile service station	1 m³ water per vehicle (Mazumder and Mukherjee, 2011)	1 m³ of wastewater per vehicle (Mazumder and Mukherjee, 2011)
7.	Iron and steel industry	25–60 m³ of water average each ton of steel production (Sirajuddin et al., 2019)	–
8.	Malted barley brewery	2.5–6.4 liter water consumption per 1 liter of beer production (Donoghue et al., 2012)	Average 2.7 liters of wastewater per liter of beer produced (Donoghue et al., 2012)
9.	Vegetable oils and fats processing	30–80 m³ of water consumption per ton of raw material processing (JME, 2004)	–
10.	Slaughterhouses	-1.5–10 m³ per ton for pigs -2.5–40 m³ per ton for cattle -6–30 m³ per ton for poultry (Valta et al., 2015)	–
11.	Cheese production process	1–4 liter water consumption per liter of milk processed (Food Efficiency, 2013)	1.05–3.6 liter of wastewater per liter of milk processed (Food Efficiency, 2013)

these industries are considered as major sources of pollution (Díaz, 2004; Reineke and Knackmuss, 1988) Besides these, MSW generation is increasing and for effective management, the use of incineration is increasing for volume reduction and power generation. However, many toxic gases are released due to the burning of SW and are considered a xenobiotic in nature even at nano-concentration. Xenobiotics can enter into the environment at micropollutant concentrations (ng/L to µg/L range) or at high (ug/L to mg/L range) level (Meckenstock et al., 2015). Halogenated compounds considered as one important category of xenobiotics and generally synthesized for agricultural, industrial, and pharmaceuticals applications (Häggblom and Bossert, 2003). One report (Gribble, 2009) suggests that more than 5000 organohalogens are naturally occurring from the biogenic and geogenic sources, whereas hypochlorite, chlorine dioxide, and chlorite are used as inorganic halogenated compounds for various applications (Liebensteiner et al., 2016). The Pulp mill bleaching generates wastewater contaminated with the chlorophenols, chlorocatechols, and chloroguaiacols have negative environmental impacts due to their toxicity and persistence in the environment (Kim and Choi, 2005). Another work reports the chlorophenols generation into the environment from the industrial waste, pesticides, and insecticides, or by the degradation of complex chlorinated hydrocarbons. Chlorophenolics, thermal, and chemical degradation products are associated with public health problems such as histopathological alterations, genotoxicity, mutagenicity, and carcinogenicity among other abnormalities in humans and animals (Igbinosa et al., 2013).

Pesticides such as insecticides, herbicides, and fungicides have been used for a long time mainly for agricultural activities and are known for their xenobiotic characteristics (Schwarzenbach et al., 2010). The broad spectrum of chlorinated organophosphate chlorpyrifos (Arias-Estévez et al., 2008), the herbicide atrazine (John and Shaike, 2015), and a herbicide such as 2,4-dichlorophenoxyacetic acid (Bradberry et al., 2000) are the most extensively used pesticides worldwide. The chances of pesticide contamination are from point sources, mainly accidental release at manufacturing plants, spills on farmyards and wastewater treatment plants, and agricultural runoff as non-point sources (Vandermaesen et al., 2016). Sometimes, pesticides can contaminate due to transformation products. It was observed that 2,6-dichlorobenzamide is a highly mobile and persistent ground water pollutant that originates from the herbicide dichlobenil (Horemans et al., 2017). The exposure of pesticides could be by various routes, i.e. inhalation, dermal, and ocular routs, with spraying pesticides on protected skin/eyes of agricultural workers and worker exposed in the pesticide industry (Damalas and Eleftherohorinos, 2011). One study showed the important effects on the gut microbiome primarily by decreasing the concentrations of *Proteobacteria* by oral exposure to PCBs (Choi et al., 2013).

Organochlorine and organophosphate pesticides are widely used and associated with wide health impacts, such as endocrine disorders (Gasnier et al., 2009; Lemaire et al., 2004; McKinlay et al., 2008; Mnif et al., 2011), embryonic developments (Tiemann, 2008), lipid metabolism (Karami-Mohajeri and Abdollahi, 2010), and hematological disorders (Freire et al., 2015), which are also concerned with the organochlorine pesticides. Glyphosate and carbamates pesticides can also display endocrine-disrupting activity and possible reproductive disorder (Goad et al., 2004; Swanson et al., 2014; Thongprakaisang et al., 2013) affecting human erythrocytes *in vitro* (Kwiatkowska et al., 2014) and promoting carcinogenicity in mouse skin (George et al., 2010).

The pharmaceutical industry is fast growing and waste generated from the production of veterinary medications and anthropoid is a well-known source of prolonged environmental contamination. The nature of pharmaceutical chemicals is significantly affecting our environment due to toxicity and cause several effects on the aquatic flora and fauna compared to other industrial chemicals. One report showed the low-risk acute toxicity in the environment due to contamination of pharmaceutical compounds (Fent et al., 2006). Canadian scientist Rogers in 1986 tested and identified the concentration presence of the ibuprofen and naproxen in the wastewater. Another report is related to the presence of Diclofenac (as pain killer) which was commonly in use for veterinarian purposes for cattle treatment. It was observed in the drastic population reduction in the Indian vultures and Asian white-backed vultures (9 from 150 in 1997 to 25 in 2010). Various steroids, insect repellants, phthalates, and traces of different drugs were reported in the water supply study conducted by the Geological Survey Department (United States). One recent work identified the role of active pharmaceuticals ingredients in environmental degradation in certain areas due to their bulk drugs production (Fick et al., 2010; Gunnarsson et al., 2009). However, recently, the concern is increasing globally on pharmaceuticals residues present in the water resources that have potential effects on the aquatic flora and fauna (Bonjoko, 2014). Various antibiotics and several pharmaceutical compounds have been reported in wastewater treatment plants in pharmaceutical industries. Previous estimation from the EU has shown that around 3000 various chemical substances were identified in the wastewater that were used in human medicine such as beta-blockers, antibiotics, analgesics, anti-inflammatory drugs, and several others. Similarly, veterinary medicines are also used in a large number of the pharmaceuticals sector, mainly anti-inflammatories and antibiotics (Fent et al., 2006). Thus, in the present paragraph, xenobiotic compound generation and their effects on the surrounding environment and public health have been summarized.

1.2.4 WASTE BIOMASS GENERATION

Agricultural, forestry, and industrial practices produce a large amount of residues and wastes. The organic fraction of municipal waste, bio-solids, and animal residues are also generated in a substantial amount globally. Accumulation of these wastes presents a disposal and management problems and can have negative environmental impacts (Iakovou et al., 2010; Tripathi et al., 2019). Agricultural processes generate different residues such as stems, stalks, leaves, seedpods, husks, seeds, roots, and bagasse from different crops. Various crop residues such as rice straw, wheat straw, sugarcane bagasse, rice husk, reed, corn cob, sunflower stalks, sunflower hulls, empty fruit bunches of oil palm, Bermuda grass, and rapeseed stover are produced from agricultural activities. The cumulated potential of agricultural residues production is dominated by China, United States, and India with 716,682, and 605 million tons of fresh weight, respectively. After these countries, Europe, Brazil, Argentina, and Canada also have the potential of 580, 451, 148, and 105 million tons of fresh weight, respectively (Bentsen and Felby, 2010). The juice industries generate a huge amount of peel and pomace wastes such as orange peels, lemon peel, and apple pomace. The coffee industry generates a large quantity of coffee pulp as waste (Abraham et al., 2015; Kumar et al., 2016b; Sadh et al., 2018). The wastes of animal origin that are produced in significant quantities include tannery SWs, cow, and other ruminant's dung, chicken feathers, and fish-related wastes (Iakovou et al., 2010). Food processing industries such as juices, chips, meat, confectionery, and fruit industries generate a huge quantity of organic residues and related effluents (Sadh et al., 2018). If these waste residues are released without proper treatment into the environment, the result may be environmental deterioration that in turn affects the public health and animal's depending upon their exposure. Most of these wastes are untreated and many reports show that these agro-industrial SWs are disposed of by dumping, open burning, or unsystematic landfilling (Sadh et al., 2018). The agricultural waste residue burning practice not only destroys the soil biomass resources but also produces GHG and pollutes the environment (Kumar et al., 2016a).

1.3 NANOTECHNOLOGY FOR SUSTAINABLE ENVIRONMENTAL PROCESSES

According to one estimation, the fast-growing world's population reaching around 10 billion by 2050 (United Nations, 2019). Currently, the global environment is also facing major challenges to meet the rising necessary demands of things like water, food, and energy along with finished products (e.g. cars, airplanes, cell phones) and services (e.g. healthcare, shelter, employment) by considering the impacts on the reduction on environment and climate (Diallo et al., 2013). Thus, the nanotechnology branch emerged as a wide platform that could provide cost-effective and environmentally sustainable solutions to meet the global challenges facing society. Nanotechnology is the preparation of engineered materials manipulated at the nanoscale (1–100 nm); it offers potential applications in environmental remediation of contamination by toxic metals, organics, inorganic solutes, and microorganisms (Theron et al., 2008). Further, the nanotechnology branch involves the measuring, imaging, modeling, and manipulating of materials at the nanoscale level. The National Nanotechnology Initiative (NNI) was started for research and development in this area (NNI, 2010). The present section deals with the achievement, improvements, and new development in the area of nanotechnology for future sustainable development goals. The Brundtland Commission of United Nations defined 'sustainable development' as 'that which meets the need of the present without compromising the ability of future generations to meet their own needs' (Brundtland, 1987). In the current scenario, society is facing global environmental challenges; it was envisioned by the NNI that nanotechnology could provide better solutions to our society. Besides many challenges, the role of nanotechnology in the convergence of knowledge, technology, and society in achieving the sustainable goals provides the societal perspectives (Diallo et al., 2013). Nanotechnology is considered as the multi-trillion-dollar industry of the future. Another estimation has shown that total revenue from the nanomaterial products is expected to be higher in the future and sometimes observed slower than the original due to global economic slowdown (Dhingra et al., 2010). Thus, the growing field of nanotechnology must evaluate the life cycle approach to their manufactured products from environmental and human health impacts at every stage of development (Dhingra et al., 2010). Figure 1.2 shows some of the selected nanomaterial applications that are presently used to build materials for the future generation of sustainable materials and technologies in various challenging areas.

This section presents the recent advances and opportunities of nanotechnology to address environmental issues in the following areas:

1. Nano based filters or nanomembranes for drinking water treatment (Theron et al., 2008);
2. Site remediation and wastewater treatment (Watlington, 2005), clean energy technologies;

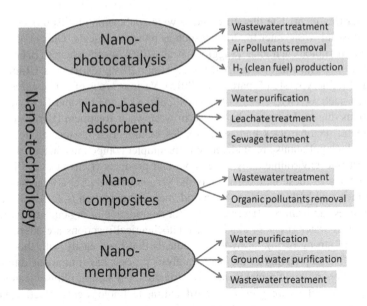

FIGURE 1.2   Nanotechnology for sustainable environment.

3. Use of nanophotocatalyst (NP) for air filtration (Sinha et al., 2007) GHG control;
4. Materials supply and utilization; and
5. Green manufacturing and chemistry.

1.3.1   NANO-PHOTOCATALYSIS FOR WASTEWATER TREATMENT

In the present scenario, water pollution and a limited safe water supply are the global concerns associated with the environment. As we are aware, with the increasing population, waste generation into water bodies is growing creating more pressure on the drinking water resources for the future. Thus, a drinking water shortage will lead to more emphasis on the wastewater treatment and recycling of treated water (Zekić et al., 2018). Wastewater treatment involves conventional methods such as physical, chemical, and biological processes. The conventional methods have some drawbacks such as too high an investment and poor treatment efficiency against xenobiotic or bio-recalcitrant compounds available in wastewater. Therefore, new approaches such as nanomaterial-based photocatalytic wastewater treatments are continuously being checked and tested in the traditional wastewater purification (Sharma and Sharma, 2013).

Photocatalysis means the degradation of organic compounds in the presence of light. However, it is a well-established process to stimulate or activate the semi-conductor materials by using UV/visible/solar light or change the rate of reaction without any involvement (Yaqoob et al., 2020). Photocatalysis is being considered due to advantages over the xenobiotic pollutants removal and zero secondary waste sludge generation. Another advantage of nanotechnology is the possibility of developing new and environmentally sound technology for drinking water purification in wastewater treatment. The literature cites a wide application of NPs process in different industrial wastewater treatments such as pharmaceuticals (Dhir et al., 2016), endocrine disruptor chemicals (Saggioro et al., 2019), 4-ethylphenol (Brüninghoff et al., 2019), textile wastewater (Al-Mamun et al., 2019), textile dyes (Ram et al., 2012), and refinery waste (Topare et al., 2015). Various types of nanocatalysts have been used such as TiO_2, ZnO, N_2O, AgO, CuO, SiO_2, Al_2O_3, and CeO_2 for their applications due to their remarkable use in wastewater treatment (Berekaa, 2016). These metal nanoparticles and metal oxides exhibit high photocatalytic activity through which pollutants are oxidized to form simpler compounds that are less toxic or convert into environmentally acceptable end products (Lens et al., 2013). The main property of these nanoparticles is the very small size, i.e. a large surface to volume ratio and high reactivity. NP-based chemical oxidation is an advanced oxidation process and can be effectively used for degradation of organic and inorganic contaminated water (Lens et al., 2013). Photochemical reactions are based on the formation of strong oxidizing hydroxyl radicals that react easily with the organic pollutants. The application of the photocatalysis process has limitations due to the high cost of energy requirements, mainly ozonators, UV lamps, ultrasonicators, etc. Moreover, researchers are going on toward making technology cost-effective; i.e. the use of solar energy in the tropical regions of the world.

1.3.2 NANOMEMBRANE FOR WATER PURIFICATION

According to the United Nations prediction, with the present population growth, in the coming decade 50% of the world's geographical locations will drastically face water scarcity (Goh and Ismail, 2018). Desalination techniques have been widely used more and more around the world to provide people with clean and safe water, especially in water-stressed countries such as Qatar, the United Arab Emirates, and Israel. Since 1980, nanoporous membrane-equipped filtration systems have been commercialized and commonly used at the domestic level. Membrane separation methods are become rapidly emerging areas for several industrial applications, for example the petroleum industry, food industry, chemical processing industry, pulp and paper mills, pharmaceuticals, and the electronics industry (Bhattacharyya et al., 2019; Lively and Sholl, 2017; Hansen et al., 2018; Tortora et al., 2018; Wen et al., 2016; Yamjala et al., 2016). The membrane separation technology is an essential process in various industries; it is based on the particle size of the various retained impurities. The water purification system was introduced globally, which involves reverse osmosis (RO), nanofiltration (NF), ultrafiltration (UF), and microfiltration (MF) (Davis, 2019; Mohammad et al., 2015; Moslehyani and Goh, 2019; Zhao and Yu, 2015). However, NF applications are increasing exponentially in the areas of water and wastewater purification, separation of solute, bio-materials production, pharmaceuticals, and preferred over the MF and UF membrane system. Moreover, the research and development in the area of RO membranes received significant attention; membranes were developed from the various materials for desalination applications (Ali et al., 2019). Other research also supports findings about the membrane filtration system that plays a significant role in the removal of pollutants and improves the water purification to the highest quality. Thus, the major objective of

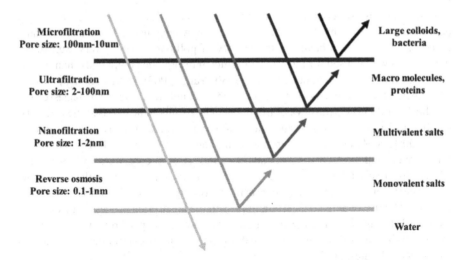

FIGURE 1.3 Classification of membranes for water purification in terms of pore size and retained species. (Yang et al., 2019.)

this review is to illustrate the NF applications that will, for example, increase in water recycling and reuse, wastewater treatment as tertiary options, and soft water use in industrial applications. (Abdel-Fatah, 2018).

The major disadvantages of nanomembrane technology are the high capital investment (estimated that around 70% of the total cost refers to the membranes) and a decrease in filtration efficiency with time, which requires further replacement costs. Membrane wastewater purification systems will become more and more popular in the market due to the reduced price and cost. The process has advantages for the removal of solid content, monovalent, divalent ions, and various pathogens, with high efficiency. The features of the types of membrane used with their corresponding pore space and retained impurities are shown in Figure 1.3.

1.3.3 NANO-BASED ADSORBENTS AND NANOCOMPOSITES FOR WASTEWATER TREATMENT

Adsorption is one of the most promising methods applied for the removal of several pollutants such as heavy metals, dyes, and others from the water and wastewater. Recently, several research works have been performed on the removal of various types of toxic pollutants from water and wastewater using nanoadsorbents (Deliyanni et al., 2009; Kurniawan et al., 2012). Further, adsorbent use is increasing due to its various benefits, i.e. cost-effective, non-toxic, easily available, and can be easily regenerated (Gupta and Bhattacharyya, 2006). The selection of adsorbent is based upon the removal of water pollutants and further depends on the type and concentration of pollutant, efficiency, and adsorption capacity against pollutants. Several studies have been investigated on the use of adsorbents for water purification (Schadler et al., 2014). Other advantages of adsorbents over the complex materials in water and wastewater treatment are the simplicity of design, convenience, ease of operation, and affordability (Gautam et al., 2014). Among nanoadsorbents, activated carbon is used as adsorbents for the variety of pollutants; this is very well documented and widely used for the efficient removal of organic compounds than metals and other inorganic pollutants (Hassler, 1980; Yang, 2003). In this context, authors investigated the sorption capacity of corn cob, coconut shell, sugarcane bagasse, and rice husk for domestic wastewater treatment. The results indicated that the turbidity, biochemical oxygen demand, and chemical oxygen demand for rice husk removal were 100%, 89%, and 100%; for sugarcane bagasse were 89%, 100% and 100%; for coconut shell were 78%, 97%, and 70%; and for corn cobs were 78%, 97%, and 63%, respectively (Waziri et al., 2017). Another study reported synthesis of nano zero-valent iron (nZVI) nanoparticles using activated carbon (AC-nZVI) to decrease the cost. The use of activated carbon improved the adsorption process. The effect of AC-nZVI dosage, pH, and the initial concentration for copper removal were studied and 96% removal rate was observed for 200 mg/L concentration in the case of AC-nZVI (Altuntas, 2017).

The reuse potential of nanoadsorbents has improved in wastewater treatment by impregnating into other matrices, especially polymers, to synthesize the nanocomposite adsorbents (Pan et al., 2009). The polymer nanocomposites are considered much better adsorbents. Two or more materials in combination are known as composites where one material is used for the reinforcing, in the form of fibers, sheets, or particles, and is embedded into the other materials. The idea of the developing nanocomposites originated at the Toyota research laboratory, Japan, in 1980. An article entitled 'Nano sandwiches', stating, 'Nature is a master chemist with incredible talent' was published in the journal *Chemistry* in Britain in 1998 (Oriakhi, 1998). Nanocomposites have multifunctional properties that depend upon various other parameters (Gupta et al., 2006; Hussain et al., 2006; Ramanathan et al., 2005; Schadler et al., 2007). Nanocomposites are classified into two major classes: (1) nonpolymer based nanocomposite and (2) polymer-based nanocomposite. Further, non-polymer is classified into the composites of three types: (1) metal-metal nanocomposites, (2) metal-ceramic nanocomposites, and (3) ceramic-ceramic nanocomposites (CCNC). The capability of removing heavy metals like arsenic is investigated by using iron oxide nanomaterials (Fe_2O_3 and Fe_3O_4) as cost-effective adsorbents by various researchers (Lai and Chen, 2001; Oliveira et al., 2004). Polymer nanocomposites are well-known and rigorously investigated for a wide range of applications among all the nanomaterials. Polymer nanocomposites have advantages in terms of low production costs and high specific strength. Hence, nanocomposites exhibit excellent advantages as compared to other peer composites in several areas.

1.4 BIOTECHNOLOGY FOR SUSTAINABLE ENVIRONMENTAL PROCESSES

Biotechnology has played an essential role in society, life, and economics since ancient times. Biotechnology potentially addressed global problems such as environmental pollution, global warming, and energy security by assisting in the renewable production of fuels and chemicals (Liang et al., 2016). Bioremediation is a process that involves the degradation of organic contaminants by indigenous or inoculated microorganisms such as fungi and bacteria. Organic contaminants or pollutants present in soil and water are converted into innocuous end products by microbial treatment (Jakovljević and Vrvić, 2018). Microbial enzymes are proved effective for bioremediation of environmental pollutants from industrial wastes due to their high specificity to a broad range of substrates (pollutants). Diverse microbial enzymes are used for degradation and detoxification of a variety of organic and inorganic pollutants (Saxena et al., 2020).

1.4.1 MICROBIAL BIOREMEDIATION

Anthropogenic activities result in generating hazardous wastes into the environment. These wastes represent various types of pollutants such as pesticides, heavy metals, polyaromatic hydrocarbons (PAHs), PCB, antibiotics, dyes, disinfectants, and detergents. Most of them are toxic and can cause diseases or even death. Soil, surface water, groundwater, and air become contaminated by these pollutants. Therefore, remediation of contaminated environments with cost-effective methods becomes an essential task. The degradation and removal of these pollutants can be attained by bioremediation. Bioremediation, although not a recent term, describes a natural process that uses biological agents (microorganisms or their products) for pollutant mineralization and recovery of the contaminated site (Dellagnezze et al., 2018; Verma and Jaiswal, 2016). Petroleum-based hydrocarbons are one of the major pollutants released into the environment either by human activities or accidentally. The nature and amount of hydrocarbon affects their biodegradation. Microbial degradation of petroleum hydrocarbon is the ultimate natural mechanism for their cleanup from the environment (Das and Chandran, 2011; Lal and Khanna, 1996). Fungi, bacteria, and yeasts are able to degrade hydrocarbons. *Arthrobacter, Burkholderia, Mycobacterium, Pseudomonas, Sphingomonas*, and *Rhodococcus* were involved in the degradation of alkyl aromatic compounds (Jones et al., 1983). Nine bacterial strains, namely, *Pseudomonas fluorescens, Pseudomonas aeruginosa, Bacillus subtilis, Bacillus* sp., *Alcaligenes* sp., *Acinetobacter lwoffii, Flavobacterium* sp., *Micrococcus roseus*, and *Corynebacterium* sp. were isolated from the polluted stream that could degrade petroleum hydrocarbons (Adebusoye et al., 2007; Das and Chandran, 2011). Pesticides are commonly used for agriculture, forestry, parks, industrial sites, sports grounds, and educational institutions. The use of pesticides degrades soil quality and contaminates the water table. Hence, it enters aquatic environments and the remediation of areas polluted with pesticides is a complex process (Marican and Durán-Lara, 2018; Uqab et al., 2016). Pesticides are classified into insecticides, herbicides, bactericides, fungicides, nematicides, wood preservatives, and rodenticides, based on the target action (Foo and Hameed, 2010; Marican and Durán-Lara, 2018). The complete biodegradation of pesticides involves the oxidation of parent compounds into CO_2 and water to generate energy for microbial growth. Several bacterial species from the genera *Flavobacterium, Arthobacter, Azotobacter, Burkholderia*, and *Pseudomonas* have been found to degrade pesticides. Parameters such as temperature, pH, and moisture content affect the degradation of pesticides by microbial action. Some white-rot and other fungi have also been reported in the degradation of pesticides. Fungi perform minor structural changes and convert pesticides into non-toxic substances that become susceptible to further degradation (Uqab et al., 2016). Heavy metals are also one of the contaminants of soil and water. Some metals such as nickel, copper, iron, and arsenic are beneficial for the human body at very low amounts, but they are cytotoxic, mutagenic, and carcinogenic at high concentrations. High density holding heavy metals is found to be hazardous at minimum concentrations (Valko et al., 2016; Verma and Kuila, 2019). Several microorganisms such as *Flavobacterium, Pseudomonas, Bacillus, Arthrobacter, Corynebacterium, Methosinus, Rhodococcus, Mycobacterium, Stereum hirsutum, Nocardia, Metahnogens, Aspergillus niger, Pleurotus astreatus, Rhizopus arrhizus, Azotobacter, Alcaligenes, Phormidium valderium, Ganoderma applantus*, etc. are involved in the bioremediation of heavy metals (Verma and Kuila, 2019).

1.4.2 Microbial Enzymes for Pollutants Degradation

Biocatalysis offers the green and clean solution to chemical processes and it is emerging as a sustainable alternative to chemical technology. A variety of chemical processes are now carried out by biocatalysts (enzymes) (Gautam et al., 2017; Prakash et al., 2013). Enzymes are biocatalysts that facilitate the conversion of substrates into products by lowering their required activation energy. Recent developments in technology showed that biocatalysis through isolated enzyme is more economical compared to whole-cell utilization. Enzymes are not restrained by inhibitors of microbial metabolism and are also applicable in extreme conditions, limiting microbial activity. Moreover, enzymes are more effective at low pollutant concentrations and can work in the presence of microbial predators or antagonists. The small size of enzymes makes them more mobile compared to microorganisms. All of these properties render enzymes very useful for bioremediation (Rao et al., 2010; Saxena et al., 2020). Bioremediation is carried out by the enzymes that mainly belong to oxidoreductases, hydrolases, and transferases. The main producers of these enzymes are bacteria, fungi (especially white-rot fungi), plants, and microbe-plant association (Rao et al., 2010).

1.4.2.1 Microbial Oxidoreductases

The oxidoreductases perform humification of different phenolic compounds that are generated from the decomposition of lignin in the soil environment. Similarly, oxidoreductases can also detoxify toxic xenobiotic substances such as phenolic or anilinic compounds polymerization, copolymerization with other substrates, or binding to humic substances (Karigar and Rao, 2011; Park et al., 2006). Oxygenases are the oxidoreductases that perform the oxidation of reduced substances by utilizing oxygen from molecular oxygen and FAD/NADH, NADPH as a co-substrate. Oxygenases are categorized into two groups (monooxygenases and dioxygenases) based on the number of oxygen atoms used for oxygenation (Arora et al., 2009; Karigar and Rao, 2011). Mono-oxygenases incorporate one atom of oxygen into the substrate; whereas, dioxygenases add both atoms of oxygen to a substrate. Mono-oxygenases are grouped into flavin-dependent mono-oxygenases and P450 mono-oxygenases, based on the co-factor requirement (Arora et al., 2010) (Table 1.2). Dioxygenases

TABLE 1.2

Microbial Enzymes for Pollutants Degradation

Enzyme	Enzyme Action	References
Microbial oxidoreductases		
Mono-oxygenases	• Desulfurization, dehalogination, denitrification, biodegradation, amminification, hydroxylation and biotranformation of aromatic and aliphatic compounds • Methane monooxygenases performs the degradation of methanes, alkanes, cycloalkanes, alkenes, haloalkenes, ethers, and aromatic and heterocyclic hydrocarbons	Arora et al. (2010); Grosse et al. (1999); Karigar and Rao (2011)
Dioxygenases	• Degrade aromatic compounds into aliphatic products	Krzmarzick et al. (2018); Sharma et al. (2019)
Laccases	• Laccases catalyze the oxidation of phenolic compounds such as polyphenols, amino phenols, and methoxy phenols • Diamines, aromatic amines and related substances, N-heterocyles, phenothiazines, thio groups, etc. are oxidized	Barrios-Estrada et al. (2018); Yadav et al. (2018)
Microbial peroxidases		
Lignin peroxidases	• Non-phenolic lignin derivatives are degraded to homologous ketones or aldehydes • Involves in hydroxylation of benzylic methylene groups of aromatic ring cleavages • Catalyzes oxidation of phenolic compounds such as vanillyl alcohol, catechol, acetosyringone, syringic acid, and guaicol etc. preferentially at a much faster rate compared to non-phenolic compounds	Ikehata et al. (2004); Kumar and Chandra (2020); Wong (2009)
Manganese peroxidases	• It oxidizes both phenolic and non-phenolic compounds • Phenolic compounds such as phenol containing dyes, amines, and lignin derivatives	Kumar and Chandra (2020)
Versatile peroxidases	• Are able to oxidize phenolic, non-phenolic and lignin derivatives • They do not require any mediator for oxidation of compounds	Knop et al. (2016); Kumar and Chandra (2020)
Hydrolytic enzymes		
Lipases	• Catalyze the hydrolysis and synthesis of long chain acylglycerols • Perform hydrolysis of triglycerides and reverse reaction (interesterification and esterification)	Hassan et al. (2018); Lajis (2018)
Proteases	• Conduct proteolysis by hydrolysis of peptide bonds that link amino acids together in polypeptide chain forming the protein	Wanyonyi and Mulaa (2019)
Cellulases and hemicellulases	• Cellulases catalyze the hydrolysis of cellulose into simple glucose units • Hemicellulases perform the hydrolysis of hemicelluloses to release simple fermentable sugars	Kumar et al. (2019a)

are a multicomponent enzyme system that incorporates oxygen into the substrate. Dioxygenases primarily oxidize aromatic compounds and therefore are applicable in environmental remediation (Karigar and Rao, 2011).

Laccases are oxidoreductases that contain a copper active center and are widely used for environmental protection. Laccases are broadly practiced for the treatment of effluents from dye, textile, leather, pulp and paper, and petrochemical industries. The efficiency of enzymes can be improved by using it in combination with redox mediator molecules. Laccases are involved in dye degradation, especially azo dyes, that are very difficult to degrade as they contain polyphenolic components (Abadulla et al., 2000; Kanagaraj et al., 2015; Le et al., 2016). Lignin peroxidase (LiP) belongs to the class oxidoreductase that degrades lignin and its derivatives in the presence of H_2O_2. These are heme-containing enzymes that are mainly produced by fungi and bacteria and degrade the polymer via the oxidative process (Kumar and Chandra, 2020). Manganese peroxidase (MnP) is also a heme-containing enzyme that is crucial for lignin degradation. It is a glycoprotein dependent on H_2O_2 that requires Mn^{++} ion for the oxidation of monoaromatic phenols (Kumar et al., 2017; Kumar and Chandra, 2020). The processing of olive oil, distillery, and pulp and paper industry produces several billion liters of colored, toxic, and harmful wastewater over the world annually. Pulp and paper effluents have been treated with several fungi such as *Ceriporiapsis subvermispora*, *Phenerocheate chrysosporium*, *Trametes versicolor*, *Rhizopus oryzae*, etc. (Yadav and Yadav, 2015). Versatile peroxidases (VP) are known as hybrid enzymes that belong to oxidoreductase family. They have combined catalytic activities of both LiP and MnP, hence designated as hybrid enzymes. VPs are able to oxidize the compounds with low to high redox potential due to two additional active sites via a mechanism similar to that described for LiPs (Knop et al., 2016; Kumar and Chandra, 2020; Ravichandran et al., 2019).

1.4.2.2 Hydrolytic Enzymes

Hydrolytic enzymes such as lipase, protease, cellulases, xylanase, and amylase play an important role in the waste treatment and sustainable industrial processing. Lipases are the ubiquitous enzymes that act at the interface between hydrophobic lipid substrates and hydrophilic aqueous medium to catalyze the hydrolysis of ester bonds in triglycerides. Lipases act on a broad range of substrates in harsh reaction conditions without any requirement of expensive co-factors (Gupta et al., 2004; Wang et al., 2019). Lipases are explored as the promising alternatives to assist the degradation of effluents rich in lipids, especially in milk and meat industries. The lipase pretreatment improves the hydrolysis and dissolution of fats available in the effluent from milk processing and reduces the time consumed in later treatments (Adulkar and Rathod, 2014; Golunski et al., 2017). Lipase assists the degradation of slop oil, a by-product of oil refining, and also generates during cleanup of oil tanks and filters. The high content of hydrocarbons (C3–C40) and other organic compounds make this waste difficult to eliminate from the environment. The combination of *Bacillus cereus* EN18 and lipase degraded slop oil efficiently (Marchut-Mikolajczyk et al., 2020).

Proteases are enzymes of ubiquitous nature that catalyze the breakdown of proteins into peptides and amino acids. Based on catalytic action, proteases are divided into exopeptidases and endopeptidases. Proteases are exploited in food, pharmaceutical, leather, and detergent industry (Sharma et al., 2017). Proteases catalyze the breakdown of proteinaceous substances that enter the environment due to the shedding and moulting of appendages and animal carcasses. Industries such as poultry, fishery, and leather also generate proteinaceous pollutants. Proteases hydrolyze the peptide bond in an aqueous environment (Karigar and Rao, 2011). Proteases can increase the degradation rated of biodegradable substances such as activated sludge, allowing more efficient treatment processes (Karn and Kumar, 2015). Proteases efficiently removes the vanes from the shaft of the detached bird feather and dehairs the goat skin. Protease treatment also reduces the BOD, COD, and pH of tannery waste effluent (Majumder et al., 2015).

A huge quantity of lignocellulosic biomass is generated through forestry, agricultural practices, timber, pulp and paper, and many agro-industries that create an environmental pollution problem. This large amount of residual plant biomass considered as 'waste', can be converted into value-added products such as bioethanol, biomethane, biohydrogen, biobutanol, organic acids, microbial polysaccharides, xylitol, etc. (Kumar et al., 2016b). The bioconversion of lignocellulosic wastes into valuable products is completed in three steps: pretreatment, hydrolysis of carbohydrates, and fermentation of monomer sugars. The hydrolysis of cellulose and hemicelluloses into fermentable sugars is carried out by cellulases and xylanases. Cellulases, hemicellulases, and pectinases are the hydrolytic enzymes that decompose the cellulose, hemicelluloses, and pectin, respectively. Cellulases are a multi-component enzymatic system that consists of three major groups of enzymes: endo-β-1, 4-glucanases, exo-β-1, 4-glucanase, and β-glucosidases. Endo-β-1, 4-glucanases randomly cleave β-1, 4-glycosidic linkages in amorphous part of cellulose away from chain ends. Exo-β-1, 4-glucanase, or cellobiohydrolase produces cellobiose by attacking cellulose from reducing and non-reducing chain ends. β-glucosidases that converts cellobiose and soluble oligosaccharides into glucose (Bansal et al., 2009; de Castro and de Castro, 2012; Kumar et al., 2019b). The enzymes that break down the hemicelluloses are collectively called hemicellulases. Endoxylanases randomly cleave the backbone of the xylan chain, producing a mixture of xylooligosaccharide (Kumar et al., 2019b).

1.4.3 PHYTOREMEDIATION FOR POLLUTANTS DEGRADATION AND HEAVY METALS REMOVAL

Phytoremediation is the in-situ technology that exploits plants and their rhizosphere to remove the contaminants or lower their bioavailability in soil and water. It results in the ecological rehabilitation of contaminated sites (DalCorso et al., 2019). Various organic pollutants such as polynuclear aromatic hydrocarbons, PCBs, and pesticides are degraded by phytoremediation. It is also able to remove inorganic pollutants like heavy metals (Rezania et al., 2015). Phytoremediation is considered more suitable when pollutants cover a wide area and are present within the root zone of the plants (Chibuike and Obiora, 2014). Plants and rhizosphere, i.e. soil and microorganisms associated with roots, perform the mechanisms that altogether are responsible for the rehabilitation of contaminated soil. The phytoremediation system includes phytodegradation, phytoextraction, phytovolatilization, phyto(rhizo)satabilization, and phyto(rhizo)filtration (DalCorso et al., 2019). Phytoextraction is the process of absorption of contaminants and their translocation to aerial parts of plants, mainly shoots and sometimes leaves. The removal of heavy metals and other pollutants from contaminated water by plant roots is known as rhizofiltration (Khan and Faisal, 2018; Martínez-Alcalá et al., 2012). A wide variety of plants have been tested for phytoremediation purposes. The plants belonging to multiple species of families such as *Brassicaceae, Poaceae, Fabaceae, Asteraceae, Salicaceae, Chenopodiaceae,* and *Careophylaceae* have shown phytoremediation potential. Some individual species from families including *Cyperaceae, Amaranthaceae, Cannabaceae, Cannaceae, Typhaceae,* and *Pontederiaceae* have also been used for phytoremediation. Every plant shows certain advantages and suffers from some limitations for phytoremediation applications (Gawronski and Gawronska, 2007).

1.5 CONCLUSION

This chapter described an overview of current environmental problems created by MSWs, wastewaters, and different types of industrial xenobiotic pollutants. Several nanobiotechnological approaches have shown great potential to address these problems. Microorganisms and their enzymes can be utilized to maintain the quality, health, and sustainability of the environment. Nanotechnology has also emerged as an effective and sustainable way for the treatment of several types of pollutants.

REFERENCES

Saadia, A., AshfAq, A., 2010. Environmental management in pulp and paper industry. *Jr. Ind. Pollut. Control* 26, 71–77.

Abadulla, E., Tzanov, T., Costa, S., Robra, K.-H., Cavaco-Paulo, A., Gübitz, G.M., 2000. Decolorization and detoxification of textile dyes with a laccase from *Trametes hirsuta. Appl. Environ. Microbiol.* 66, 3357–3362.

Abbasi, S.A., 2018. The myth and the reality of energy recovery from municipal solid waste. *Energy. Sustain. Soc.* 8, 36. https://doi.org/10.1186/s13705-018-0175-y

Abdel-Fatah, M.A., 2018. Nanofiltration systems and applications in wastewater treatment: Review article. *Ain Shams Eng. J.* 9, 3077–3092. https://doi.org/https://doi.org/10.1016/j.asej.2018.08.001

Abraham, J., Cerda, A., Barrena, R., Ponsá, S., Gea, T., Sánchez, A., Abraham, J., Cerda, A., Barrena, R., Ponsá, S., Gea, T., Sánchez, A., 2015. From wastes to high value added products : Novel aspects of SSF in the production of enzymes. *Crit. Rev. Environ. Sci. Technol.* 45, 1999–2045. https://doi.org/10.1080/10643389.2015.1010423

Adebusoye, S.A., Ilori, M.O., Amund, O.O., Teniola, O.D., Olatope, S.O., 2007. Microbial degradation of petroleum hydrocarbons in a polluted tropical stream. *World J. Microbiol. Biotechnol.* 23, 1149–1159.

Adulkar, T.V, Rathod, V.K., 2014. Ultrasound assisted enzymatic pre-treatment of high fat content dairy wastewater. *Ultrason. Sonochem.* 21, 1083–1089.

Akimoto, H., 2003. Global air quality and pollution. *Science.* 302, 1716–1719. https://doi.org/10.1126/science.1092666

Alam, P., Ahmade, K., 2013. Impact of solid waste on health and the environment. *Spec. Issue Int. J. Sustain. Dev. Green Econ.* 2, 165–168.

Albert, S., Leon, J.X., Grinham, A.R., Church, J.A., Gibbes, B.R., Woodroffe, C.D., 2016. Interactions between sea-level rise and wave exposure on reef island dynamics in the

Solomon Islands. *Environ. Res. Lett.* 11, 54011. https://doi.org/10.1088/1748-9326/11/5/054011

Ali, T.M., 2009. Assessment of solid waste management system in Khartoum locality. *Master Thesis.* 1–71.

Ali, Z., Al Sunbul, Y., Pacheco, F., Ogieglo, W., Wang, Y., Genduso, G., Pinnau, I., 2019. Defect-free highly selective polyamide thin-film composite membranes for desalination and boron removal. *J. Memb. Sci.* 578, 85–94. https://doi.org/https://doi.org/10.1016/j.memsci.2019.02.032

Al-Mamun, M.R., Kader, S., Islam, M.S., Khan, M.Z.H., 2019. Photocatalytic activity improvement and application of UV-TiO$_2$ photocatalysis in textile wastewater treatment: A review. *J. Environ. Chem. Eng.* 7, 103248. https://doi.org/https://doi.org/10.1016/j.jece.2019.103248

Altuntas, K., 2017. Adsorption of copper metal ion from aqueous solution by nanoscale zero valent iron (nZVI) supported on activated carbon 5, 3–6. https://doi.org/10.21533/pen.v5i1.77

Anand, S.V., 2013. Global environmental issues 2. Open Access Scientific Reports. 2, 1–9. https://doi.org/10.4172/scientificreports.

Ansari, F., Pandey, Y.K., Kumar, P., Pandey, P., 2013. Performance evaluation of effluent treatment plant for automobile industry. *Int. J. Energy Environ.* 4, 1079–1086.

Arias-Estévez, M., López-Periago, E., Martínez-Carballo, E., Simal-Gándara, J., Mejuto, J.-C., García-Río, L., 2008. The mobility and degradation of pesticides in soils and the pollution of groundwater resources. *Agric. Ecosyst. Environ.* 123, 247–260. https://doi.org/https://doi.org/10.1016/j.agee.2007.07.011

Arora, P.K., Kumar, M., Chauhan, A., Raghava, G.P.S., Jain, R.K., 2009. OxDBase: A database of oxygenases involved in biodegradation. *BMC Res. Notes* 2, 67. https://doi.org/10.1186/1756-0500-2-67

Arora, P.K., Srivastava, A., Singh, V.P., 2010. Application of monooxygenases in dehalogenation, desulphurization, denitrification and hydroxylation of aromatic compounds. *J. Bioremed. Biodegrad.* 1, 1–8.

Atashgahi, S., Sánchez-Andrea, I., Heipieper, H.J., van der Meer, J.R., Stams, A.J.M., Smidt, H., 2018. Prospects for harnessing biocide resistance for bioremediation and detoxification. *Science.* 360, 743–746. https://doi.org/10.1126/science.aar3778

Badar, S., Farooqi, I.H., 2012. Pulp and paper industry—Manufacturing process, wastewater generation and treatment, in: Malik, A., Grohmann, E. (Eds.), *Environmental Protection Strategies for Sustainable Development.* Strategies for Sustainability. Springer: Dordrecht, Netherlands, pp. 397–436. https://doi.org/10.1007/978-94-007-1591-2_13

Bansal, P., Hall, M., Realff, M.J., Lee, J.H., Bommarius, A.S., 2009. Modeling cellulase kinetics on lignocellulosic substrates. *Biotechnol. Adv.* 27, 833–848. https://doi.org/http://dx.doi.org/10.1016/j.biotechadv.2009.06.005

Barrios-Estrada, C., de Jesús Rostro-Alanis, M., Muñoz-Gutiérrez, B.D., Iqbal, H.M.N., Kannan, S., Parra-Saldívar, R., 2018. Emergent contaminants: Endocrine disruptors and their laccase-assisted degradation – A review. *Sci. Total Environ.* 612, 1516–1531. https://doi.org/https://doi.org/10.1016/j.scitotenv.2017.09.013

Bentsen, N.S., Felby, C., 2010. Technical potentials of biomass for energy services from current agriculture and forestry in selected countries in Europe, The Americas and Asia. *Forest & Landscape Working Papers.* 55.

Berekaa, M.M., 2016. Nanotechnology in wastewater treatment; influence of nanomaterials on microbial systems. *Int. J. Curr. Microbiol. Appl. Sci.* 5, 713–726.

Bernstein, L., Bosch, P., Canziani, O., Chen, Z., Christ, R., Riahi, K., 2008. IPCC, 2007: Climate Change 2007: Synthesis Report. Geneva: IPCC. ISBN 2-9169-122-4.

Bhardwaj, S., Ruhela, M., Bhutiani, R., Ahamad, F., 2019. Distillery spent wash (DSW) treatment methodologies and challenges with special reference to incineration : An overview. *Environ. Conserv. J.* 20, 135–144. https://doi.org/10.36953/ECJ.2019.20318

Bhattacharyya, S., Das, P., Datta, S., 2019. Removal of ranitidine from pharmaceutical waste water using activated carbon (AC) prepared from waste lemon peel, in: Ghosh, S.K. (Ed.), *Waste Water Recycling and Management*; Springer: Singapore, pp. 123–141.

Biermann, F., 2014. Governance in the anthropocene: Towards planetary stewardship. At the 4th Interactive Dialogue of the United Nations General Assembly on Harmony with Nature. New York City. 1–5.

Bisgrove, R., Hadley, P., 2002. Gardening in the global greenhouse : The impacts of climate change on gardens in the UK. The UK Climate Impacts Programme.

Bogner, J.E., Spokas, K.A., Chanton, J.P., 2011. Seasonal greenhouse gas emissions (methane, carbon dioxide, nitrous oxide) from engineered landfills: Daily, intermediate, and final California cover soils. *J. Environ. Qual.* 40, 1010–1020. https://doi.org/10.2134/jeq2010.0407

Bonjoko, B., 2014. Environmental pharmacology: An overview, *Pharmacol. Ther..* InTech Open, 1, pp. 133–178.

Bradberry, S.M., Watt, B.E., Proudfoot, A.T., Vale, J.A., 2000. Mechanisms of toxicity, clinical features, and management of acute chlorophenoxy herbicide poisoning: A review. *J. Toxicol. Clin. Toxicol.* 38, 111–122. https://doi.org/10.1081/CLT-100100925

Brundtland, H., 1987. Report of the world commission on environment and development: Our common future. https://sustainabledevelopment.un.org/content/documents/5987our-common-future.pdf.

Brüninghoff, R., van Duijne, A.K., Braakhuis, L., Saha, P., Jeremiasse, A.W., Mei, B., Mul, G., 2019. Comparative analysis of photocatalytic and electrochemical degradation of 4-ethylphenol in saline conditions. *Environ. Sci. Technol.* 53, 8725–8735. https://doi.org/10.1021/acs.est.9b01244

Chibuike, G.U., Obiora, S.C., 2014. Heavy metal polluted soils: Effect on plants and bioremediation methods. *Appl. Environ. Soil Sci.* 2014.

Choi, J.J., Eum, S.Y., Rampersaud, E., Daunert, S., Abreu, M.T., Toborek, M., 2013. Exercise attenuates PCB-induced changes in the mouse gut microbiome. *Environ. Health Perspect.* 121, 725–730. https://doi.org/10.1289/ehp.1306534

CSS, 2019. US water supply and distribution factsheet. Center for Sustainable Systems, University of Michigan, Pub. No. CSS05-17.

DalCorso, G., Fasani, E., Manara, A., Visioli, G., Furini, A., 2019. Heavy metal pollutions: State of the art and innovation in phytoremediation. *Int. J. Mol. Sci.* 20, 3412.

Damalas, C.A., Eleftherohorinos, I.G., 2011. Pesticide exposure, safety issues, and risk assessment indicators. *Int. J. Environ. Res. Public Health* 8, 1402–1419. https://doi.org/10.3390/ijerph8051402

Damtew, T., Desta, N., 2015. Micro and small enterprises in solid waste management : Experience of selected cities and towns in Ethiopia : A review. *Pollution* 1, 461–472.

Dan, A., Oka, M., Fujii, Y., Soda, S., Ishigaki, T., Machimura, T., Ike, M., 2017. Removal of heavy metals from synthetic landfill leachate in lab-scale vertical flow constructed wetlands. *Sci. Total Environ.* 584–585, 742–750. https://doi.org/https://doi.org/10.1016/j.scitotenv.2017.01.112

Mansuy, D., 2013. Metabolism of xenobiotics: beneficial and adverse effects. *Biol Aujourdhui.* 207, 33–47.

Das, N., Chandran, P., 2011. Microbial degradation of petroleum hydrocarbon contaminants: An overview. *Biotechnol. Res. Int.* 2011, 1–14.

Davis, R.H., 2019. Microfiltration in pharmaceutics and biotechnology, in: Basile, A. & Ghasemzadeh, K. (Ed.), *Current Trends and Future Developments on (Bio-) Membranes.* Elsevier: Amsterdam, the Netherlands, pp. 29–67.

de Castro, S., de Castro, A., 2012. Assessment of the Brazilian potential for the production of enzymes for biofuels from agroindustrial materials. *Biomass Convers. Biorefinery.* 2, 87–107. https://doi.org/10.1007/s13399-012-0031-9

Deliyanni, E.A., Peleka, E.N., Matis, K.A., 2009. Modeling the sorption of metal ions from aqueous solution by iron-based adsorbents. *J. Hazard. Mater.* 172, 550–558. https://doi. org/https://doi.org/10.1016/j.jhazmat.2009.07.130

Dellagnezze, B.M., Gomes, M.B., de Oliveira, V.M., 2018. Microbes and petroleum bioremediation, in: *Microbial Action on Hydrocarbons*. Springer: Switzerland AG, 97–123.

Dhingra, R., Naidu, S., Upreti, G. and Sawhney, R., 2010. Sustainable nanotechnology: through green methods and life-cycle thinking. *Sustainability.* 2, 3323–3338. https:// doi.org/10.3390/su2103323

Dhir, A., Kamboj, M., Ram, C., 2016. Studies on the use of calcium hypochlorite in the TiO_2 mediated degradation of pharmaceutical wastewater. *Environ. Eng. Manag. J.* 15, 1713–1720.

Diallo, M.S., Fromer, N.A., Jhon, M.S., 2013. Nanotechnology for sustainable development: Retrospective and outlook. *J. Nanoparticle Res.* 15, 2044. https://doi.org/10.1007/ s11051-013-2044-0

Diez-roux, A.V., Nieto, F.J., Caulfield, L., Tyroler, H.A., Watson, R.L., Szklo, M., 1999. Neighbourhood differences in diet : the Atherosclerosis Risk in Communities (ARIC) study. *J Epidemiol Community Health* 53, 55–63. doi: 10.1136/jech.53.1.55

Donoghue, C., Jackson, G., Koop, J.H., Heuven, A.J.M., 2012. The environmental performance of the European brewing sector. https://brewersofeurope.org/uploads/mycms-files/documents/archives/publications/2012/envi_report_2012_web.pdf

Díaz, E., 2004. Bacterial degradation of aromatic pollutants: A paradigm of metabolic versatility. *Int. Microbiol.* 7, 173–180.

Egen, J.G., Ouyang, W., 2010. Even neurons are excited by Th17 cells. *Immunity* 33, 298–300. https://doi.org/https://doi.org/10.1016/j.immuni.2010.09.004

Epule, E.T., Ford, D.J., Lwasa, S., Lepage, L., 2017. Vulnerability of maize yields to droughts in Uganda. *Water.* 9, 181. https://doi.org/10.3390/w9030181

Fent, K., Weston, A.A., Caminada, D., 2006. Ecotoxicology of human pharmaceuticals. *Aquat. Toxicol.* 76, 122–159. https://doi.org/https://doi.org/10.1016/j.aquatox.2005.09.009

Ferreira, C., Barbi, F., 2016. The challenge of global environmental change in the anthropocene: An analysis of Brazil and China. *Chinese Polit. Sci. Rev.* 1, 685–697. https://doi. org/10.1007/s41111-016-0028-9

Ferronato, N., 2019. Waste mismanagement in developing countries : A review of global issues. *Int. J. Environ. Res. Public Health.* https://doi.org/10.3390/ijerph16061060

Fetzner, S., 2002.. Biodegradation of xenobiotics. Encycl. Life Support Syst. Publ. Dev. under Auspices UNESCO. Biotechnol. Ed. by Doelle Da Silva. https://www.eolss.net/ Sample-Chapters/C17/E6-58-09-08.pdf

Fick, J., Lindberg, R.H., Tysklind, M., Larsson, D.G.J., 2010. Predicted critical environmental concentrations for 500 pharmaceuticals. *Regul. Toxicol. Pharmacol.* 58, 516–523. https://doi.org/https://doi.org/10.1016/j.yrtph.2010.08.025

Foo, K.Y., Hameed, B.H., 2010. Detoxification of pesticide waste via activated carbon adsorption process. *J. Hazard. Mater.* 175, 1–11.

Food Efficiency, 2013. Water usage, waste water and biogas potential within the dairy manufacturing sector. http://www.foodefficiency.eu/system/resources/BAhbBlsHOg ZmSSJFMjAxMy8wNC8xMS8xNS8wMS8yNC81MzYvV2F0ZXJfV2FzdGV3YXR R lcl9hbmRfQmlvZ2FzX0ZlYl8yMDEzLnBkZgY6BkVU/Water%20Wastewater%20 and%20Biogas%20Feb%202013.pdf

Forster, J., 2014. Cooling for electricity production dominates water use in industry, Catalogue number: KS-SF-14-014-EN-N. *Eurostat Stat. Eur. Comm.* 2011, 1–12.

Freire, C., Koifman, R.J., Koifman, S., 2015. Hematological and hepatic alterations in Brazilian population heavily exposed to organochlorine pesticides. *J. Toxicol. Environ. Health* 78, 534–548.

Friedrich, E., Trois, C., 2011. Quantification of greenhouse gas emissions from waste management processes for municipalities – A comparative review focusing on Africa. *Waste Manag.* 31, 1585–1596. https://doi.org/10.1016/j.wasman.2011.02.028

Gasnier, C., Dumont, C., Benachour, N., Clair, E., Chagnon, M.-C., Séralini, G.-E., 2009. Glyphosate-based herbicides are toxic and endocrine disruptors in human cell lines. *Toxicology* 262, 184–191. https://doi.org/https://doi.org/10.1016/j.tox.2009.06.006

Gautam, A., Kumar, A., Dutt, D., 2017. Production and characterization of cellulase-free xylanase by aspergillus flavus ARC-12 and its application in pre-bleaching of ethanol-soda pulp of eulaliopsis binata. *Res. J. Biotechnol.* 12, 63–71.

Gautam, R.K., Mudhoo, A., Lofrano, G., Chattopadhyaya, M.C., 2014. Biomass-derived biosorbents for metal ions sequestration: Adsorbent modification and activation methods and adsorbent regeneration. *J. Environ. Chem. Eng.* 2, 239–259. https://doi.org/https://doi.org/10.1016/j.jece.2013.12.019

Gawronski, S.W., Gawronska, H., 2007. Plant taxonomy for phytoremediation, in: *Advanced Science and Technology for Biological Decontamination of Sites Affected by Chemical and Radiological Nuclear Agents*. Springer, Dordrecht, 79–88.

George, J., Prasad, S., Mahmood, Z., Shukla, Y., 2010. Studies on glyphosate-induced carcinogenicity in mouse skin: A proteomic approach. *J. Proteomics* 73, 951–964. https://doi.org/https://doi.org/10.1016/j.jprot.2009.12.008

Goad, R.T., Goad, J.T., Atieh, B.H., Gupta, R.C., 2004. Carbofuran-induced endocrine disruption in adult male rats. *Toxicol. Mech. Methods* 14, 233–239. https://doi.org/10.1080/15376520490434476

Goh, P.S., Ismail, A.F., 2018. A review on inorganic membranes for desalination and wastewater treatment. *Desalination* 434, 60–80. https://doi.org/https://doi.org/10.1016/j.desal.2017.07.023

Golunski, S.M., Mulinari, J., Camargo, A.F., Venturin, B., Baldissarelli, D.P., Marques, C.T., Vargas, G.D.L.P., Colla, L.M., Mossi, A., Treichel, H., 2017. Ultrasound effects on the activity of *Aspergillus niger* lipases in their application in dairy wastewater treatment. *Environ. Qual. Manag.* 27, 95–101.

Gribble, G.W., 2009. *Naturally Occurring Organohalogen Compounds – A Comprehensive Update*, 91st ed. Springer-Verlag: Vienna.

Grosse, S., Laramee, L., Wendlandt, K.-D., McDonald, I.R., Miguez, C.B., Kleber, H.-P., 1999. Purification and characterization of the soluble methane monooxygenase of the Type II methanotrophic Bacterium Methylocystis sp. strain WI 14. *Appl. Environ. Microbiol.* 65, 3929–3935.

Gunnarsson, L., Kristiansson, E., Rutgersson, C., Sturve, J., Fick, J., Förlin, L., Larsson, D.G.J., 2009. Pharmaceutical industry effluent diluted 1:500 affects global gene expression, cytochrome P450 1A activity, and plasma phosphate in fish. *Environ. Toxicol. Chem.* 28, 2639–2647. https://doi.org/10.1897/09-120.1

Gupta, N., Yadav, K.K., Kumar, V., 2015. A review on current status of municipal solid waste management in India. *J. Envrionmental Sci.* 37, 206–217. https://doi.org/10.1016/j.jes.2015.01.034

Gupta, R., Gupta, N., Rathi, P., 2004. Bacterial lipases: an overview of production, purification and biochemical properties. *Appl. Microbiol. Biotechnol.* 64, 763–781.

Gupta, S.S., Bhattacharyya, K.G., 2006. Adsorption of Ni(II) on clays. *J. Colloid Interface Sci.* 295, 21–32. https://doi.org/https://doi.org/10.1016/j.jcis.2005.07.073

Gupta, S., Zhang, Q., Emrick, T., Balazs, A.C., Russell, T.P., 2006. Entropy-driven segregation of nanoparticles to cracks in multilayered composite polymer structures. *Nat. Mater.* 5, 229–233. https://doi.org/10.1038/nmat1582

Gurdak, J.J., Qi, S.L., 2012. Vulnerability of recently recharged groundwater in principle aquifers of the United States to nitrate contamination. *Environ. Sci. Technol.* 46, 6004–6012. https://doi.org/10.1021/es300688b

Häggblom, M.M., Bossert, I.D., 2003. *Dehalogenation: Microbial Processes and Environmental Applications*. Kluwer Academic Publisher Group: Boston, MA.

Hansen, É., Rodrigues, M.A.S., Aragão, M.E., de Aquim, P., 2018. Water and wastewater minimization in a petrochemical industry through mathematical programming. *J. Clean. Prod.* 172, 1814–1822. https://doi.org/10.1016/j.jclepro.2017.12.005

Hassan, S.W.M., Abd El Latif, H.H., Ali, S.M., 2018. Production of cold-active lipase by free and immobilized marine *Bacillus cereus* HSS: Application in wastewater treatment. *Front. Microbiol.* 9, 2377.

Hassler, J.W., 1980. *Carbon Adsorption Handbook*, in: Cheremisinoff, P.N. ed. Ann Arbor Science, Ann Arbor, MI.

Hoornweg, D., Bhada-Tata, P., 2012. What a waste: A global review of solid waste management. World Bank Report. 1–116. http://hdl.handle.net/10986/17388

Horemans, B., Raes, B., Vandermaesen, J., Simanjuntak, Y., Brocatus, H., T'Syen, J., Degryse, J., Boonen, J., Wittebol, J., Lapanje, A., Sørensen, S.R., Springael, D., 2017. Biocarriers improve bioaugmentation efficiency of a rapid sand filter for the treatment of 2,6-dichlorobenzamide-contaminated drinking water. *Environ. Sci. Technol.* 51, 1616–1625. https://doi.org/10.1021/acs.est.6b05027

Husain, Q., 2006. Potential applications of the oxidoreductive enzymes in the decolorization and detoxification of textile and other synthetic dyes from polluted water: A review. *Crit. Rev. Biotechnol.* 26, 201–221. https://doi.org/10.1080/07388550600969936

Hussain, F., Hojjati, M., Okamoto, M., Gorga, R.E., 2006. Review article: Polymer-matrix nanocomposites, processing, manufacturing, and application: An overview. *J. Compos. Mater.* 40, 1511–1575. https://doi.org/10.1177/0021998306067321

Iakovou, E., Karagiannidis, A., Vlachos, D., Toka, A., Malamakis, A., 2010. Waste biomass-to-energy supply chain management: A critical synthesis. *Waste Manag.* 30, 1860–1870. https://doi.org/https://doi.org/10.1016/j.wasman.2010.02.030

Igbinosa, E.O., Odjadjare, E.E., Chigor, V.N., Igbinosa, I.H., Emoghene, A.O., Ekhaise, F.O., Igiehon, N.O., Idemudia, O.G., 2013. Toxicological profile of chlorophenols and their derivatives in the environment: The public health perspective. *Sci. World J.* 2013, 1–11. https://doi.org/10.1155/2013/460215

Ikehata, K., Buchanan, I.D., Smith, D.W., 2004. Recent developments in the production of extracellular fungal peroxidases and laccases for waste treatment. *J. Environ. Eng. Sci.* 3, 1–19. https://doi.org/10.1139/S03-077

Indarto, A., Choi, J.-W., Lee, H., Song, H.K., 2008. Decomposition of greenhouse gases by plasma. *Environ. Chem. Lett.* 6, 215–222. https://doi.org/10.1007/s10311-008-0160-3

Jakovljević, V.D., Vrvić, M.M., 2018. Potential of pure and mixed cultures of Cladosporium cladosporioides and *Geotrichum candidum* for application in bioremediation and detergent industry. *Saudi J. Biol. Sci.* 25, 529–536.

JME, 2004. Examples of food processing wastewater treatment, Japanese Ministry of the Environment. https://www.env.go.jp/earth/coop/coop/document/male2_e/007.pdf

John, E.M., Shaike, J.M., 2015. Chlorpyrifos: pollution and remediation. *Environ. Chem. Lett.* 13, 269–291. https://doi.org/10.1007/s10311-015-0513-7

Jones, D.M., Douglas, A.G., Parkes, R.J., Taylor, J., Giger, W., Schaffner, C., 1983. The recognition of biodegraded petroleum-derived aromatic hydrocarbons in recent marine sediments. *Mar. Pollut. Bull.* 14, 103–108.

Joseph, N., Ryu, D., Malano, H.M., George, B., Sudheer, K.P., 2019. Estimation of industrial water demand in India using census-based statistical data. *Resour. Conserv. Recycl.* 149, 31–44. https://doi.org/10.1016/j.resconrec.2019.05.036

Kanagaraj, J., Senthilvelan, T., Panda, R.C., 2015. Degradation of azo dyes by laccase: Biological method to reduce pollution load in dye wastewater. *Clean Technol. Environ. Policy* 17, 1443–1456. https://doi.org/10.1007/s10098-014-0869-6

Karami-Mohajeri, S., Abdollahi, M., 2010. Toxic influence of organophosphate, carbamate, and organochlorine pesticides on cellular metabolism of lipids, proteins, and carbohydrates: A systematic review. *Hum. Exp. Toxicol.* 30, 1119–1140. https://doi.org/10.1177/0960327110388959

Karigar, C.S., Rao, S.S., 2011. Role of microbial enzymes in the bioremediation of pollutants: A review. *Enzyme Res.* 2011, 805187. https://doi.org/10.4061/2011/805187

Karn, S.K., Kumar, A., 2015. Hydrolytic enzyme protease in sludge: Recovery and its application. *Biotechnol. Bioprocess Eng.* 20, 652–661.

Kaur, H., 2017. Sustainable environmental biotechnology, in: Kumar, R., Sharma, A.K., Ahluwalia, S.S. (Eds.), *Advances in Environmental Biotechnology.* Springer: Singapore, pp. 145–154. https://doi.org/10.1007/978-981-10-4041-2_8

Khan, H.N., Faisal, M., 2018. Phytoremediation of industrial wastewater by hydrophytes, in: *Phytoremediation.* Springer International Publishing, Cham, Switzerland, 179–200.

Khan, S.H., 2020. Green nanotechnology for the environment and sustainable development, in: Naushad, M., Lichtfouse, E. (Eds.), *Green Materials for Wastewater Treatment.* Springer International Publishing, Cham, Switzerland, 13–46. https://doi.org/10.1007/978-3-030-17724-9_2

Kim, S., Choi, W., 2005. Visible-light-induced photocatalytic degradation of 4-chlorophenol and phenolic compounds in aqueous suspension of pure titania: Demonstrating the existence of a surface-complex-mediated path. *J. Phys. Chem. B.* 109, 5143–5149. https://doi.org/10.1021/jp045806q

Knop, D., Levinson, D., Makovitzki, A., Agami, A., Lerer, E., Mimran, A., Yarden, O., Hadar, Y., 2016. Limits of versatility of versatile peroxidase. *Appl. Environ. Microbiol.* 82, 4070–4080. https://doi.org/10.1128/AEM.00743-16

Krzmarzick, M.J., Taylor, D.K., Fu, X., McCutchan, A.L., 2018. Diversity and niche of archaea in bioremediation. *Archaea.* 2018. https://doi.org/10.1155/2018/3194108

Kumar, A., Aggarwal, D., Yadav, M., Kumar, P., Kumar, V., 2019a. Biotechnological conversion of plant biomass into value-added products, in: Yadav, M., Kumar, V., Sehrawat, N. (Eds.), *Industrial Biotechnology.* De Gruyter, 51–72. https://doi.org/10.1515/9783110563337-003

Kumar, A., Chandra, R., 2020. Ligninolytic enzymes and its mechanisms for degradation of lignocellulosic waste in environment. *Heliyon* 6, e03170. https://doi.org/https://doi.org/10.1016/j.heliyon.2020.e03170

Kumar, A., Dutt, D., Gautam, A., 2016a. Production of crude enzyme from *Aspergillus nidulans* AKB-25 using black gram residue as the substrate and its industrial applications. *J. Genet. Eng. Biotechnol.* 14, 107–118. https://doi.org/10.1016/j.jgeb.2016.06.004

Kumar, A., Gautam, A., Dutt, D., 2016b. Biotechnological transformation of lignocellulosic biomass in to industrial products: An overview. *Adv. Biosci. Biotechnol.* 7, 149–168. https://doi.org/10.4236/abb.2016.73014

Kumar, A., Gautam, A., Dutt, D., Yadav, M., Sehrawat, N., Kumar, P., 2017. Applications of microbial technology in the pulp and paper industry, in: Kumar, V., Singh, G., Aggarwal, N. (Eds.), *Microbiology and Biotechnology for a Sustainable Environment.* Nova Science Publishers, Inc.: New York, NY, pp. 185–206.

Kumar, A., Gautam, A., Minuye, N., Bezie, Y., Bharti, A.K., 2019b. Food industry based application of cellulases, xylanases and pectinases in: Kumar, A., Yadav, M., Sehrawat, N. (Eds.), *Microbial Enzymes and Additives for the Food Industry.* Nova Science Publishers, Inc.: New York, NY, pp. 170–200.

Kumar, A., Samadder, S.R., 2017. A review on technological options of waste to energy for effective management of municipal solid waste. *Waste Manag.* 69, 407–422. https://doi.org/https://doi.org/10.1016/j.wasman.2017.08.046

Kurniawan, T.A., Sillanpää, M.E.T., Sillanpää, M., 2012. Nanoadsorbents for remediation of aquatic environment: Local and practical solutions for global water pollution problems.

Crit. Rev. Environ. Sci. Technol. 42, 1233–1295. https://doi.org/10.1080/10643389.201
	1.556553

Kwiatkowska, M., Huras, B., Bukowska, B., 2014. The effect of metabolites and impurities
	of glyphosate on human erythrocytes (in vitro). *Pestic. Biochem. Physiol.* 109, 34–43.
	https://doi.org/https://doi.org/10.1016/j.pestbp.2014.01.003

Lai, C.H., Chen, C.Y., 2001. Removal of metal ions and humic acid from water by iron-coated
	filter media. *Chemosphere.* 44, 1177–1184. https://doi.org/https://doi.org/10.1016/
	S0045-6535(00)00307-6

Lajis, A.F.B., 2018. Realm of thermoalkaline lipases in bioprocess commodities. *J. Lipids.*
	2018. https://doi.org/10.1155/2018/5659683

Lal, B., Khanna, S., 1996. Degradation of crude oil by *Acinetobacter calcoaceticus* and
	Alcaligenes odorans. J. Appl. Bacteriol. 81, 355–362.

Le, T.T., Murugesan, K., Lee, C.-S., Vu, C.H., Chang, Y.-S., Jeon, J.-R., 2016. Degradation of
	synthetic pollutants in real wastewater using laccase encapsulated in core–shell mag-
	netic copper alginate beads. *Bioresour. Technol.* 216, 203–210. https://doi.org/https://
	doi.org/10.1016/j.biortech.2016.05.077

Lee, C.C., Lin, S., 2000. *Handbook of Environmental Engineering Calculations.* McGraw-
	Hill Professional: New York, NY.

Lemaire, G., Terouanne, B., Mauvais, P., Michel, S., Rahmani, R., 2004. Effect of
	organochlorine pesticides on human androgen receptor activation in vitro. *Toxicol.
	Appl. Pharmacol.* 196, 235–246. https://doi.org/https://doi.org/10.1016/j.taap.2003.
	12.011

Lens, P., Virkutyte, J., Jegatheesan, V., Al-Abed, S., Kim, S.-H., 2013. *Nanotechnology for
	water and wastewater treatment.* IWA Publishing: London, UK.

Liang, M., Zhou, X., Xu, C., 2016. Systems biology in biofuel. *Phys. Sci. Rev.* 1, 1–10. https://
	doi.org/https://doi.org/10.1515/psr-2016-0047

Liao, Z., Zhi, G., Zhou, Y., Xu, Z., Rink, K., 2016. To analyze the urban water pollution dis-
	charge system using the tracking and tracing approach. *Environ. Earth Sci.* 75, 1–10.
	https://doi.org/10.1007/s12665-016-5881-1

Liebensteiner, M.G., Oosterkamp, M.J., Stams, A.J.M., 2016. Microbial respiration with chlo-
	rine oxyanions: Diversity and physiological and biochemical properties of chlorate-
	and perchlorate-reducing microorganisms. *Ann. N. Y. Acad. Sci.* 1365, 59–72. https://
	doi.org/10.1111/nyas.12806

Lively, R.P., Sholl, D.S., 2017. From water to organics in membrane separations. *Nat. Mater.*
	16, 276–279. https://doi.org/10.1038/nmat4860

Mackenbach, J.P., 2007. Global environmental change and human health: A public health
	research agenda. *J. Epidemiol. Community Heal.* 61, 92–94.

Maenpaa, K.A., 2007. *The Toxicity of Xenobiotics in an Aquatic Environment: Connecting
	Body Residues with Adverse Effects.* University of Joensuu: Finland.

Majumder, R., Banik, S.P., Ramrakhiani, L., Khowala, S., 2015. Bioremediation by alka-
	line protease (AkP) from edible mushroom *Termitomyces clypeatus*: Optimization
	approach based on statistical design and characterization for diverse applications.
	J. Chem. Technol. Biotechnol. 90, 1886–1896.

Marchut-Mikolajczyk, O., Drożdżyński, P., Struszczyk-Świta, K., 2020. Biodegradation of
	slop oil by endophytic *Bacillus cereus* EN18 coupled with lipase from *Rhizomucor
	miehei* (Palatase®). *Chemosphere.* 250, 126203.

Marican, A., Durán-Lara, E.F., 2018. A review on pesticide removal through different pro-
	cesses. *Environ. Sci. Pollut. Res.* 25, 2051–2064.

Martínez, J., Ortiz, A., Ortiz, I., 2017. State-of-the-art and perspectives of the catalytic and
	electrocatalytic reduction of aqueous nitrates. *Appl. Catal. B Environ.* 207, 42–59.
	https://doi.org/https://doi.org/10.1016/j.apcatb.2017.02.016

Martínez-Alcalá, I., Clemente, R., Bernal, M.P., 2012. Efficiency of a phytoimmobilisation strategy for heavy metal contaminated soils using white lupin. *J. Geochemical Explor.* 123, 95–100.

Mason, L.R., Shires, M.K., Arwood, C., Borst, A., 2017. Social work research and global environmental change. *J. Soc. Social Work Res.* 8, 645–672. https://doi.org/10.1086/694789

Mateo-Sagasta, J., Zadeh, S.M., Turral, H., Burke, J., 2017. *Water Pollution from Agriculture: A Global Review. Executive Summary.* The Food and Agriculture Organization of the United Nations Rome, 2017 and the International Water Management Institute, Colombo.

Mazumder, D., Mukherjee, S., 2011. Treatment of automobile service station wastewater by coagulation and activated sludge process. *Int. J. Environ. Sci. Dev.* 2, 64–69.

McDonald, R.I., Green, P., Balk, D., Fekete, B.M., Revenga, C., Todd, M., Montgomery, M., 2011. Urban growth, climate change, and freshwater availability. *Proc. Natl. Acad. Sci.* 108, 6312–6317. https://doi.org/10.1073/pnas.1011615108

McKinlay, R., Plant, J.A., Bell, J.N.B., Voulvoulis, N., 2008. Endocrine disrupting pesticides: Implications for risk assessment. *Environ. Int.* 34, 168–183. https://doi.org/https://doi.org/10.1016/j.envint.2007.07.013

Meckenstock, R.U., Elsner, M., Griebler, C., Lueders, T., Stumpp, C., Aamand, J., Agathos, S.N., Albrechtsen, H.-J., Bastiaens, L., Bjerg, P.L., Boon, N., Dejonghe, W., Huang, W.E., Schmidt, S.I., Smolders, E., Sørensen, S.R., Springael, D., van Breukelen, B.M., 2015. Biodegradation: Updating the concepts of control for microbial cleanup in contaminated aquifers. *Environ. Sci. Technol.* 49, 7073–7081. https://doi.org/10.1021/acs.est.5b00715

Meyer, A.M., Klein, C., Fünfrocken, E., Kautenburger, R., Beck, H.P., 2019. Real-time monitoring of water quality to identify pollution pathways in small and middle scale rivers. *Sci. Total Environ.* 651, 2323–2333. https://doi.org/https://doi.org/10.1016/j.scitotenv.2018.10.069

Mnif, W., Hassine, A.I.H., Bouaziz, A., Bartegi, A., Thomas, O., Roig, B., 2011. Effect of endocrine disruptor pesticides: A review. *Int. J. Environ. Res. Public Health* 8, 2265–2303. https://doi.org/10.3390/ijerph8062265

Mohammad, A.W., Teow, Y.H., Ang, W.L., Chung, Y.T., Oatley-Radcliffe, D.L., Hilal, N., 2015. Nanofiltration membranes review: Recent advances and future prospects. *Desalination.* 356, 226–254. https://doi.org/https://doi.org/10.1016/j.desal.2014.10.043

Monavari, S.M., Omrani, G.A., Karbassi, A., Raof, F.F., 2012. The effects of socioeconomic parameters on household solid-waste generation and composition in developing countries (a case study: Ahvaz, Iran). *Environ. Monit. Assess.* 184, 1841–1846. https://doi.org/10.1007/s10661-011-2082-y

Moodley, P., Roos, J.C., Rossouw, N., Heath, R., Coleman, T., 2011. Planning level review of water quality in South Africa, Sub-series No. WQP 2.0. Pretoria, South Africa.

Moslehyani, A., Goh, P.S., 2019. Recent progresses of ultrafiltration (UF) membranes and processes in water treatment, in: *Membrane Separation Principles and Applications.* Elsevier, Amsterdam, the Netherlands, 85–110.

Mouhoun-Chouaki, S., Derridj, A., Tazdaït, D., Salah-Tazdaït, R., 2019. A study of the impact of municipal solid waste on some soil physicochemical properties: The case of the landfill of Ain-El-Hammam Municipality, Algeria. *Appl. Environ. Soil Sci.* 2019, 3560456. https://doi.org/10.1155/2019/3560456

Shaikh, M. A., 2009. Water conservation in textile industry. *PTJ.* 58, 48–51.

Newman, M., 1998. *Fundamentals of Ecotoxicology.* Sleeping Bear/Ann Arbor Press, Chelsea.

Ngusale, G.K., Oloko, M., Agong, S., Nyakinya, B., 2017. Energy recovery from municipal solid waste. Energy Sources, Part A recover. *Util. Environ. Eff.* 39, 1807–1814. https://doi.org/10.1080/15567036.2017.1376007

NNI, 2010. Nanotech facts: What is nanotechnology? National Nanotechnology Initiative. https://www.nano.gov/

Oliveira, L.C.A., Petkowicz, D.I., Smaniotto, A., Pergher, S.B.C., 2004. Magnetic zeolites: A new adsorbent for removal of metallic contaminants from water. *Water Res.* 38, 3699–3704. https://doi.org/https://doi.org/10.1016/j.watres.2004.06.008

Oriakhi, C.O., 1998. Nano sandwiches. *Chem. Br.* 34, 59–62.

Pan, B., Pan, B., Zhang, W., Lv, L., Zhang, Q., Zheng, S., 2009. Development of polymeric and polymer-based hybrid adsorbents for pollutants removal from waters. *Chem. Eng. J.* 151, 19–29. https://doi.org/https://doi.org/10.1016/j.cej.2009.02.036

Park, J.-W., Park, B.-K., Kim, J.-E., 2006. Remediation of soil contaminated with 2,4-dichlorophenol by treatment of minced shepherd's purse roots. *Arch. Environ. Contam. Toxicol.* 50, 191–195. https://doi.org/10.1007/s00244-004-0119-8

Parvez, F., Wasserman, G.A., Factor-Litvak, P., Liu, X., Slavkovich, V., Siddique, A.B., Sultana, R., Sultana, R., Islam, T., Levy, D., Mey, J.L., van Geen, A., Khan, K., Kline, J., Ahsan, H., Graziano, J.H., 2011. Arsenic exposure and motor function among children in Bangladesh. *Environ. Health Perspect.* 119, 1665–1670. https://doi.org/10.1289/ehp.1103548

Prakash, D., Nawani, N., Prakash, M., Bodas, M., Mandal, A., Khetmalas, M., Kapadnis, B., 2013. Actinomycetes: A repertory of green catalysts with a potential revenue resource. *Biomed Res. Int.* 2013, 8. https://doi.org/10.1155/2013/264020

Przydatek, G., Kanownik, W., 2019. Impact of small municipal solid waste landfill on groundwater quality. *Environ. Monit. Assess.* 191, 169. https://doi.org/10.1007/s10661-019-7279-5

Rahman, A., Persson, L.Å., Nermell, B., Arifeen, S.E., Ekström, E.C., Smith, A.H., Vahter, M., 2010. Arsenic exposure and risk of spontaneous abortion, stillbirth, and infant mortality. *Epidemiology* 21, 797–804.

Ram, C., Pareek, R.K., Singh, V., 2012. Photocatalytic degradation of textile dye by using titanium dioxide nanocatalyst. *Int. J. Theoretical Appl. Sci.* 4, 82–88.

Ramanathan, T., Liu, H., Brinson, L.C., 2005. Functionalized SWNT/polymer nanocomposites for dramatic property improvement. *J. Polym. Sci. Part B Polym. Phys.* 43, 2269–2279. https://doi.org/10.1002/polb.20510

Rao, M.A., Scelza, R., Scotti, R., Gianfreda, L., 2010. Role of enzymes in the remediation of polluted environments. *J. Soil Sci. Plant Nutr.* 10, 333–353.

Reineke, W., Knackmuss, H.J., 1988. Microbial degradation of haloaromatics. *Annu. Rev. Microbiol.* 42, 263–287. https://doi.org/10.1146/annurev.mi.42.100188.001403

Rezania, S., Ponraj, M., Talaiekhozani, A., Mohamad, S.E., Din, M.F.M., Taib, S.M., Sabbagh, F., Sairan, F.M., 2015. Perspectives of phytoremediation using water hyacinth for removal of heavy metals, organic and inorganic pollutants in wastewater. *J. Environ. Manage.* 163, 125–133.

Ravichandran, A., Rao, R.G., Gopinath S., M., Sridhar, M., 2019. Purification and characterization of versatile peroxidase from *Lentinus squarrosulus* and its application in biodegradation of lignocellulosics. *J. Appl. Biotechnol. Bioeng.* 6, 280–286. https://doi.org/10.15406/jabb.2019.06.00205

Sadh, P.K., Duhan, S., Duhan, J.S., 2018. Agro-industrial wastes and their utilization using solid state fermentation: A review. *Bioresour. Bioprocess.* 5, 1. https://doi.org/10.1186/s40643-017-0187-z

Saggioro, E.M., Chaves, F.P., Felix, L.C., Gomes, G., Bila, D.M., 2019. Endocrine disruptor degradation by UV/chlorine and the impact of their removal on estrogenic activity and toxicity. *Int. J. Photoenergy.* 2019, 1–9.

Sahu, O.P., Chaudhari, P.K., 2015. Electrochemical treatment of sugar industry wastewater: COD and color removal. *J. Electroanal. Chem.* 739, 122–129. https://doi.org/https://doi.org/10.1016/j.jelechem.2014.11.037

Sakhel, S.R., Geissen, S.-U., and Vogelpohl, A., 2013. Virtual industrial water usage and wastewater generation in the Middle East/North African region. *Hydrol. Earth Syst. Sci. Discuss.* 10, 999–1039. https://doi.org/10.5194/hessd-10-999-2013

Sánchez-Porro, C., Martín, S., Mellado, E., Ventosa, A., 2003. Diversity of moderately halophilic bacteria producing extracellular hydrolytic enzymes. *Journal of Applied Microbiology.* John Wiley & Sons, Ltd, pp. 295–300. https://doi.org/10.1046/j.1365-2672.2003.01834.x

Sarayu, K., Sandhya, S., 2012. Current technologies for biological treatment of textile wastewater: A review. *Appl. Biochem. Biotechnol.* 167, 645–661. https://doi.org/10.1007/s12010-012-9716-6

Saxena, G., Kishor, R., Bharagava, R.N., 2020. Application of microbial enzymes in degradation and detoxification of organic and inorganic pollutants, in: Saxena, G., Bharagava, R.N. (Eds.), *Bioremediation of Industrial Waste for Environmental Safety: Volume I: Industrial Waste and Its Management.* Springer: Singapore, pp. 41–51. https://doi.org/10.1007/978-981-13-1891-7_3

Schadler, L., Kumar, S., Benicewicz, B., Lewis, S., & Harton, S., 2014. Clay–polymer nanocomposites (CPNs): Adsorbents of the future for water treatment. *Appl. Clay Sci.* 99, 83–92. https://doi.org/https://doi.org/10.1016/j.clay.2014.06.016

Schadler, L.S., Kumar, S.K., Benicewicz, B.C., Lewis, S.L., Harton, S.E., 2007. Designed interfaces in polymer nanocomposites: A fundamental viewpoint. *MRS Bull.* 32, 335–340. https://doi.org/DOI: 10.1557/mrs2007.232

Schwarzenbach, R.P., Egli, T., Hofstetter, T.B., von Gunten, U., Wehrli, B., 2010. Global water pollution and human health. *Annu. Rev. Environ. Resour.* 35, 109–136. https://doi.org/10.1146/annurev-environ-100809-125342

SDG Target 6.3. 2017. Water quality and wastewater. United Nations. https://www.sdg6monitoring.org/indicators/target-63/#:~:text=What%20and%20why%3F,of%20pollution%20into%20water%20bodies

Sharma, A., Sharma, T., Sharma, T., Sharma, S., Kanwar, S.S., 2019. Role of microbial hydrolases in bioremediation, in: Kumar, A., Sharma, S. (Eds.), *Microbes and Enzymes in Soil Health and Bioremediation.* Springer: Singapore, pp. 149–164. https://doi.org/10.1007/978-981-13-9117-0_7

Sharma, K.M., Kumar, R., Panwar, S., Kumar, A., 2017. Microbial alkaline proteases: Optimization of production parameters and their properties. *J. Genet. Eng. Biotechnol.* 15, 115–126. https://doi.org/https://doi.org/10.1016/j.jgeb.2017.02.001

Sharma, V., Sharma, A., 2013. Nanotechnology: An emerging future trend in wastewater treatment with its innovative products and processes. *Int. J. Enhanc. Res. Sci. Technol. Eng.* 2, 1–8.

Shekdar, A.V, 2009. Sustainable solid waste management : An integrated approach for Asian countries. *Waste Manag.* 29, 1438–1448. https://doi.org/10.1016/j.wasman.2008.08.025

Singh, S., 2015. Study of waste water effluent characteristics generated from paper industries. *J. Basic Appl. Eng. Res.* 2, 1505–1509.

Singh, C.K., Kumar, A., Roy, S.S., 2018. Quantitative analysis of the methane gas emissions from municipal solid waste in India. *Sci. Rep.* 8, 2913. https://doi.org/10.1038/s41598-018-21326-9

Singh, R.L., Singh, P.K., 2017. Global environmental problems, in: Singh, R.L. (Ed.), *Principles and Applications of Environmental Biotechnology for a Sustainable Future.* Springer: Singapore, pp. 13–41. https://doi.org/10.1007/978-981-10-1866-4_2

Sinha, A.K., Suzuki, K., Takahara, M., Azuma, H., Nonaka, T., Fukumoto, K., 2007. Mesostructured manganese oxide/gold nanoparticle composites for extensive air purification. *Angew. Chemie Int. Ed.* 46, 2891–2894. https://doi.org/10.1002/anie.200605048

Sirajuddin, A., Rathi, R.K., Chandra, U., 2019. Wastewater treatment technologies commonly practised in major steel Industries in India. Theme: Global and national policy processes on sustainable development. Proceedings of the 16th Annual International Sustainable Development Research Conference. 1–17.

Stocker, T.F., Qin, D., Plattner, G.K., Tignor, M., Allen, S.K., Boschung, J., Nauels, A., Xia, Y., Bex, V., Midgley, P.M., 2013. Climate change 2013: The physical science basis. Working Group I Contribution to the Fifth Assessment Report of the Intergovernmental Panel on Climate Change.

Streit, B., 1992. Bioaccumulation processes in ecosystems. *Experientia* 48, 955–970. https://doi.org/10.1007/BF01919142

Swanson, N.L., Leu, A., Abrahamson, J., Wallet, B., 2014. Genetically engineered crops, glyphosate and the deterioration of health in the United States of America. *J. Org. Syst.* 9, 6–37.

Theron, J., Walker, J.A., Cloete, T.E., 2008. Nanotechnology and water treatment: Applications and emerging opportunities. *Crit. Rev. Microbiol.* 34, 43–69. https://doi.org/10.1080/10408410701710442

Thongprakaisang, S., Thiantanawat, A., Rangkadilok, N., Suriyo, T., Satayavivad, J., 2013. Glyphosate induces human breast cancer cells growth via estrogen receptors. *Food Chem. Toxicol.* 59, 129–136. https://doi.org/https://doi.org/10.1016/j.fct.2013.05.057

Tiemann, U., 2008. In vivo and in vitro effects of the organochlorine pesticides DDT, TCPM, methoxychlor, and lindane on the female reproductive tract of mammals: A review. *Reprod. Toxicol.* 25, 316–326. https://doi.org/https://doi.org/10.1016/j.reprotox.2008.03.002

Topare, N.S., Joy, M., Joshi, R.R., Jadhav, P.B., Kshirsagar, L.K., 2015. Treatment of petroleum industry wastewater using TiO2/UV photocatalytic process Treatment of petroleum industry wastewater using TiO_2/UV photocatalytic process. *J. Indian Chem. Soc.* 92, 219–222.

Tortora, F., Innocenzi, V., Prisciandaro, M., De Michelis, I., Vegliò, F., Mazziotti di Celso, G., 2018. Removal of tetramethyl ammonium hydroxide from synthetic liquid wastes of electronic industry through micellar enhanced ultrafiltration. *J. Dispers. Sci. Technol.* 39, 207–213. https://doi.org/10.1080/01932691.2017.1307760

Toufexi, E., Tsarpali, V., Efthimiou, I., Vidali, M.-S., Vlastos, D., Dailianis, S., 2013. Environmental and human risk assessment of landfill leachate: An integrated approach with the use of cytotoxic and genotoxic stress indices in mussel and human cells. *J. Hazard. Mater.* 260, 593–601. https://doi.org/10.1016/j.jhazmat.2013.05.054

Tripathi, N., Hills, C.D., Singh, R.S., Atkinson, C.J., 2019. Biomass waste utilisation in low-carbon products: harnessing a major potential resource. *NPJ Clim. Atmos. Sci.* 2, 35. https://doi.org/10.1038/s41612-019-0093-5

UNEP, 2016. *A Snapshot of the World's Water Quality: Towards a Global Assessment.* United Nations Environment Programme (UNEP): Nairobi.

UNESCO, 2017. World Water Assessment Programme (UNESCO WWAP). http://www.unesco.org/new/en/natural-sciences/environment/water/wwap/

United Nations, 2014. Department of Economic and Social Affairs, Population Division. World Urbanization Prospects: The 2014 Revision, Highlights (ST/ESA/SER.A/352). https://population.un.org/wup/

United Nations, 2019. Department of Economic and Social Affairs, Population Division. World Population Prospects 2019: Highlights (ST/ESA/SER.A/423). https://population.un.org/wpp/

Uqab, B., Mudasir, S., Nazir, R., 2016. Review on bioremediation of pesticides. *J. Bioremediat. Biodegrad.* 7, 2, 343.

Vaish, B., Srivastava, V., Singh, P., Singh, A., Singh, P.K., Singh, R.P., 2016. Exploring untapped energy potential of urban solid waste. *Energy Ecol. Environ.* 1, 323–342. https://doi.org/10.1007/s40974-016-0023-x

Valko, M., Jomova, K., Rhodes, C.J., Kuča, K., Musílek, K., 2016. Redox-and non-redox-metal-induced formation of free radicals and their role in human disease. *Arch. Toxicol.* 90, 1–37.

Valta, K., Kosanovic, T., Malamis, D., Moustakas, K., Loizidou, M., 2015. Overview of water usage and wastewater management in the food and beverage industry. *Desalin. Water Treat.* 53, 3335–3347. https://doi.org/10.1080/19443994.2014.934100

Vandermaesen, J., Horemans, B., Bers, K., Vandermeeren, P., Herrmann, S., Sekhar, A., Seuntjens, P., Springael, D., 2016. Application of biodegradation in mitigating and remediating pesticide contamination of freshwater resources: State of the art and challenges for optimization. *Appl. Microbiol. Biotechnol.* 100, 7361–7376. https://doi.org/10.1007/s00253-016-7709-z

Vasanthi, P., Srinivasaraghavan, R., Prasad, P., 2013. Ground water contamination fue to solid waste disposal: A solute transport model based on Perungudi Dumpyard, Chennai, India BT – On a sustainable future of the Earth's natural resources, in: Ramkumar, M. (Ed.) Springer: Berlin, Heidelberg, pp. 425–434. https://doi.org/10.1007/978-3-642-32917-3_24

Verma, J.P., Jaiswal, D.K., 2016. Book review: Advances in biodegradation and bioremediation of industrial waste. *Front. Microbiol.* 6, 1555.

Verma, S., Kuila, A., 2019. Bioremediation of heavy metals by microbial process. *Environ. Technol. Innov.* 14, 100369.

Verstraeten, W.W., Neu, J.L., Williams, J.E., Bowman, K.W., Worden, J.R., Boersma, K.F., 2015. Rapid increases in tropospheric ozone production and export from China. *Nat. Geosci.* 8, 690–695. https://doi.org/10.1038/ngeo2493

Vitorino de Souza Melaré, A., Montenegro González, S., Faceli, K., Casadei, V., 2017. Technologies and decision support systems to aid solid-waste management: A systematic review. *Waste Manag.* 59, 567–584. https://doi.org/https://doi.org/10.1016/j.wasman.2016.10.045

Wang, H., Liu, W., Hao, R., 2019. Lipase from Bacillius cereus: A potential solution to alleviate dietary oil pollution, in: *IOP Conference Series: Earth and Environmental Science.* IOP Publishing: UK, p. 52067.

Wang, Y., Bian, Y., Xu, H., 2015. Water use efficiency and related pollutants' abatement costs of regional industrial systems in China: a slacks-based measure approach. *J. Clean. Prod.* 101, 301–310. https://doi.org/https://doi.org/10.1016/j.jclepro.2015.03.092

Wanyonyi, W.C., Mulaa, F.J., 2019. Alkaliphilic enzymes and their application in novel leather processing technology for next-generation tanneries. *Adv. Biochem. Eng. Biotechnol.* 172, 195–220.

Wasserman, G.A., Liu, X., Parvez, F., Ahsan, H., Factor-Litvak, P., van Geen, A., Slavkovich, V., LoIacono, N.J., Cheng, Z., Hussain, I., Momotaj, H., Graziano, J.H., 2004. Water arsenic exposure and children's intellectual function in Araihazar, Bangladesh. *Environ. Health Perspect.* 112, 1329–1333. https://doi.org/10.1289/ehp.6964

Watlington, K., 2005. Emerging Nanotechnologies for Site Remediation and Wastewater Treatment. Washington DC: Environmental Protection Agency.

Waziri, S.A., Nain, K.S., Ram, C., 2017. Application of activated carbon in the treatment of domestic effluent : A comparative analysis. *Int. J. Earth Sci. Eng.* 10, 435–441. https://doi.org/10.21276/ijee.2017.10.0246

Wen, Z., Di, J., Zhang, X., 2016. Uncertainty analysis of primary water pollutant control in China's pulp and paper industry. *J. Environ. Manage.* 169, 67–77. https://doi.org/https://doi.org/10.1016/j.jenvman.2015.11.061

White, S.S., Birnbaum, L.S., 2009. An overview of the effects of dioxins and dioxin-like compounds on vertebrates, as documented in human and ecological epidemiology. *J. Environ. Sci. Health. C. Environ. Carcinog. Ecotoxicol. Rev.* 27, 197–211. https://doi.org/10.1080/10590500903310047

WHO/UNICEF, 2015. Joint Water Supply, Sanitation Monitoring Programme, and World Health Organization. *Progress on sanitation and drinking water: 2015 update and MDG assessment.* World Health Organization, Geneva, Switzerland.

Wong, D.W.S., 2009. Structure and action mechanism of ligninolytic enzymes. *Appl. Biochem. Biotechnol.* 157, 174–209. https://doi.org/10.1007/s12010-008-8279-z

WWAP, 2017. Wastewater, the untapped resource. United Nations World Water Assessment Programme (WWAP), Paris.

Yadav, M., Sehrawat, N., Kumar, A., 2018. Microbial laccases in food processing industry: Current status and future Perspectives. *Research Journal of Biotechnology.* 13, 108–113.

Yadav, M., Yadav, H.S., 2015. Applications of ligninolytic enzymes to pollutants, wastewater, dyes, soil, coal, paper and polymers. *Environ. Chem. Lett.* 13, 309–318. https://doi.org/10.1007/s10311-015-0516-4

Yamjala, K., Nainar, M.S., Ramisetti, N.R., 2016. Methods for the analysis of azo dyes employed in food industry – A review. *Food Chem.* 192, 813–824. https://doi.org/https://doi.org/10.1016/j.foodchem.2015.07.085

Yang, R.T., 2003. *Adsorbents: Fundamentals and Applications.* John Wiley & Sons Inc.: Hoboken, NJ.

Yang, Z., Zhou, Y., Feng, Z., Rui, X., Zhang, T., Zhang, Z., 2019. A review on reverse osmosis and nanofiltration membranes for water purification. *Polymers (Basel).* 11, 1252. https://doi.org/10.3390/polym11081252

Yaqoob, A.A., Parveen, T., Umar, Khalid, M.N.M.I., 2020. Role of nanomaterials in the treatment of wastewater: A review. *Water* 12, 495–525.

Yunus, I.S., Harwin, Kurniawan, A., Adityawarman, D., Indarto, A., 2012. Nanotechnologies in water and air pollution treatment. *Environ. Technol. Rev.* 1, 136–148. https://doi.org/10.1080/21622515.2012.733966

Zekić, E., Vuković, Ž., Halkijević, I., 2018. Application of nanotechnology in wastewater treatment. *GRAĐEVINAR* 70, 315–323.

Zhang, C., Xu, T., Feng, H., Chen, S., 2019. Greenhouse gas emissions from landfills: A review and bibliometric analysis. 11, 1–15.

Zhao, D., Yu, S., 2015. A review of recent advance in fouling mitigation of NF/RO membranes in water treatment: Pretreatment, membrane modification, and chemical cleaning. Desalin. *Water Treat.* 55, 870–891. https://doi.org/10.1080/19443994.2014.928804

2 Municipal Solid Waste Management
Recent Practices

Chhotu Ram
Adigrat University

Amit Kumar
Debre Markos University

CONTENTS

2.1 INTRODUCTION

Worldwide waste is an issue. Municipal solid waste (MSW) management issue is more prominent in middle and low-income countries where population and urbanization are increasing rapidly (Pervez and Kafeel, 2013; Tefera and Negussie, 2015). Solid waste (SW) includes materials such as bottles, packages, newspapers, leftovers, food residue, batteries, equipment, and dyes, etc., which are produced as a result of everyday activities (Ekmekçioğlu et al., 2010). MSW management includes all functional elements from SW generation to final disposal (Parvez et al., 2019). For instance, as per the World Bank report (2019), the MSW generated globally is expected to increase by 2.01 billion tons in2016 to 3.40 billion tons in 2050. One recent (2016) estimation shows that world's cities generated around 2.01 billion tons of SW, which further accounts 0.74 kilograms per capita per day (World Bank Data, 2019). Yang et al. (2018) reported that the limited financial resources of local municipalities to combat with the advanced solid waste management (SWM) provisions especially in the developing world. However, the quantity of SW generation increases with the technological advancements, increase in population, and public life styles (Ali, 2015; Bello, 2018; Monavari et al., 2012; Sankoh et al., 2013). Ferronato and Torretta (2019) show comparisons that the developed world generates more MSW than the developing countries. Thus, the developing countries are facing more problems and challenges in MSW management. Maximum per capita SW-generating countries include Canada, the U.S.,. Switzerland, Denmark, Australia, Ireland, Israel, and others (OECD, 2013). The World Bank report about waste generation prediction in 2050 indicates that East Asian and Pacific regions show the highest whereas the Middle East and North Africa show the lowest waste generation (Pervez and Kafeel, 2013). India being a populated country generates nearly 62 million tons (MT) of MSW annually, which creates significant environmental problems without proper waste management. One latest report (Lahiry, 2019) indicates that out of a total of 62 MT per annum of MSW generated, only 43 MT (69.35%) of the waste has been collected. Out of the 43 MT of MSW, 11.9 MT (27.67%) has been treated and 31 MT (72.09%) has been dumped in the sites. Thus, the developed and developing worlds are required to manage and dispose of the MSW properly without affecting our surrounding environment. MSW generation and its management have become a major issue in many developing countries (Maria et al., 2018). The final disposal of waste is still greatly based on uncontrolled dumping and/or littering, and burning it through domestic ways in developing towns or cities. This mismanagement leads to serious health and environmental risks (Guerrero et al., 2013). Few reports (Lohri et al., 2014; Memon, 2010) suggest the scenario from developing countries that, maximum municipalities spend 20–50% of their financial budget on SW manage- ment, which covers less than the 50% of total population. Thus, the overall goal of SW management is the proper collection, segregation, treatment, and final disposal of SW generated.

Recent studies presented from Ethiopia (Africa) indicated that a significant part of (~30–60%) of the urban SWs generated is uncollected with less than 50% of the total population served (Monyoncho, 2013). Kassa (2010) also indicated that SW management is the key issue in urban areas and affects the public health and envi- ronmental concern; only 2% of the total population received waste collection ser- vices. One of the earlier estimation from the U.S. showed that the annual amount of MSW increased between 1960 and 2015, approximately 262.4 MT of SW was gener- ated in 2015 (Wang, 2019). A report from Canada indicates about 965 kg per capita SW generation rate in the 2010. Thus, Canada is considered one of the maximum SW producing countries in the world with 32,947,000 tons of total wastes (Wang et al., 2016). The large amount of SW has the potential to contaminate water resources, increase disease transmission and greenhouse gas emissions, damage ecosystem, and discourage tourism and other business activities in the world (Chinasho, 2015; Miezah et al., 2015; Pervez and Kafeel, 2013; Thanh, et al., 2011).

In several countries, waste management is the municipalities' major responsibility and sometimes, due to various reasons, results in the inadequate service provisions (Bewketu, 2013; Hagos et al., 2012). The U.S. Public Health Service department found that around 22 human diseases are connected with the improper SW management systems. Furthermore, it can correlate with the poor sanitation and waste management systems in the developing countries, which neither adequate nor acceptable levels of practice in the waste handling and disposal systems are done (USAID, 2015). The current techniques in practice are pyrolysis, incineration, bio-refining composting, biogas plants, recycling, and secured landfill facilities for the treatment, processing, and final disposal of MSW. Therefore, this chapter has been planned to provide a systematic and in-depth study on the status and recent practices used for MSW generation to the final disposal. Further, this book will be helpful in the conservation of natural resources, environment, and public health.

2.2   GENERATION OF MUNICIPAL SOLID WASTE

MSW generation is growing with the increase in population and change in public lifestyle. Several factors affecting the SW generation rate have been suggested and reported by various authors. Any waste which is produced by the commercial, household, and institutional activities is well defined as MSW (Agarwal et al., 2015). SW is categorized mainly into the three types: household or residential levels that arise from domestic areas from individual houses, commercial and institutional wastes that arise from individually larger sources like schools, hotels, office buildings, etc. The waste coming from area sources like streets, parks, etc. is considered municipal services waste. In general, SW includes mainly food waste, plastics, textiles, paper, glass, cardboard, metals, street sweeping, wood, metals, landscape and tree trimmings, waste from recreational activities and other sources (Abas and Wee, 2014). Table 2.1 represents the sources and different types of components in

TABLE 2.1

Sources, Types and Components of Municipal Solid Waste (Daniel and Thomas, 1999)

Sources	Typical Waste Generators	Components of Solid Waste
Residential	Single or multifamily dwellings	Food wastes, paper, cardboard, plastics, textiles, glass, metals, ashes, special wastes (bulky items, consumer electronics, batteries, oil, tires), and household hazardous wastes
Commercial	Stores, hotels, restaurants markets office buildings	Paper, cardboard, plastics, wood, food, wastes, glass, metals, and hazardous wastes
Municipal services	Street cleaning, landscaping parks, beaches, recreational areas	Street sweepings, landscape and tree trimmings, general wastes from parks, beaches, and other recreational areas
Institutional	Schools, hospitals, prisons, government center	Paper, plastics, cardboard, wood, food waste, metals, hazardous wastes

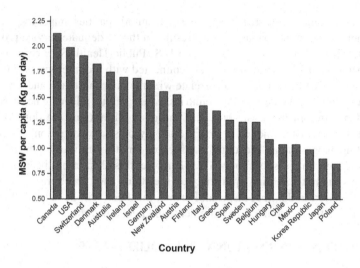

FIGURE 2.1   MSW generation per capita in OECD member countries. (OECD, 2013.)

MSW generation in various sectors (Abas and Wee, 2014). In this connection, Kawai and Tasaki's (2016) report presents global data on MSW per capita generation in various nations. Another study presented the SW generation in the Organization for Economic Co-operation and Development (OECD) member countries that are shown in Figure 2.1 (OECD, 2013). The OECD countries report indicates that Canada generates the highest per capita waste generation whereas Poland generates the least SW. A gradual increase in the amount of MSW leads to various problems in transportation, storage, and disposal of MSW and thus, making inefficient waste management practices. The important aspect of MSW management is the proper waste collection. Thus, Figure 2.2 shows the collection efficiency of SW in different cities in the world (UN-Habitat, 2010). It is also observed that low waste collection efficiencies were observed in the low-income countries. In fact, the observation shows the variations in waste collection efficiency within the nations in developing countries. The metro cities achieve a higher collection rate, whereas smaller cities are behind in the collection rate (Talyan et al., 2008). Present waste generation data shows that approximately 1.3 billion tons per year of SW is produced from major cities globally and data further predicts the increase in volume of waste to 2.2 billion tons by 2025. In low-income countries, the waste production rate will be more than double in the next two decades. An estimation indicates the management of SW costs will increase from 2020's annual $205.4 billion to about $375.5 billion in 2025. However, the low-income countries will be more pronounced in cost increases (more than 5-fold increases) as compared to lower-middle-income countries (more than 4-fold increases) (Hoornweg and Bhada-Tata, 2012). One study conducted in Ethiopia revealed that the present system of MSW management relied entirely on the municipality and involved the waste collection, transportation, and disposal service.

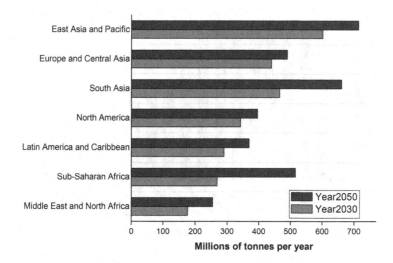

FIGURE 2.2 Collection efficiency of MSW in 20 municipalities around the world. (UN-Habitat, 2010.)

Around 23% of the total waste generation is collected and disposed at the town disposal site (Cheru, 2011).

Kumar and Pandit (2013) presented that waste generation in small cities is lesser than larger cities. The Tata Energy Research Institute, New Delhi, India, investigation depicts that higher income groups generate more SW than middle- and lower-income groups; hence, the economic status of families. The lower-income groups in New Delhi generate less than one-third of the SW than their higher-income counterparts. Thus, it can be concluded that the SW generated quantity is influenced by the urbanization, economic conditions, living standards (Liu and Wu, 2010; Saeed et al., 2009), and population (Chiemchaisri et al., 2007). Asia and African continents are evidence of the dramatic increase in population in urban areas (United Nations (undated), 2009) As a consequence, the waste quantity generated is also growing rapidly. In the current scenario, urban areas constitute 54% of the total world population, and, according to one estimation, it will increase to 66% or more by year 2050. Further, this increasing trend will add 2.5 billion people to the urban population in the world (United Nations, 2014). One report regarding waste management in urban centers reports that more than 70% of the daily waste generated is left near the houses, in markets, streets, or drainage channels (Boateng et al., 2019). Another study indicates that around 42% of total waste generated is illegally dumped at the roadsides and open areas in Ethiopia (Erasu, 2018). With increasing waste problems the existing waste infrastructure is not suitable for its proper management. Figure 2.3 indicates the latest World Bank data (2019) waste generation prediction up to 2050 for different regions in the world. East Asian and Pacific regions show highest; whereas, the Middle East and North Africa shows the lowest generation.

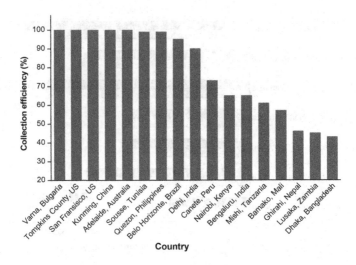

FIGURE 2.3   Projected waste generated by different regions in the world. (Silpa et al., 2018.)

2.3   COMPOSITION OF MUNICIPAL SOLID WASTE

Total MSW generation, per capita generation and its composition differ not only from country to country but also from region to region and even within the city. The MSW generation rate in terms of quantity is provided in the above section.

The current section reviews the composition of the MSW in different countries. The SW composition plays a significant role in waste management practices. It was observed from various studies from different regions that more than 50% of the generated SW is organic in nature, except from Europe, Central Asia, and North America, where waste contains higher amounts of dry waste as shown in Figure 2.4.

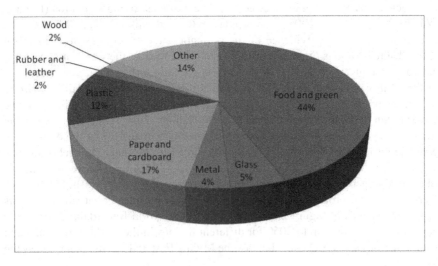

FIGURE 2.4   Global waste composition. (Silpa et al., 2018.)

The lowest waste generations indicated are rubber, leather, and wood (Figure 2.4). Furthermore, Table 2.2 shows the relative types of waste composition in current and future predictions by income level.

2.3.1 Physical Composition of Waste

The scientific study correlated to the composition of MSW includes physical and chemical characteristics which plays a significant role in the deciding of suitable waste management technique (Katiyar et al., 2013). To analyze the proper composition of SW, well-controlled designated disposal sites were used to estimate the weight and unloading of collected waste samples. MSW compositions were analyzed and performed as per the American Society of Testing and Materials (ASTM) standard procedure (Hla and Roberts, 2015). The routine used is to collect representative samples (100–200 kg) during the week from a residential area. Another method, i.e. reduction method, was used for large waste generation sources to get a representative sample of approximately 200 kg and if the sample volume is on the lower side, direct segregation was used. The SW sample is mixed thoroughly to prepare a homogenous and representative sample. Furthermore, the sample is divided into four piles of equal volume followed by the removal of two parts of waste at the diagonal opposite ends and mixing the remaining amount. One research (John et al., 2006) analyzed the organic content for the macronutrient and physical constituents of the samples taken in Nigeria. The waste was sorted to know the physical components were organic materials (73.7%), polythene/plastics (12.9%), metals/cans (4.3%), glasses/bottles (4.2%), shoes/clothes (4.3%), and ceramics (0.6%).

Authors categorize MSW into six categories, textiles, paper, food residue, plastics, wood waste, and rubber (Zhou et al., 2014). Another study (Srivastava et al., 2014) shows the characterization of SW from Varanasi, India. MSW compositions are food waste (31.9%), plastic waste (22%), textile (10.6%), paper (9.6%), glass (6.7%), cardboard (6.2%), leather (5.7%), ash (5.3%), and metal waste (2.8%), which is minimum. The physical composition of MSW produced from several countries is shown in Table 2.3.

2.3.2 Chemical Composition of Waste

As far as the role of the physical constituents of SW is important, similarly, chemical characterization also has a great concern in deciding the management strategy. Thus, various studies have been reported so far to evaluate the chemical composition of SW in several countries and among several cities within the country. The study investigated the chemical composition for moisture content, carbon, nitrogen, and a three-component analysis. A total of 36 samples were analyzed for different parameters; moisture content, C:N ratio, average value of combustible content, and average ash content were reported as 54.8%, 109.55%, 86.2%, and 13.5%, respectively (Nadeem et al., 2016). Another laboratory analysis (John et al., 2006) from the African continent (Nigeria) shows that the organic constituents of SW indicate potassium, nitrogen, and phosphorous content 10.7, 11, and 3.2 g kg^{-1}, respectively. Other content elements investigated were calcium (87.7 g kg^{-1}), sulfur (2.3 g kg^{-1}),

TABLE 2.2
Relative Types of Waste Composition in Current and Future Predictions By Income Level

Income Level	Organic (%)		Paper (%)		Plastic (%)		Glass (%)		Metal (%)		Other (%)	
	Current	2025	Current	2025	Current	2025	Current	2025	Current	2025	Current	2025
Low income	64	62	5	6	8	9	3	3	3	3	17	17
Lower middle income	59	55	9	10	12	13	3	4	2	3	15	15
Upper middle income	54	50	14	15	11	12	5	4	3	4	13	15
High income	28	28	31	30	11	11	7	7	6	6	17	18

Source: Hoornweg and Bhada-Tata (2012).

TABLE 2.3
Physical Composition of Municipal Solid Waste in Different Countries

Composition (%)	Kenya (Ngau and Kahiu, 2009; Njoroge, et al., 2014)	U.S. (U.S. Environmental Protection Agency, 2013)	Egypt (Ibrahim and Mohamed, 2016)	Jordan (MMA, 2015; Saidan et al., 2017)	China (Zhou et al., 2014)	Canada (Assamoi and Lawryshyn, 2012)	Mexico (Moreno et al., 2013)	Nigeria (Ogwueleka, 2009)	India (Srivastava et al., 2014)
Organics	50.9	–	79	84.77	81.63	65	49.5	49.78	–
Food	–	15	56	52.7	55.86	27	–	–	31.9
Wood	–	–			2.94	2	0.45	–	–
Textile	–	–	10	10.22	3.16	2	3.64	–	10.6
Paper	17.5	27	13	8	8.52	26	5.89	12.79	9.6
Plastic	16.1	13	13	12.85	11.15	8	13.16	8.4	22
Glass	2	4	4	1.25	18.37	35	2.65	3.75	–
Metals	2	9	2	4.82			–	5.48	2.8
Ash and others	11.4	3	15	9.16			4.19	18.04	5.3

Source: MMA: Ministry of Municipal Affairs (2015).

TABLE 2.4

Chemical Composition of Municipal Solid Waste in Different Countries

Constituents	Bangladesh (Sarkar and Bhuyan, 2018)	Ireland (Yusuff et al., 2014)	India (Thitame et al., 2010)	Indonesia (Pasek et al., 2013)	Ethiopia (Kebede et al., 2018)	South Africa (Ayeleru, et al., 2016)	China (Zhou et al., 2014)
Moisture content	91.21	16.36	38.5	8.04	41.9	63.47	–
Fixed residue	87.13	–	–	18.5	14.56	–	0–26.93
Organic carbon	20.83	38.7	40.2	42.2	–	45.03	–
Organic nitrogen	0.39	1.2	0.73	–	–	1.98	–
Phosphorus	0.03	–	0.93	–	–	–	–
Potassium	Nil	–	0.35	–	–	–	–
Hydrogen	–	7.2	–	5.79	–	–	4.1–12.7
Oxygen	–	48.1	–	43.69	–	41.16	1–56.9
Ash	–	4.6	–	7.21	16.9	5.56	0–18.44
Sulfur	–	0.2	–	0.25	–	0	0–0.87

and sodium (18.4 g kg^{-1}). One report investigated the analysis related to the chemical characterization of the SW which showed that polyethylene (PE), polypropylene (PP), and polystyrene (PS) has the highest volatile matter with almost no ash and fixed carbon. Plastic, especially polyethylene terephthalate (PET), contain low hydrogen and high carbon content. Polyvinyl chloride has the highest value of chlorine content at around 55%. PP, PS, and PE have the highest values, followed by chemical products such as rubber and chemical fiber (Zhou et al., 2014). Table 2.4 shows the chemical composition of municipal SW in different countries.

2.4 MUNICIPAL SOLID WASTE MANAGEMENT RECENT STRATEGIES

MSW generation rate is continuously increasing throughout the world and the environmental deterioration is becoming a serious issue. Various researchers are going in this direction to combat the pollution problem (Henry et al., 2006; Huisman et al., 2016). The latest World Bank Data (2019) shows that the worldwide SW generation was around 2.01 billion tons in year 2016. This further calculated as 0.74 kilograms per capita/per day waste generation. It can also conclude that the problems in developing countries are more severe than the in developed countries because of unsustainable waste management practices. The previous report from low-income countries has also shown that more than 90% of SW is often disposed in an unregulated manner or openly burned (World Bank, 2019). Several studies have been reported from the developing world's MSW management challenges. For example, waste collection in Kenya has always been on the low side (less than 40%), on average, during a period of the total waste generated in Nairobi. Thus, MSW is disposed of in open dumping areas that cause environmental pollution due to the lack of pollution control measures and poor SW disposal systems, which were found to be the major source of environmental pollution in the cities (Henry et al., 2006). One study from Nepal shows that 35–45% of the total waste generated is collected (Gurung and Oh, 2012). Various studies were performed in other countries and similar collection efficiency by local municipalities were observed in other geographical locations such as Ethiopia (Getahun et al., 2012), Vietnam (Omran and Gavrilescu, 2008), Iran (Ghafour et al., 2013), and India (Vishwakarma et al., 2012). The reason suggested is mainly due to the lack of technology, poor infrastructure, and financial capacity; these are the leading challenges for an inadequate waste management system. These practices serve as a breeding ground for the disease vectors, contribute to greenhouse gases generation, and can even promote urban violence (World Bank, 2019). Further, the proper SW management system is essential for building the sustainable and livable cities, but it will remain a major challenge for the developing world and cities. Another important aspect is the finance for effective SW management which is expensive and often comprising 20–50% of municipal budgets. Therefore, municipal services require an integrated system that is well efficient, socially supported, and sustainable. MSW management requires appropriate financial resources, infrastructure, and technology (Emmanuel and Jiquan, 2019). Thus, in this connection, several methods and techniques are proposed, such as informal and formal recycling, biological treatment, aerobic composting, small scale biomethanation, waste-to-energy (WtE) combustion, refuse derived fuel (RDF) and landfill. WtE is a very attractive. which has advantages in waste volume reduction and power generation as a by-product.

2.4.1 GLOBAL WASTE TREATMENT AND DISPOSAL SCENARIO

The landfill used in high- and upper-middle-income countries operated more stringently whereas open dumping is more preferred in low-income countries. Low-income countries dump approximately 93% of the SW, whereas high-income countries dump only 2% of their total SW generated.. World is producing a huge amount SW which can fill a line of trash truck 5,000 kilometers long every day (Cohen et al., 2015). Three regions, mainly the Middle East and North Africa, Sub-Saharan Africa, and South Asia, dump their half of the total SW generated. The estimation has shown that 1.6 billion tons of carbon dioxide was produced from the total SW management system in 2016 which is approximately 5% of the global emissions (Cohen et al., 2015). In upper-middle-income countries, 54% of waste goes into landfills, which is considered the highest compared to high- and low-income countries. In comparison with high-income countries, this rate decreases to 39%, with the diversion of 17% of waste to recycling, composting, and 22% to incineration. The emissions resulted from SW are anticipated to increase to 2.38 billion tons of CO_2-equivalent per year by 2050 if no improvements are made in the sector (Lardinios and Klundert, 1997). Thus, the composition of waste describes the waste

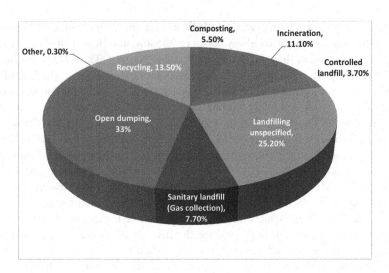

FIGURE 2.5   Global solid waste processing and disposal. (World Bank Data, 2019.)

constituents, and this varied widely from place to place, as is evident from Tables 2.3 and 2.4. Figure 2.5 shows the overall SW processing and disposal throughout the world. The current scenario indicates that SW is mostly openly dumped (33%) and disposed of in the form of a landfill (~37%) along with the sanitary landfill having a gas collection system. Around 13.5% of waste is recycled, 5.5% is recovered through composting, and the remaining 11% is incinerated for the final disposal (World Bank, 2019).

2.4.2   Resource Recovery

Waste reduction, reuse, and recycling are the three methods that have always been preferred options for effective waste management. By following these options, we have various environmental advantages, such as reduction in greenhouse gas emission and contaminants, reduction in landfill areas, preservation of natural resources, energy saving, and reduction in the demand for waste treatment technology. Therefore, various techniques should be adopted by any municipality and integrated into the waste management planning depending upon waste types as described in Table 2.5. The sustainable waste management hierarchy developed by the Earth Engineering Center at Columbia University is widely accepted and used as a reference for the sustainable SW management and disposal (Annepu et al., 2012). One study presented that open burning and unscientific landfilling were not considered as the original hierarchy of SW management that should be end up with the sanitary landfills. Many developing countries including India are facing the same issue regarding waste management, such as unsanitary landfilling and open burning. It is also shown that most are environmentally friendly approaches, i.e. reduction in the materials use and reusing them into the hierarchy of waste management. The source reduction begins with the decrease in the amount of waste generated and reusing the waste further prevent materials entering into the waste stream (Bhada and Themelis, 2008). Hence, waste

TABLE 2.5

Waste Streams and Potential Treatment/Management Options

Waste Stream/ Treatment Options	Organic: Kitchen Waste, Market Waste, Garden Waste	Plastic and Rubber: Film, Rigid, Other Related Plastic Waste	Paper: Paper and Paperboard	Glass	Metal: Steel, Aluminum	Others: Tetrapacks, Diapers, Non-Ferrous Metal
Composting vermicomposting	+	-	-	-	-	-
Recycling	-	+	+	+	+	-
Incineration with energy recovery	-	+	+	-	-	+
Pyro-gasification: pyrolysis, gasification, anaerobic digestion	+	-	-	-	-	-
Landfilling: sanitary bioreactor	+	+	+	+	-	+

Source: (Emmanuel and Jiquan, 2019).

is not generated until the end of the reusing phase. For example, in Hong Kong in the year 2006, the government provided the financial incentive of recycling financial incentives for the construction and demolition waste generators in order to reduce waste and encourage recycling. The estimation shows the reduction in the construction waste sent to an average landfill of 7,890 tons (1999) every day was reduced to 3,300 tons (2013) per day (Yau, 2010).

The MSW materials recovery is enhanced by using the most effective way of recycling and composting process (Abdel-Shafy and Mansour, 2018). In Indian scenario, most of the SW generated ends up in the landfills. This is mainly due to some technical and economical limitations in recycling, product design, inadequate source separation, and lack of proper markets for sorting and insufficient capacity. The improvement can be done by awareness among the local authorities and working for the separation of waste at source. This could be achieved by increasing recycling and provisions should be made to handle the non-recyclable waste. Thus, energy recovery potential from non-recyclable waste is a sustainable way and this option falls under the materials recovery category. The use of landfilling for MSW is considered a misuse of natural resources which could be used as source of energy or as secondary raw materials. However, in current scenario, a small fraction of MSW could be utilized to be landfilled. In fact, open dumping or unsanitary landfilling of waste are not considered a sustainable options to handle the MSW and its not recommended anywhere.

2.4.2.1   Recycling of Non-Biodegradable Waste

Reuse and reduction in waste generation are considered the most effective ways to reduce the amount of SW generation. Recycling is an important aspect at the source of waste generation and collection, and the best way to handle this is, the materials need to undergo a chemical transformation. Another way leading to a reduction in waste generation is reusing at the source. Just after collection by the traders of the materials from households, they further re-shape and repair them and sell in the open market. It is well established that 95% of a product's environmental impact occurs before it is discarded, mainly during extraction of raw materials, processing and manufacturing. Therefore, the concept of recycling is important in decreasing the overall impact during the life cycle of a product on the environment and public health. However, recycling is better taken place due to the source segregation of wastes and further waste mixing at the source makes process difficult to separate waste components. Recycling materials from MSW can be separated manually in low- and middle-income countries. Thus, the resource recovery from mixed MSW generates employment and provides the livelihood for the marginalized urban population. In developed countries, well-developed machinery systems are used to solve the purpose. The recyclable waste needs to be collected as a separate dry waste without mixing with the biodegradable food waste. Hence, the segregated items such as paper, glass, plastics, and metal could be further recycled. Further, it is very difficult to separate 100% of the components of MSW, which is highly time-consuming and energy-intensive thus, usually not carried out. Another issue associated with mixed waste is that it always produces some waste fraction residues, which can neither be composted nor recycled. These are combusted in WtE plants where refuse-derived fuel can be prepared for energy generation and landfilling can be avoided. For example, in New York (U.S.), a total of 3.8 MT of SW collected yearly and 76% of that is sent to landfills, while 14% is recycled, and 10% of is sent to energy facility for energy conversion (Citizens Budget Commission, 2014). Another example from Hong Kong shows that construction and demolition waste generators reduce the waste from an average of 7,890 tons to 3,300 tons per day (Yau, 2010). One study (Mian et al., 2016) cites the increase in the recycling of MSW generated from China where waste management strategy is dominated by the 60.16% landfilling, followed by incineration, untreated discharge, and other waste treatment at 29.84%, 8.21%, and 1.79%, respectively. Another recent report from China shows that they have planned to launch 100 new large-scale recycling platforms by the end of next year for improvement and better use of waste as a resource after a ban on trash imports (Stanway, 2019). America generated about 251 MT of waste in 2012 and 87 MT of SW is recycled and composted which is equivalent to 34.5% of the waste. The per capita waste generation is around 4.38 pounds per day from U.S. and, on average per person, 1.51 pounds of that waste is recycled and composted (USEPA, 2012). One report from Hong Kong indicated the importance and set out a series of waste reduction initiatives. One report in 2013 indicated that approximately 70% of MSW was disposed at landfills, and only 30% was recovered from recycling (Environmental Protection Department (EPD), 2013).

2.4.2.2 Biodegradable Waste Processing

The composting method is one of the most feasible and economical among the various options available. The techniques available for waste management are minimization, recycling, sanitary landfilling, incineration, and composting (Mohammed et al., 2018). Composting is most suitable for biodegradable waste, which is waste capable of undergoing anaerobic or aerobic degradation, such as biodegradable food waste, garden waste, paper including paper board, etc. (Council Directive 1999/31/EC, 1999). The Energy Information Administration (EIA) of the U.S. defines biogenic waste as the waste produced by the biological process of living organisms. The UNEP data shows that outdoor decay of biogenic organic matter of MSW contributes to about 5% of global greenhouse gas (GHG) emissions annually (UNEP, 2011). Thus, to combat this global issue, a number of technological developments for waste treatment have taken place. Mainly the composting process is to make organic fertilizer, anaerobic digestion to generate biogas, and thermal treatment to generate energy have all been developed (Laner et al., 2012). The major drawback of the implementation of some of these techniques at field levels is due to the high implementation costs and environmental concerns.

Organic waste could be utilized in composting and the anaerobic digestion process, which is used as an agricultural fertilizer. The produced natural compost is rich in micronutrients and macro-nutrients (like phosphorous, nitrogen, potassium). The biodegradable waste has the potential to produce more than 50% of the energy content of MSW in both developed and developing countries (Beneduci et al., 2012). Another process to convert organic waste into compost is anaerobic composting. The next section is discussed about both the aerobic and anaerobic processes.

2.4.2.2.1 Aerobic Composting

Biological decomposition of MSW degradable materials carried out under the predominantly aerobic conditions. The produced compost is sufficiently stable for nuisance free handling, storage, and satisfactorily matured to increase the agricultural productivity (UNEP, 2005). The microbes play a significant role in the biological decomposition process that involves the oxidation of organic carbon in the organic waste. The microbe-prepared compost is a good soil amendment and can be used as fertilizer. Care must be taken for pollutants such as heavy metals in the organic waste. During the composting process, energy is released which is responsible for the rise in the temperature during the composting. Energy loss and aerobic composting converted into the anaerobic composting in the hierarchy of the waste management system. Waste handling in environmentally friendly manner materials recovery is considered for extracting important raw materials and new products manufacturing by helping the life cycle assessment. The mixture of SW is used to form compost by aerobic composting technique and it can be contaminated by mainly heavy metals. Thus, MSW compost has the chance of heavy metal contamination and can cause environmental degradation as well as harm to public health. The major environmental concern is the restriction to use in agricultural practices as manure (Lasat, 2000; Pinamonti et al., 1997). Thus, mixed SW composting might not be a preferred option for the composting; thereby, it is not considered a sustainable waste management option. However, mixed waste composting is widely used (Ravi, 2011) in various countries including India where more than 91% of MSW is landfilled and there are no other options.

Micro- and macro-flora along with the pH, temperature, aeration, and moisture content, as well as the chemical and physical availability of the nutrients, are key factors in controlling the composting processing (Beneduci et al., 2012). It is very important to consider the public health issue and environmental impacts without the unsanitary landfilling that are more firmly formed by research. Thus, it can reduce the heavy metal contamination to groundwater than the open dumping of wastes. Previously, industrial waste generators disposed of their waste into the environment without any treatment and now regulation is very strict to treat and manage their waste as per regulation (Musee et al., 2008; Thandavamoorthy, 2016). One investigation (Ball et al., 2017) shows the effect composting to bio-stabilize the biodegradable fraction of MSW from an advanced waste treatment system in Australia. The aerobic system process observed a reduction in oxygen consumption of 30% in immature compost and 45% in mature compost when compared with the input material. The aerobic composting technique is the most used of MSW technology in India. The previous estimation has shown that 6% of the total MSW collected in composts in various membrane bioreactor facilities (DEA, Ministry of Finance, Government of India, 2009).

2.4.2.2.2 Anaerobic Digestion

Anaerobic digestion of MSW on a large scale has a challenge in terms of source segregation of organic waste stream. However, for smaller scales e.g. hotels or restaurants, for vegetables and meat markets and at household levels anaerobic digestion is successful. The anaerobic process recovers both energy and compost and is positioned at the higher levels in the hierarchy of waste management, as compared to other options i.e. energy recovery. The compost quality further depends upon the quality of the input of MSW. The composting of MSW results in a low-quality of compost that is less productive to crops and has the potential to introduce toxic heavy metals into the food chain. Pellera et al. (2016) evaluated the aerobic and anaerobic conditions for the combined treatment of an organic fraction of SW in the laboratory-scale landfill bioreactors. The initial operating system under anaerobic conditions could allow energetic exploitation of the substrates, while the implementation of a leachate treatment system ultimately aimed at the nutrient recovery. The organic fraction of MSW for the composting process takes around 37 days followed by a curing phase of 79 days. Various process parameters were tested and their optimum range (oxygen>140 mL l^{-1}, moisture content > 500 g kg^{-1}, temperature < 65°C w.w.). The processing of the sample is of utmost importance to increase biogas production from the production of organic acids, which, further reduces the pH and interfere with the biogas generation process (Adani et al., 2000). Another advantage is the considerable reduction of greenhouse gas emissions by diverting the biodegradable SW from the landfill (Ball et al., 2017). One interesting example from Biotech Company, Kerala (India), who installed twenty thousand household biogas units to divert about 2.5% of biodegradable waste from landfill that ultimately saves up to 4.5 million USD every year in terms of transportation cost.

Further, biogas production from these units avoids the generation of 7,000 tons of carbon dioxide equivalent to (TCO_2) emissions annually. The mixed waste composting is a better option as compared to landfilling or open burning due to environmental constraints (Ravi, 2011).

2.4.3 THERMAL TREATMENT

Thermal treatment is any waste treatment method which involves high-temperature processing of the SW materials. High-temperature processing of waste leads to significant volume reduction and renders it innocuous by destroying harmful pathogens. Thermal treatment, mainly WtE recovery, from waste is a desirable and viable option often used in the industrialized countries. The process commonly is based upon the combustion of SW that is usually considered to be thermal treatment and includes pyrolysis, incineration, and gasification. The combustibles waste constitutes an average of about 60% of the weight of SW and results in average heating value of about 3,000–6,000 Btu/lb. Thus, it makes MSW attractive for thermal treatment (McIntyre and Papic, 1974). Incineration thermal system generates dioxin, a toxic gas which has strict regulation releasing their emissions into the environment. Thus, activated carbon is one of the simplest and widely used methods due to its' large surface area and it strong adsorption capacity for the removal of dioxin gas. In the present section, important technologies such as incineration, gasification, and other options are discussed.

2.4.3.1 Incineration

The incineration is preferred in high-capacity, high income, and land-constrained countries. Previously waste is disposed of in open dumping areas and landfills without the landfill gas collection systems or unscientific manner. In the U.S., anaerobic digestion and incineration have a long history as a management strategy for SW with landfilling as an alternative (Perrot and Subiantoro, 2018). The anaerobic digestion process applies to only biodegradable organic waste in SW, whereas incineration works for all combustible waste materials. Both methods require the separation of recyclables waste to achieve optimum resource recovery. Anaerobic digestion and incineration can produce heat, electricity, or both. As the waste quantity is increasing, the world is shifting toward the adoption of sustainable and effective techniques to tackle the global waste crisis. In this context, innovative approach, design, and forward-thinking is needed in the waste-to- energy sector in the world. Modern waste treatment plants where as much as 99% of the energy will be utilized for electricity and heating and 95% of water is recovered. If SW contains some metal components, such as aluminum packaging, even 90% of the metals will be recovered. Slag produced could be recycled, the flue gas is 90–99% clean, and unwanted materials are removed from the circulation. It is essential that the municipal services integrated system should be self-sufficient, socially supported, and sustainable.

In several years, the advancements in the WtE technology will change to dual-purpose technology, ridding urban areas of their growing waste while generating electricity as a by-product. Developed nations have an important role as they are technically more sound compared to developing counties. Thus, they can support in terms of technical assistance in the waste management areas which can be solved with the help of mechanization (Lardinios and Klundert, 1997). Another important viewpoint shows that the 'blind technology transfer' from developed nations to developing nations also leads to subsequent failures. It indicates the need for appropriate technology to suit the local conditions like the type of waste, composition, and its treatment (Beukering et al., 1999). According to UN estimation, it is predicted that world's population will reach around 9.8 billion in 2050. Further, 68% of this total population will be living in urban areas; thus, effectively removing the SW and generating energy will attract the investors. Another estimation from the World Energy Council indicates that WtE production technologies are rapidly growing and are going to increase with the global market to 40 billion USD by 2023. China has the largest installed WtE capacity in the world and it is increasing at a rate of 26% annually. OECD countries observed to have on average a 4% growth rate in the capacity of WtE. One report by the World Economic Forum (WEF) shows that China is going to be the world's largest trash generator in 2030 and the waste volume is projected to be double American's volume of MSW. Chinese authorities are planning more incinerators for the MSW disposal to divert the waste from landfills, rivers, or oceans. Hence, in this connection, one Chinese company, Shenzhen Energy, is working to develop innovative WtE incinerator plants with the most advanced technologies. The world's largest WtE plant (with a capacity of 5,000 tons of waste per day) started its commissioning in 2016 and it will be nearly fully commissioned in early 2020. This plant will generate power approximately 550 million kWh each year and it is a simple, clean, and compact design with improved efficiency (The Global Power and Energy Elites, 2020). One case study from the U.K. made an effort to innovate the thermal-treatment options after recycling, the residual MSW has usually burnt in the WtE incinerator plants (Breeze, 2014). The European Union (EU) is also faced with several challenges during the handling of large quantities of SW such as landfill reduction, along with the renewable energy supply, taxes, greenhouse gas reduction, and SW recycling. Thus, it needs to re-think and broaden the opportunities for better alternatives to the landfill (European Commission, 1999). The European countries' view point indicated and stimulated a shift of WtE facilities upper side in the waste hierarchy than the landfill or incineration without energy recovery. The European Commission (2008) discussed the waste hierarchy on the statutory basis given by the Waste Framework Directive. The importance should be given to the recovery of important materials and its preference over the disposal system. It is usually considered that for a waste combustion plant it must generate sufficient energy to fulfill the 65% efficiency threshold (Levidow and Upham, 2017).

The U.S. has more than 200 incinerators in operation for MSW management in the early 1990s and reduced to 97 in 2001 with a further decrease to 77 in 2016 (EPA, 2017; Reitze and Davis, 1993). The major reason behind the declining number of incinerators is more stringent air pollution control regulation requirements (Clean Air Act amendments of 1990), low electricity prices, which is the main source of revenues for the survival of these plants. The first new plant was constructed and commissioned in 2015 in Florida and is considered the most advanced and cleanest WtE plant in North America with air pollution control measures and advanced combustion.

However, there are certain challenges associated with the incineration plant that involves the high initial investment, higher operating costs, and requires both local and international currency during its operation (National Research Council, 2000). Another challenge is the release of the toxic pollutants from the waste burning in the incineration process. Air pollution control devices such as equipped scrubbers with all the precautions, many chemicals and substances, which are toxic in nature, can be removed before emitted into the atmosphere. Some of them are toxic which are still released. In developed countries like the U.S., the incinerator system needs to improve its efficiency continuously due to the emission targets of the Clean Air Act regulation (Solid Waste Incineration, 2019).

2.4.3.2 Pyrolysis

Pyrolysis is a fast growing biomass thermal energy conversion technique and getting much attention globally. Pyrolysis has further advantages due to its high efficiency and eco-friendly nature. This technique is widely recommended for the conversion of MSW, scrap tires, agricultural residues, and non-recyclables plastics into renewable energy. Pyrolysis offers an attractive way of converting SW into the generation of heat, electricity, and certain chemicals (Zafar, 2019b).The process consists of the simultaneous and successive reactions when the carbon-rich organic waste is heated in a non-reactive atmosphere under high temperature in the absence of oxygen. Thus, the temperature has an important role in the thermal degradation of organic waste and it starts at the 350–550°C and goes up to 700–800°C in the absence of oxygen. On comparison with gasification, the pyrolysis usually takes place at a lower temperature. This results in the less volatilization of carbon and certain other pollutants such as heavy metals and dioxin formation.

Several studies have been reported so far related to the pyrolysis of SW. In this connection, one review article (Chen et al., 2014) shows that pyrolysis has been an attractive option to the incineration of SW disposal that recovers the energy and resources. Another important aspect discussed is the influence of operating parameters (heating rate, final temperature, and residence time). Thus, the pyrolysis reactors and technologies adopted the yields and main properties of the process end-products from SW constituents, and refuse-derived fuel from prepared from SW. Chen et al. (2014) showed the release emissions of HCl, SO_2, and NH_3 pollutants from pyrolytic processes and also included in the products polychlorinated dibenzo-p-dioxins and dibenzofurans, and heavy metals. This is further reviewed and available measures are taken to improve the environmental impacts of the pyrolysis process. Hammoodi and Almukhtar (2019) investigated the present thermal pyrolysis of two types of plastic: PET and PS in a batch tubular reactor. The yield investigated was maximum thermal pyrolysis of PS, 10% PET+ PS, 20% PET+ PS, and 30% PET + PS was 98.4% at 450°C, 72.5% at 450°C, and 55.8% at 400°C, respectively. Previous study (Sipra et al., 2018) has shown that MSW composition is a type of biomass and it has been widely used for biofuel production, thus considered as a renewable energy source. The study focused on the pyrolysis of SW by using its various components as a feedstock material with varying composition. The influence of interaction between MSW contents and their heating values have been systematically reviewed.

2.4.3.3 Gasification

The gasification method has various potential benefits over the traditional combustion of SW. Gasification uses only a fraction of the oxygen that would be needed to burn the waste material. The major product of the gasification system is the syngas, and also contains hydrogen, carbon monoxide, and methane. The syngas has a net calorific value of 4–10 MJ/Nm^3 (OECD, 2013). Another advantage of the gasification process is a gas fuel which is burnt in a conventional burner or in a gas engine to utilize the heat and thus produce the electricity. Further, the technique could be utilized as a building block for the production of valuable products such as chemicals and other forms of fuel energy (Blasi, 2000; Klinghoffer and Castaldi, 2013). The solid residue of noncombustible materials is the product of the gasification system that contains relatively low levels of carbon content. Korea developed the most efficient and effective pathway to utilize the MSW gasification technique to generate power with a newly developed system that has a decrease in tar and pollutants emissions (Seo et al., 2018). However, numerous other SW gasification systems are operating or under development around the world (Hauserman et al., 1997). SW gasification has various advantages over the traditional combustion processes for waste treatment. The process takes place in a low oxygen environment that has further advantages over the formation of dioxins and a large quantity of SOx and NOx (Baykara and Bilgen, 1981). Further calculations indicate the requirement of just a fraction of the stoichiometric amount of oxygen necessary for the combustion. The major disadvantages of the process is the gas resulting from gasification of MSW that contains various tars, halogens, particulates, heavy metals, and alkaline compounds. It could result in agglomeration in the gasification system, which can lead to clogging of fluidized beds and increased tar formation (Zafar, 2019a). Generally, steam produced is used in various applications for the indirect gasification system and it also increases the hydrogen content in the produced gas (Hauserman et al., 1997). Indirect and direct gasification processes are classified in the gasification system. The indirect gasification process is performed without air or oxygen injection. Nitrogen gas is present in the syngas produced and significantly affects the heating value. Thus, without the nitrogen contamination in the indirect gasification process, the heating value and the volumetric efficiency of syngas both increase (Belgiorno et al., 2003; Paisley and Anson, 1998). Hence, the indirect gasification system decreases the gas cleaning cost and energy could be recovered by the gas generation rate. In fact, the gas generation process is quite complex and initial investment is slightly higher (Hauserman et al., 1997). The various reactions and steps involved in the gasification system are shown in Figure 2.6. The SW gasification has involved both exothermic and endothermic reactions, which are ultimately successful and repetitive (Knoef, 2005; Souza-Santos, 2004). Figure 2.6 has shown the main reactants and the process of gasification is carried out in two main steps. The first step is the heating of biomass at around 600°C and the volatile contents, such as hydrogen, hydrocarbon gases, carbon monoxide, carbon dioxide, water, and tar, vaporize by various reactions. The first step is basically the endothermic reaction, oxygen is not required and the remaining by-products are the char and ash (Roos, 2010). This produced char is gasified in the second step by reacting with oxygen, steam, and hydrogen at a very high temperature. Unburnt char is formed due to combustion and is used to provide

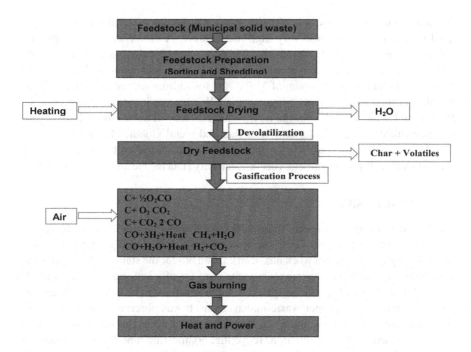

FIGURE 2.6 The main reactions and steps of the gasification process. (Modified from Seo et al., 2018, published under CC BY 3.0 license.)

the heat which is required for the endothermic reactions. The vapor production leads to the formation of char and gas due to thermal cracking (Roos, 2010). The composition, quantities, and nature of chemical species released due to the devolatilization process are dependent upon various factors such as SW composition, pressure, temperature, and heating rate (Kawaguchi et al., 2002; Souza-Santos, 2004). In the case of MSW treatment, the gasification technology has been updated from present status with an improvement. It is distinguished between the current commercial gasifier and the advanced gasifier with more benefits. Previous studies have analyzed the technological expectations for the gasification for the homogenous biomass (Kirkels, 2014) but not for homogenous SW. The major aim of this is to highlight the socio-cognitive aspects of expectations.

One research has shown that the plasma-assisted gasification technique for MSW has received much attention recently. In this plasma torch system, an instrument used to convert heat into power by-passing the current through a gas stream (Seo et al., 2018). WtE process involves the plasma-assisted gasification which leads to the partial oxidation hydrocarbons present in the MSW and the use of plasma. However, in this process, an electric arc is created by passing high current and high voltage between the two electrodes. The inert gas is used during the process under a very high pressure through the arc and is transferred to a closed container of MSW. All this process leads to an increase the maximum temperature to around 13,900°C in the heated arc and it can increase the torch temperature from 2760 to 4427°C. This high temperature has an important role in the decomposition of MSW which thermally converts into the gaseous elements and complex molecules that are separated into atoms at a specific very high temperature. A device known as a plasma converter is used for the decomposition MSW into a molecular gas and slag as SW. Further, the net electricity generation process is more dependent upon the SW composition, and thereby the amount of MSW sent to landfills will be decreased. Thus, the plasma torch can be used to gasify the SW generated and the gasification of MSW could be done by the plasma torch, which thereby dissolves the volatile gases, and electrification leads to formation of slag and metal globules from ash content. Finally, a syngas could be a viable option for the gas engine or to run the turbine generators by using as a synthetic fuel or electricity (Ducharme and Themelis, 2010).

2.5 CONCLUSION

The World Bank report depicts that global MSW generation will be increased to 2.2 billion tons annually in 2025 due to urbanization, increasing population, and public lifestyle. The present work has been carried out for the status of MSW generation, properties, and the various techniques used for efficient proper management. Physical and chemical properties play an important role in the selection of appropriate technology and proper waste management. It was observed that waste problems faced by various municipalities needed a systematic plan that integrated several techniques such as waste reduction, recycling, composting, and thermal treatment process. Globally, around 50% of waste is biodegradable, and hence, composting could be a preferred option for resource recovery that leads to the generation of good quality compost which can replace the chemical fertilizers on large scale. Another technique recently increasing much attention as thermal treatment is WtE, which converts the combustible and non-recyclables portion of the SW to power, reduce the MSW materials sent to the landfills, preventing air or water pollution, recycling improvement, and less dependency upon fossil fuel for electricity. The estimation shows that every year 2.3 billion tons of MSW will be generated, and this is equivalent to 2.58×10^{23} MJ of energy. Thus, WtE is a promising alternative energy option in the present or future because with this amount of energy 10% of global energy demand can be satisfied.

ACKNOWLEDGMENTS

We are very much thankful to our concerned department and universities for all kinds of support to the smooth finishing of our work.

REFERENCES

Abas, M.A. and Wee, S.T. 2014. Municipal solid waste management in Malaysia: An insight towards sustainability. *4th International Conference on Human Habitat & Environment* 4:192–206.
Abdel-Shafy, H.I. and Mansour, M.S.M. 2018. Solid waste issue: Sources, composition, disposal, recycling, and valorization. *Egyptian Journal of Petroleum* 27: 1275–1290.
Adani, F., Scatigna, L. and Genevini, P. 2000. Biostabilization of mechanically separated municipal solid waste fraction. *Waste Management & Research* 18 (5): 471–477.

Agarwal, R., Chaudhary, M. and Singh, J. 2015. Waste management initiatives in India for human well-being. *European Scientific Journal* 11: 105–127.

Ali, T.M. 2015. *Assessment of solid waste management system in Khartoum locality.* Khartoum: UOFK.

Annepu, R.K., Themelis, N.J. and Stanley-Thompson, T. 2012. Sustainable solid waste management in India, MSc. thesis, Columbia University in the City of New York.

Assamoi, B. and Lawryshyn, Y. 2012. The environmental comparison of landfilling vs. incineration of MSW accounting for waste diversion. *Waste Management* 32 (5): 1019–1030.

Ayeleru, O., Ntuli, F. and Mbohwa, C. 2016. Utilization of organic fraction of municipal solid waste (OFMSW) as compost: A case study of Florida, South Africa, Proceedings of the World Congress on Engineering and Computer Science, 1–6.

Ball, A.A.S., Shahsavari, E., Aburto-Medina, A., Amer, K.K.K., Shaiban, A.J. and Stewart, R.J. 2017. Biostabilization of municipal solid waste fractions from an advanced waste treatment plant. *Journal of King Saud University –Science* 29 (2): 145–150.

Baykara, S.Z. and Bilgen, E. 1981. *A feasibility study on solar gasification of Albertan coal. Alternative Energy Sources IV.* Vol. 6. New York: Ann Arbor Science.

Belgiorno, V., Feo, G.D., Rodolfo M.A. and Napoli, R.M.A. 2003. Energy from gasification of solid wastes. *Waste Management* 23 (1):1–15.

Bello, H. 2018. Impact of changing lifestyle on municipal solid waste generation in residential areas: Case study of Qatar. *International Journal Waste Resources* 8: 1–7.

Beneduci, A., Costa, I. and Chidichimo, G. 2012. Use of iron (II) salts and complexes for the production of soil amendments from organic solid wastes. *International Journal of Chemical Engineering* Article ID 701728, 2012: 1–9.

Beukering, P., Sehker, M., Gerlagh, R. and Kumar, V. 1999. Analyzing urban solid waste in developing countries: A perspective on Bangalore, India, Working paper 24, CREED, India.

Bewketu, E. 2013. *Assessment of the sustainability of solid waste collection and transport service delivery by MSEs: The case of Bahir Dar city, Ethiopia.* Rotterdam: International Institute of Urban Management.

Bhada, P. and Themelis, N. 2008. *Feasibility analysis of waste-to-energy as a key component of integrated solid waste management in Mumbai, India.* New York: Earth Engineering Center, Waste-to-Energy Research and Technology Council.

Blasi, C.D. 2000. Dynamic behavior of stratified downdraft gasifier. *Chemical Engineering Science* 55: 2931–2944.

Boateng, K.S., Agyei-Baffour P., Daniel, B., Rockson, G.N.K., Mensah, K.A. and Edusei, A.K. 2019. Household willingness-to-pay for improved solid waste management services in four major metropolitan cities in Ghana. *Journal of Environmental and Public Health* 2019: 1–9.

Breeze, P. (ed.) 2014. *Power from waste in power generation technologies.* Boston, MA: Newnes.:335–352.

Chen, D., Yin, L., Wang, H. and He, P. 2014. Pyrolysis technologies for municipal solid waste: A review. *Waste Management* 34 (12): 2466–2486.

Cheru, S. 2011. *Assessment of municipal solid waste management service in Dessie town.* Addis Ababa: Addis Ababa University.

Chiemchaisri, C., Juanga, J.P. and Visvanathan, C. 2007. Municipal solid waste management in Thailand and disposal emission inventory. *Environmental Monitoring and Assessment* 135: 13–20.

Chinasho, A. 2015. Review on community based municipal solid waste management and its implication for climate change mitigation. *American Journal of Scientific and Industrial Research* 6: 41–46.

Citizens Budget Commission. 2014. 12 Things New Yorkers should know about their garbage. http://www.cbcny.org/sites/default/files/REPORT_GarbageFacts_05222014.pdf.

Cohen, S., Martinez, H. and Schroder, A. 2015. *Waste management practices in New York City*. Hong Kong and Beijing. ALEP Waste Manage. https://www.swsu.ru/sbornik-statey/pdf/ALEP%20Waste%20Managent%20 FINAL.pdf

Council Directive 1999/31/EC. 1999. On the landfill of waste. *Official Journal of the European Communities* 182: 1–19. http://eurlex. europa.eu/LexUriServ/LexUriServ. do?uri=OJ:L:1999:182:0001:0019:EN:PDF.

Daniel, H. and Thomas, L. 1999. What A waste: Solid waste management in Asia. Working paper series No. 1. Urban development sector unit. East Asia and Pacific Region. p. 5.2019.

DEA, Ministry of Finance, Government of India. 2009. Position paper on the solid waste management sector in India. Public Private Partnerships in India.

Ducharme, C. and Themelis N. 2010. Analysis of thermal plasma-assisted waste to energy processes. In: Proceedings of the 18th annual North American Waste-to-Energy Conference, Orland: 11–13.

European Commission. 1999. Council directive online. http://eur-lex.europa.eu/legalcontent/EN/TXT/? uri¼celex:31999 L0031.

European Commission, 2008. Directive 2008/98/EC on waste (Waste Framework Directive). https://ec.europa.eu/environment/waste/framework/

Ekmekçioğlu, M., Kaya, T. and Kahraman, C. 2010. Fuzzy multicriteria disposal method and site selection for municipal solid waste. *Waste Management* 30: 1729–1736.

Emmanuel, K. and Jiquan, Z. 2019. Analyzing municipal solid waste treatment scenarios in rapidly urbanizing cities in developing countries: The case of Dar es Salaam, Tanzania. *International Journal of Environmental Research and Public Health* 16 (11): 2035–2056.

Environmental Protection Department (EPD). 2013. Monitoring of solid waste in Hong Kong: Waste statistics. https://www.wastereduction.gov.hk/sites/default/files/msw2013.pdf.

EPA. 2017. Energy recovery from the combustion of municipal solid waste (MSW). https://www.epa.gov/smm/energy-recovery-combustion-municipal-solid-waste-msw.

Erasu, D., Feye, T., Kiros, A. and Balew, A., 2018. Municipal solid waste generation and disposal in Robe town, Ethiopia. *Journal of the Air and Waste Management Association* 68 (12): 1391–1397.

Ferronato, N. and Torretta, V. 2019. Waste mismanagement in developing countries: A review of global issues. *International Journal of Environmental Research and Public Health* 16 (6): 1060–1088.

Getahun, T., Mengistie, E., Haddis, A., Wasie, F., Alemayehu, E. and Dadi, D. 2012. Municipal solid waste generation in growing urban areas in Africa: Current practices and relation to socioeconomic factors in Jimma, Ethiopia. *Environmental Monitoring and Assessment* 184: 6337–6345.

Ghafour, S., Shoresh, A., and Chiman, M. 2013. Strategic planning of factors affecting municipal solid waste management by analysis SWOT a case study: Bukan City, Kurdistan Province. *Environmental Based Territorial Planning* 6: 1–21.

Guerrero, L.A., Maas, G. and Hogland, W. 2013. Solid waste management challenges for cities in developing countries. *Waste Management* 33: 220–232.

Gurung, A. and Oh, S.E. 2012. Municipal solid waste management: Challenges and opportunities in Nepal. *Korean Journal of Soil Science and Fertility* 45: 421–427.

Hagos, D., Mekonnen, A. and Gebreegziabher, Z. 2012. Households' willingness to pay for improved urban waste management in Mekelle city, Ethiopia. Environment for Development Initiative. 1–25. http://www.jstor.com/stable/resrep14960

Hammoodi, S.I. and Almukhtar, R.S. 2019. Thermal pyrolysis of municipal solid waste (MSW). IOP Conf. Serial. *Materials Science and Engineering* 579: 1–9.

Hauserman, W.B., Giordano, N. and Recupero, V. 1997. Biomass gasifiers for fuel cells systems. *Environmental Science* 79: 199–206.

Henry, R.K., Yongsheng, Z. and Jun, D., 2006. Municipal solid waste management challenges in developing countries–Kenyan case study. *Waste management* 26 (1): 92–100.

Hla, S.S. and Roberts, D. 2015. Characterization of chemical composition and energy content of green waste and municipal solid waste from Greater Brisbane, Australia. *Waste Management* 41:12–19.

Hoornweg, D. and Bhada-Tata, P. 2012. What a waste: A global review of solid waste management. Urban development series; Knowledge papers no. 15. World Bank, Washington, DC USA.

Huisman, H., Breukelman, H. and Keesman, B. 2016. Expert mission on integrated solid waste management (ISWM) to Dar es Salaam; MetaSus. The Hague, the Netherlands.https://www.rvo.nl/sites/default/files/2016/11/Tanzania%20Report%20Expert%20Mission%20Solid%20Waste%202016.pdf

Ibrahim, M.I.M. and Mohamed, N.A.E.M. 2016. Towards sustainable management of solid waste in Egypt. *Procedia Environmental Sciences* 34: 336–347.

John, N.M., Edem, S.O., Ndaeyo, N.U. and Ndon, B.A. 2006. Physical composition of municipal solid waste and nutrient contents of its organic component in Uyo municipality, Nigeria. *Journal of Plant Nutrition* 29 (2): 189–194.

Kassa, Z. 2010. *The challenges of solid waste management in urban areas, the case of Debremarkos town.* Addis Ababa: Addis Ababa University.

Katiyar, R.B., Suresh, S. and Sharma, A.K. 2013. Characterization of municipal solid waste generated by the city of Bhopal, India. *International Journal of ChemTech Research* 5 (2): 623–628.

Kawaguchi, K., Miyakoshi, K. and Momonoi, K. 2002. Studies on the pyrolysis behavior of gasification and melting systems for municipal solid waste. *Journal of Material Cycles and Waste Management* 4 (2): 102–110.

Kawai, K. and Tasaki, T. 2016. Revisiting estimates of municipal solid waste generation per capita and their reliability. *Journal of Material Cycles Waste Management* 18: 1–13.

Kebede, G., Mekonnen, E. and Manikandan, R. 2018. Composition and characterization of municipal solid waste in landfill site: In case of Dilla city, Ethiopia. *Environmental Science: An Indian Journal* 14 (4): 171–178.

Kirkels, A. 2014. Punctuated continuity: The technological trajectory of advanced biomass gasifiers. *Energy Policy* 68: 170–82.

Klinghoffer, N.B. and Castaldi, M.J. 2013. *Gasification and pyrolysis of municipal solid waste (MSW). Book: Waste to energy conversion technology.* Woodhead Publishing: Amsterdam. 146–176.

Knoef, H. 2005. Practical aspects of biomass gasification (Chapter 3). *Handbook of biomass gasification,* Enschede, The Netherlands: BTG Biomass Technology Group, 13–37.

Kumar, V. and Pandit, R.K. 2013.Problems of solid waste management in Indian cities. *International Journal of Scientific and Research Publications* 3 (3): 1–9.

Lahiry, S. 2019. India's challenges in waste management, Down to Earth Magazine. https://www.downtoearth.org.in/blog/waste/india-s-challenges-in-waste-management-56753.

Laner, D., Crest, M., Scharff, H., Morris, J.W.F. and Barlaz, M.A. 2012. A review of approaches for the long-term management of municipal solid waste landfills. *Waste Management* 32: 498–512.

Lardinios, I. and Klundert, A.V. 1997. Integrated sustainable waste management. Paper for the Programme Policy Meeting Urban Waste Expertise Programme. 1–6.

Lasat, M.M. 2000. Phytoextraction of metals from contaminated soil: A review of plant/soil/metal interaction and assessment of pertinent agronomic issues. *Journal of Hazardous Substances Research* 2: 1–25.

Levidow, L. and Upham, P. 2017. Socio-technical change linking expectations and representations: Innovating thermal treatment of municipal solid waste. *Science and Public Policy* 44 (2): 211–224.

Liu, C. and Wu, X.W. 2010. Factors influencing municipal solid waste generation in China: a multiple statistical analysis study. *Waste Management & Research* 29: 371–378.

Lohri C.R., Camenzind E.J. and Zurbrügg C. 2014. Financial sustainability in municipal solid waste management – Costs and revenues in Bahir Dar, Ethiopia. *Waste Management* 34: 542–552.

Maria, F.D., Lovat E. and Caniato, M. 2018. Waste management in developed and developing countries: The case study of Umbria (Italy) and the West Bank (Palestine). *Detritus* 3: 171–180.

McIntyre, A.D. and Papic, M.M. 1974. Pyrolysis of municipal solid waste. *The Canadian Journal of Chemical Engineering* 52 (2): 263–272.

Memon, M.A. 2010. Integrated solid waste management based on the 3R approach. *Journal of Material Cycles and Waste Management* 12: 30–40.

Mian, M. M., Zeng, X., Nasry, A.A.N.B. and Al-Hamadani, S.M.Z.F. 2016. Municipal solid waste management in China: A comparative analysis. *Journal of Material Cycles and Waste Management* 19 (3): 1127–1135.

Miezah, K., Obiri-Danso, K., Kádár, Z., Fei-Baffoe, B. and Mensah, M.Y. 2015. Municipal solid waste characterization and quantification as a measure towards effective waste management in Ghana. *Waste Management* 46: 15–27.

Ministry of Municipal Affairs. 2015. Development of a national strategy to improve the municipal solid waste management sector in the Hashemite Kingdom of Jordan. Amman, Jordan. http://www.mma.gov.jo/Files/Docs/11102018_043930BaselineReport.pdf

Mohammed, M., Donkor, A. and Ozbay, I. 2018. Bio-drying of biodegradable waste for use as solid fuel: A sustainable approach for green waste management. *Agricultural waste and residues*. IntechOpen Science. 89–104. https://www.intechopen.com/books/agricultural-waste-and-residues/bio-drying-of-biodegradable-waste-for-use-as-solid-fuel-a-sustainable-approach-for-green-waste-manag

Monavari, S.M., Omrani, G.A., Karbassi, A. and Raof, F.F. 2012. The effects of socioeconomic parameters on household solid-waste generation and composition in developing countries (a case study: Ahvaz, Iran). *Environmental Monitoring and Assessment* 184: 1841–1846.

Monyoncho, G. 2013. *Solid waste management in urban areas Kenya: A case study of Lamu town*. Nairobi: University of Nairobi.

Moreno, A.D. et al. 2013. Mexico City's Municipal Solid Waste Characteristics and Composition Analysis. *Revista Internacional de Contaminacion Ambiental* 29 (1): 39–46.

Musee, N., Lorenzen, L. and Aldrich, C. 2008. New methodology for hazardous waste classification using fuzzy set theory – Part I. Knowledge acquisition. *Journal of Hazardous Materials* 154 (1–3): 1040–1051.

Nadeem, K., Farhan, K. and Ilyas, H. 2016. Waste amount survey and physio-chemical analysis of municipal solid waste generated in Gujranwala-Pakistan. *International Journal of Waste Resources* 6 (1): 196–204.

National Research Council. 2000. Incineration processes and environmental releases. Chapter 3. *Waste Incineration and Public Health*. Washington, DC: The National Academies Press. https://doi.org/10.17226/5803.

Ngau, P. and Kahiu. 2009. ISWM Secondary Data Report on Solid Waste Inventory in Nairobi: Report of the National Technical Taskforce (NTT) on Preparation of an Integrated Solid Waste Management Plan for Nairobi. Nairobi.

Njoroge, B.N.K., Kimani, M. and Ndunge, D. 2014. Review of municipal solid waste management: A case study of Nairobi, Kenya. *International Journal of Engineering and Science* 4 (2): 16–20.

OECD. 2013. Municipal waste. *Environment at a glance: OECD indicators*. Paris, France: OECD Publishing.

Ogwueleka, T. 2009. Municipal solid waste characteristics and management in Nigeria. *Iranian Journal of Environmental Health Science* 6: 173–180.

Omran, A. and Gavrilescu, M. 2008. Municipal solid waste management in developing countries: A perspective on Vietnam. *Environmental Engineering and Management Journal* 7: 469–478.

Paisley, M.A. and Anson, D. 1998. Biomass gasification for gas turbine-based power generation. *Journal of Engineering for Gas Turbines and Power* 120: 284–288.

Parvez, N., Agrawal, A. and Kumar, A. 2019. Solid waste management on a campus in a developing country: A study of the Indian Institute of Technology, Roorkee. *Recycling* 4: 2–22.

Pasek, A.D., Gultom K.W. and Suwono, A. 2013. Feasibility of recovering energy from municipal solid waste to generate electricity. *Journal of Engineering and Technological Sciences* 45 (3): 241–256.

Pellera, F.M., Pasparakis, E. and Gidarakos, E. 2016. Consecutive anaerobic-aerobic treatment of the organic fraction of municipal solid waste and lignocellulosic materials in laboratory-scale landfill-bioreactors. *Waste Management* 56: 181–189.

Perrot, J. and Subiantoro, A. 2018. Municipal waste management strategy review and waste-to-energy potentials in New Zealand. *Sustainability* 10: 3114–3126.

Pervez, A. and Kafeel, A. 2013. Impact of solid waste on health and the environment. *Special Issue of International Journal of Sustainable Development and Green Economics (IJSDGE)* 2: 165–168.

Pinamonti, F., Stringari, G., Gasperi, F. and Zorzi, G. 1997. The use of compost: Its effects on heavy metal levels in soil and plants: *Resources, Conservation and Recycling* 21: 129–143.

Ravi, K. 2011. Managing Director, Ramky Enviro Engineers Ltd. https://ramkyenviroengineers.com/management/

Reitze, A. and Davis, A. 1993. Regulating municipal solid waste incinerators under the clean air act: History, technology and risks, B.C. *Environmental Affairs Law Review* 21 (1).1–88.

Roos, C. 2010. *Clean heat and power using biomass gasification for industrial and agricultural projects*, U.S. Department of Energy: 1–9.

Saeed, M.O., Hassan, M.N. and Mujeebu, M.A. 2009. Assessment of municipal solid waste generation and recyclable materials potential in Kuala Lumpur, Malaysia. *Waste Management* 29: 2209–2213.

Saidan, M.N., Drais, A.A. and AL-Manaseer, E. 2017. Solid waste composition analysis and recycling evaluation: Zaatari Syrian Refugees Camp, Jordan. *Waste Management* 61: 58–66.

Sankoh, F.P., Yan, X. and Tran, Q. 2013. Environmental and health impact of solid waste disposal in developing cities: A case study of Granville Brook Dumpsite, Freetown, Sierra Leone. *Journal of Environmental Protection* 4: 665–670.

Sarkar, M.S.I. and Bhuyan, M.D.S. 2018. Analysis of physical and chemical composition of the solid waste in Chittagong city. *Journal of Industrial Pollution Control* 34 (1): 1984–1990.

Seo, Y.C., Alam M.T. and Yang, W.-S. (2018). Gasification of municipal solid waste (Chapter 7). *Gasification for low-grade feedstock*, Yongseung Yun (ed.). IntechOpen, 115–141. https://www.intechopen.com/books/gasification-for-low-grade-feedstock

Silpa, K., Yao, L., Bhada-Tata, P. and Woerden, F.V. 2018. What a waste 2.0: A global snapshot of solid waste management to 2050. *Urban development*. Washington, DC: World Bank.

Sipra, A.T., Gao, N. and Sarwar, H. 2018. Municipal solid waste pyrolysis for bio-fuel production: A review of effects of MSW components and catalysts. *Fuel Processing Technology* 175: 131–147.

Solid Waste Incineration. 2019. https://www.encyclopedia.com/environment/encyclopedias-almanacs-transcripts-and-maps/solid-waste-incineration.

Souza-Santos, M.L.D. 2004. *Solid fuels combustion and gasification: modeling, simulation, and equipment operation.* New York: Marcel Dekker, Series: 174.

Srivastava R., Krishna V. and Sonkar I. 2014. Characterization and management of municipal solid waste: a case study of Varanasi city, India. *International Journal of Current Research Academic Review* 2 (8): 10–16.

Stanway, D. 2019. *China starts new recycling drive as foreign trash ban widens.* Sustainable Business. https://www.ejinsight.com/eji/article/id/2039568/20190116-China-starts-new-recycling-drive-as-foreign-trash-ban-widens

Talyan, V., Dahiya, R.P. and Sreekrishnan, T.R. 2008. State of municipal solid waste management in Delhi, the capital of India. *Waste Management* 28: 1276–1287.

Tefera, D.Y. and Negussie, D.B. 2015.Micro and small enterprises in solid waste management: Experience of selected cities and towns in Ethiopia: A review. *Pollution* 1: 461–472.

Thandavamoorthy, T.S. 2016. Wood waste as coarse aggregate in the production of concrete, European. *Journal of Environmental and Civil Engineering* 20 (2): 125–141.

Thanh, N.P., Matsui, Y., and Fujiwara, T. 2011. Assessment of plastic waste generation and its potential recycling of household solid waste in Can Tho city, Vietnam. *Environmental Monitoring and Assessment* 175: 23–35.

The Global Power and Energy Elites. 2020. How China is turning up the heat of waste. https://www.global-energy-elites.com/articles-2020/how-china-is-turning-up-the-heat-of-waste.

Thitame, S.N., Pondhe, G.M. and Meshram, D.C. 2010. Characterisation and composition of Municipal Solid Waste (MSW) generated in Sangamner City, District Ahmednagar, Maharashtra, India, *Environmental Monitoring and Assessment* 170: 1–5.

UNEP. 2005. Chapter 8. *Composting in solid waste management* (Vol. 2, 1st ed.). United Nation Environmental Program. 19–235.

UNEP. 2011. Waste: Investing in energy and resource efficiency. In towards a green economy: Pathways to sustainable development and poverty eradication: 286–329. https://sustainabledevelopment.un.org/index.php?page=view&type=400&nr=126&menu=35

UN-Habitat. 2010. *State of the world's cities 2010/2011-cities for all: bridging the urban divide.* London: UN-Habitat.

United Nations (undated). 2009. Urban and Rural Areas. http://www.un.org/en/development/desa/population/publications/urbanization/urban-rural.shtml.

United Nations. 2014. Concise report on the world population situation in 2014. https://www.un.org/en/development/desa/population/publications/trends/concise-report2014.asp#:~:text=The%20report%20indicates%20that%20the,in%20the%20less%20developed%20regions.

USAID. 2015. *Situational analysis of urban sanitation and waste management. Strengthening Ethiopia's Urban Health Program (SEUHP).* Addis Ababa: John Snow, Inc.

U.S. Environmental Protection Agency. 2013. Wastes – Non-Hazardous Waste – Municipal Solid Waste, 1200 Pennsylvania Ave., N. W. Washington, DC.

USEPA. 2012. Municipal solid waste generation, recycling, and disposal in the United States: Facts and figures for 2012. https://www.epa.gov/sites/production/files/2015-09/documents/2012_msw_dat_tbls.pdf

Vishwakarma, A., Kulshrestha, M. and Kulshreshtha, M. 2012. Efficiency evaluation of municipal solid waste management utilities in the urban cities of the state of Madhya Pradesh, India, using stochastic frontier analysis. *Benchmarking: An International Journal* 19: 340–357.

Wang, T. 2019. U.S. municipal solid waste generation 1960-2015. https://www.statista.com/statistics/186256/us-municipal-solid-waste-generation-since-1960/.

Wang, Y., Ng, K.T.W. and Asha, A.Z. 2016. Non-hazardous waste generation characteristics and recycling practices in Saskatchewan and Manitoba, Canada. *Material Cycles and Waste Management* (18) 4: 715–724.

World Bank Data. 2019. Solid waste management. https://www.worldbank.org/en/topic/urbandevelopment/brief/solid-waste-management.

Yang R., Xu, Z. and Chai, J.A. 2018. Review of characteristics of landfilled municipal solid waste in several countries: Physical composition, unit weight, and permeability coefficient. *Polish Journal of Environmental Studies* 27: 2425–2435.

Yau, Y. 2010. Domestic waste recycling, collective action and economic incentive: The case in Hong Kong. *Waste Management* (30): 2440–2447.

Yusuff, A.S., John, W., Okoro, O. and Ajibade, A. 2014. Physico-chemical composition and energy content analysis of solid waste: A case study of Castlereagh District, Northern Ireland. *American Journal of Engineering Science and Technology Research* 2 (1): 1–9.

Zafar, S. 2019a. Gasification of municipal wastes. https://www.bioenergyconsult.com/gasification-municipal-wastes/

Zafar, S. 2019b. Pyrolysis of municipal wastes. https://www.bioenergyconsult.com/pyrolysis-of-municipal-waste/

Zhou, H., Meng, A.H., Long, Y.Q., Li, Q.H. and Zhang, Y.G. 2014. Classification and comparison of municipal solid waste based on thermochemical characteristics. *Journal of the Air & Management Association* 64 (5): 597–616.

3 Recent Advances in the Structural Modifications of Nanoparticles to Enhance Photocatalytic Activity

Anoop Kumar Verma, Steffi Talwar, and Navneet Kaur
Thapar Institute of Engineering and Technology Patiala, India

Vikas Kumar Sangal
Malaviya National Institute of Technology Jaipur, India

CONTENTS

3.1 INTRODUCTION

Amongst the different techniques used for water detoxification so far, advanced oxidation processes (AOPs) have proven themselves a number of times by successfully degrading a wide range of emerging water pollutants (Dalrymple, Yeh, and Trotz 2007). This technique also has potential to establish a green system for detoxification processes (Mahlambi, Ngila, and Mamba 2015). AOPs are based on the in-situ generation of free radicals such as hydroxyl radicals, anion radicals, and superoxide radicals, which are very reactive in nature (Kanakaraju, Glass, and Oelgemöller 2018). Since these radicals have high electrochemical oxidant potential, they attack the complex of toxins and completely mineralize them to CO_2, H_2O, inorganic ions, and acids (Buthiyappan et al. 2015; Dalrymple, Yeh, and Trotz 2007; Kanakaraju, Glass, and Oelgemöller 2018).

Various types of AOPs includes photo-Fenton, sonolysis, radiation, photocatalysis, ozonation, and electrochemical oxidation (Kanakaraju, Glass, and Oelgemöller 2018). Photocatalysis is one of the widely used AOPs for detoxification purposes. It is based on the oxidation and reduction reactions taking place on the surface of a semiconductor with the help of oxygen and UV light, which results in formation of hydroxyl radical (Hodaifa and Hodaifa 2017). In recent years, a variety of nano-photocatalysts are being studied like ZnO, ZnS, WO_3, Fe_2O_3, but still TiO_2 is considered as the best nanocatalyst semiconductor to be used as a photocatalyst because of its nontoxicity, low cost, inert nature, and high oxidizing potential (Byrne, Subramanian, and Pillai 2018). The purpose of using nanotechnology in this technique is modifying the nanoscale structured catalyst and further merging them to bulk which leads to the establishment of compounds that have novel physical and chemical properties (Mackenzie and Bescher 2007; Pradeep 2009; Zhang and Webster 2009). Moreover, the smaller the size of nanoparticles the greater is the surface energy due to an increased number of surface atoms; more surface energy ensures more removal of the contaminants (Guz and Rushchitskii 2003; Ni et al. 2007; Savage and Diallo 2005).

Hence, to alleviate the issue of water pollution, photocatalysis is one potential AOP. But, poor activity in solar light (Shekofteh-gohari, Habibi-yangjeh, and Abitorabi 2018), and high electron hole recombination (Laxma et al. 2017) is the potential drawback of heterogeneous photocatalysis.

Another AOP that has been widely investigated over several years is photo-Fenton. It includes usage of iron oxide as the catalyst and hydrogen peroxide as oxidant to generate free radicals that further target the pollutants (Antonio et al. 2008). It is a widely used technique because it requires a low dosage of oxidant and can channel solar energy to carry out the detoxification (Clarizia et al. 2017). On the other hand, it has some disadvantages as the process is disturbed at near neutral pH (Hodaifa and Hodaifa 2017). It results in a high operational cost for industrial scale up (Clarizia et al. 2017), generating huge quantity of iron sludge (Singh, Rekha, and Chand 2018). For the application of AOPs in the field scale, various literatures are focused on the utilization of natural solar radiation. The usage of natural solar light in place of artificial UV lamps is an economical as well as environmentally attractive solution for photo-Fenton and photocatalysis (Chan et al. 2003).

3.2 STRUCTURAL MODIFICATIONS IN PHOTOCATALYSIS

In countries like India, the applications of solar photocatalytic processes are the best option, as fit as immense solar light throughout the year is received. For harvesting the solar light efficiently, certain modifications in the catalyst are required. For the increased photoactivity of TiO_2 in the visible region, doping is an essential part of the study. Doping is basically the modifications in the crystal lattice of the nanoparticles for the activation of the catalyst in the visible region of light. Various studies have been reported regarding the doping of various metals like Ag (Koči et al. 2010), Cu (Fu et al. 2011), Ni (Choi, Park, and Hoffmann 2010), Fe (Yu, Xiang, and Zhou 2009), Cr (Yang et al. 2010), Au (Ayati et al. 2014), Pt (Huang et al. 2008).

It is undoubtful that unmodified semiconductor nanoparticles give efficient results for water detoxification processes as they are able to degrade even minute concentration of toxins, but due to their wide band gap their activity is confined to the UV range of solar light (Choudhury, Borah, and Choudhury 2012). But when these semiconductor nanoparticles are modified by doping with the noble metal dopant, they become more efficient as the recombination rate decreases and the spectra of the photo-catalyst gets sifted to the visible range of solar light, which results in an economical and highly efficient detoxification process as compared to the process using unmodified nanoparticles (Belver et al. 2019; Tahir 2019).

Various nano-photocatalysts, which are semiconductor in nature, have been reported in literature to carry out water detoxification such as ZnO, TiO_2, CdS, and Ag_3PO_4. But most of them are inactive in solar irradiations; therefore, extending their light response in the visible region can add an economic value to the process (Alvarez 2013; Essawy 2018). In multiple recent studies, noble metal doping of the nanoparticles has been widely reported. The Schottky barrier is created by the metal dopant on the surface of a semiconductor (Ismail and Eladl 2019; Plodinec et al. 2018). Figure 3.1 shows the structural modifications of the nanocatalyst. Photo-induced electron-hole pairs are separated well by the Schottky barrier, which results

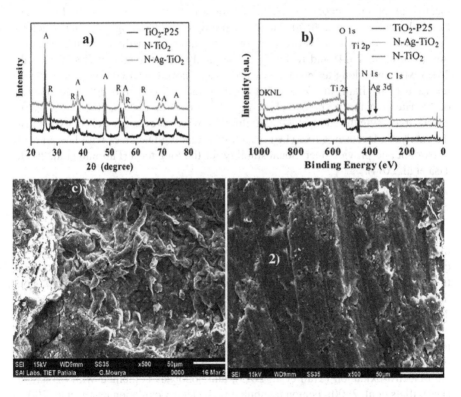

FIGURE 3.1 (a) X-ray diffraction (XRD) patterns of undoped TiO_2, nitrogen doped TiO_2 and TiO_2 doped with nitrogen and silver; (b) X-ray photoelectron spectroscopy (XPS) analysis of undoped TiO_2 (Bansal and Verma 2018c), doped TiO_2, and TiO_2 doped with nitrogen and silver; (Bansal and Verma 2018c.) (c) SEM images of (1) Fe composite and (2) TiO_2 and Fe composite. (Talwar et al. 2020.)

in reduction of electron hole recombination. Moreover, the noble dopant metal possesses the Plasmon resonance effect; this effect also enhances the visible activity of the doped photocatalyst (Koppala et al. 2019; Nanda and Rath 2020).

A huge number of different combinations of AOPs (Adish Kumar et al. 2014; Destaillats et al. 2000; Madhavan, Grieser, and Ashokkumar 2010; Rajesh Banu et al. 2018; Song et al. 2007; Varatharajan and Kanmani 2007;) based on literature review have been proven to be more efficient in comparison to individual oxidation techniques as the AOP combination leading to the generations of hydroxyl radicals in high concentration and high energy efficiency is achieved (Buthiyappan et al. 2015). Similarly, the combination of photocatalysis (using nanocatalysts) and photo-Fenton overcomes the drawbacks of each other and makes it a highly potential technique for degradation of pollutants. Additionally, 90% of operational costs of water detoxification in industry are due to post-treatment separation of the catalyst, so to overcome the economic hurdle, fixed-bed catalysis can be a lifeguard for the detoxification process. Various studies suggest that fixed bed reactors are not extensively studied for wastewater treatments; only a few researchers have discussed it (Yan, Jiang, and Zhang 2014). But in reality, this approach may lead to the development of an efficient economic technique for wastewater detoxification (Márquez, Levchuk, and Sillanpää 2018). There are multiple shortcomings of batch reactors that includes formation of various non-biodegradable intermediates during the reaction (Yan et al. 2014) and recovery of the catalyst post-treatment (Esteves et al. 2016). Moreover, long-term stability cannot be achieved due to the long duration of reactions (Yan, Jiang, and Zhang 2014). All of the above mentioned disadvantages of the batch reactor can be fixed by using a fixed bed reactor. The fixed bed reactor ensures recycling the catalyst without filtration, which prevents the excessive loss of the catalyst. Moreover, a fixed bed reactor can treat high volumes with less amount of oxidant and provide operational stability for the long term (Fatimah et al. 2009; Tisa et al. 2014).

3.3 DUAL EFFECT OF PHOTOCATALYSIS AND PHOTO-FENTON

The intensification of the advanced treatment process is a promising technique for the industrial processes aspiring for improvement in the efficiency of the process, cost-effectiveness, and environmental safety. For increasing the potential elimination of the organic pollutants present in water, recent studies have concentrated upon the development of the combined AOPs (Bejarano-Pérez, Javier and Suárez-Herrera 2007; Kanakaraju, Glass, and Oelgemöller 2018; Steter et al. 2014; Vieira Dos Santos et al. 2017). The synergistic effect of the combined process over single AOP leads to the better mineralization and degradation of pollutants. Combinations of AOPs like ozonation followed by photocatalysis (Song et al. 2007), sonolysis combined with electro-oxidation (Tran et al. 2015), and sonolysis combined with ozonation (Destaillats et al. 2000), Fenton combined with electro-oxidation (Sirés et al. 2007) have been studied. Providing higher removal efficiency than the single process, the combined processes are the most sorted option. The combined processes provide the treatment in reduced time with increased production of the hydroxyl radicals to help in faster removal of the process.

Taking into consideration various implications of the numerous coupled processes, a novel technology of combining two processes with respect to photocatalysis and photo-Fenton has been developed. The combined solar-powered AOPs possess the promising application toward the treatment of wastewater, i.e. of the capability to mineralize the toxic compounds, possessing low cost, providing a boon for the wastewater treatment.

The in-situ dual process of photocatalysis and photo-Fenton takes into consideration the shortcomings of individual processes and utilizes the advantages of both processes into one another process. The degradation rate and treatment time can be substantially improved if the two different processes (photo-Fenton and photocatalysis) are coupled together in one system which would consequently reduce the cost of the process and a high consumption of chemicals as well. For the enhancement of the degradation efficiency, few studies have suggested the basic idea of the dual effect of photo-Fenton and photocatalysis (Adish Kumar et al. 2014). The generation of more •OH has led to improved degradation efficiency of the compounds. Studies have shown the effectiveness of dual process for the treatment of various pollutants including pharmaceuticals, bacterial wastes, paints, pesticides, insecticides, and real wastewater. The reduced time for treatment along with increased degradation efficiency has proven the efficacy of the dual process.

With iron oxide and TiO_2 immobilized on PVC films lead to the synergistic effect (Mazille, Schoettl, and Pulgarin 2009). With the effect of pH and H_2O_2, hydroquinone (HQ) was degraded under simulated solar irradiation. Remarkable synergistic effects were observed for PVFf–TiO_2–Fe oxide, possibly due to Fe (II) regeneration, accelerated by electron transfer from TiO_2 to the iron oxide under light.

Utilizing the industrial waste materials as an alternative iron source has helped the dual effect gain recognition among researchers. The waste products including the fly ash and foundry sand dumped, and as such, harming the groundwater as well as the environment, have been used for the dual process. The slurry form iron source along with the nanocatalyst TiO_2 is been used for the study. The usage of the waste iron source facilitated the photo-Fenton process (Bansal and Verma 2017a). The execution of the dual effect in slurry form showed increase in the rate constant by three folds. Furthermore, the study was approached toward the fixed bed by entrapping the iron source in the cloth mesh and immobilizing on clay beads.

Bansal and Verma (2017b) studied the dual effect of photocatalysis and photo-Fenton for the degradation of cephalexin. The study presented the use of a fixed bed composed of the handmade support in the form of beads. The composite beads were made up of clay, foundry sand, and fly ash with a layer of nanocatalyst TiO_2 immobilized on the surface. The dual process takes place in-situ as the fly ash and foundry sand being the sources of iron help the photo-Fenton process to take place. The study showed the best results in the case of composite beads made from clay, foundry sand, and fly ash as compared to beads made from only clay and foundry sand, or clay and fly ash. The rate constant increased to almost 30–45% in the case of the composite beads made from clay, foundry sand, and fly ash.

Textile dye (azucryl red) was degraded using the Fe doped TiO_2 (Kerrami et al. 2019). With the help of Fe doping the rutile phase of TiO_2 was converted into the anatase leading to the synergistic effect. Complete degradation of dye was observed with 0.2% doping.

Various studies have been performed for the degradation of pharmaceutical compounds including phenazone (Talwar, Sangal, and Verma 2019), pentoxyfylline (Bansal and Verma 2017a), cephalexin (Bansal and Verma 2017b), metronidazole (Talwar, Verma, and Sangal 2019), and ofloxacin (Sharma et al. 2020). The results showed the capability of a dual effect to be applied on a large scale for the treatment of real wastewater.

Talwar, Verma, and Sangal (2019) explained the basic mechanism for the hybrid process of photocatalysis and photo-Fenton. Actually, the surface-active TiO_2 layer would lead to photocatalysis, thus leading to the generation of •OH. Besides this, there would be subsequent leaching of iron from these composite beads. The electron required for the reduction of Fe (III) comes from the conduction band of TiO_2 (Kim et al. 2012). This step helps in coverage of the shortcoming of the photocatalysis process of the electron-hole recombination, as an electron gets involved in reduction. This will eventually lead to the increase of •OH, thus increasing the reaction rate. This also helps in the oxidation of water molecules to form •OH. Moreover, Fe (II) species present in the composite are capable of changing H_2O_2 into reactive species. If the amount of Fe (II) increases in the solution, electron scavenging might take place. This could be due to the absence of H_2O_2.

Therefore, the results recommend that the binding of iron to the photo-excited TiO_2 surface might have an effect on the mechanism of the Fenton reaction to aid the •OH production (Bokare and Choi 2014).

The basic mechanism of the dual process taking place might be explained as shown in Equations 3.1–3.9. The process limitation of photocatalysis, i.e. electron-hole recombination is solved in this novel dual effect as the electrons generated in the valence band helps in the conversion of leached Fe^{3+} to Fe^{2+} (Equations 3.7 and 3.8). The scavenging effect of the H_2O_2 is also explained (Equation 3.9).

$$Fe_{aq}^{2+} + H_2O_2 \rightarrow Fe_{aq}^{3+} + OH^- + \cdot OH \tag{3.1}$$

$$Fe_{aq}^{3+} + H_2O_2 + H_2O \rightarrow Fe_{aq}^{2+} + H_3O^+ + HO_2^- \tag{3.2}$$

$$Fe_{aq}^{3+} + H_2O \overset{hv}{\rightarrow} Fe_{aq}^{2+} + H^+ + \cdot OH \tag{3.3}$$

$$TiO_2 \overset{hv}{\rightarrow} e^- + h^+ \tag{3.4}$$

$$TiO_2 + H_2O \rightarrow Ti\,OH^- + \cdot OH \tag{3.5}$$

$$Ti\,OH^- + h^+ \rightarrow Ti - HO^. \tag{3.6}$$

$$TiO_2 + O_2 + e^- \rightarrow TiO_2 + O_2^- \tag{3.7}$$

$$Fe_{aq}^{3+} + e^- + H_2O \rightarrow Fe_{aq}^{2+} + H^+ + \cdot OH \tag{3.8}$$

$$\cdot OH + H_2O_2 \rightarrow HO_2 \cdot + H_2O \tag{3.9}$$

Based upon all the credentials of the dual effect and having the potential for scale-up, the fixed bed dual effect can provide a platform for the treatment of real industrial wastewater. This technology needs to revisit all the important operating parameters and has proven to be the new emerging technology for the commercial-scale reactor.

3.4 SCALING UP OF FIXED BED REACTORS

Recent advancement in the nanoparticles has led the application of dual technologies at the larger scale. For the commercial-scale applications, successful implementation of AOPs is required at the pilot plant scale using various photo reactors. It is of exceptional scientific importance to scale up the production of nanocatalysts under non-special conditions, and at the same time the methodology should provide enough versatility to be extensible to a wide range of materials.

Besides, the complications related to the conventional reactor system, catalyst contact, mixing, flow patterns, temperature control, and mass transfer might pose hurdles in the potential scale-up. For uniform distribution of the sunlight inside the reactor and for maximizing the exposed surface area of the reactor axial and radial scale-up are also considered essential parameters (Miyawaki et al. 2016).

The intensification of the advanced treatment process is regarded as a promising technique for the industrial processes aspiring for improvement in the efficiency of the process, cost-effectiveness, and environmental safety. Plug flow reactors (PFR) being more efficient than the mixed flow reactors (MFR) are mostly preferred, as the rate of reaction is increased with the increasing concentration of the reactants. More conversion in less time makes the ideal PFR better than MFR. The major drawback of the MFR, i.e. back mixing that could be avoided in PFR, leads to a continuous change in concentration. Furthermore, other hurdles for the AOPs include patterns of flow, catalyst contact, mass transfer, reagents used, mixing, and many more which have posted the hurdle in successful scaling up of the processes (`Malato et al. 2002; Spasiano et al. 2015).

Now taking into consideration various advantages and disadvantages of PFR, a real reactor implicating the ideal PFR is mostly sorted out option. A packed bed reactor with the catalyst immobilized on the support is mostly preferred for such applications. Various obstacles like mass transfer limitations, channeling the flow and the intactness of the catalyst coating are hindering the process scale-up. The stability and durability of the catalyst are also an issue leading to a slower degradation rate along with increased time for the treatment of priority pollutants, forming a major hindrance to the potential scale-up.

The process was intensified with scale-up studies using a non-concentrating fixed-bed reactor (Figure 3.2a) incorporating the Fe–TiO$_2$ nanocomposite with a total volume handling of 5 L (Bansal and Verma 2018a). The process was controlled with proper optimization of all operating parameters like flow rate, number of baffles, catalyst, and oxidant dose. Furthermore, the studies were enhanced using other reactors. The cascade reactor system (six reactors in series) (Figure 3.2b) was used by (Bansal and Verma 2018b) for the scale up trials of dual effect utilizing Fe–TiO$_2$ nanocomposite beads. The film diffusion resistance was reduced utilizing the cascade design of the reactor system with the help of aeration. The commercial viability

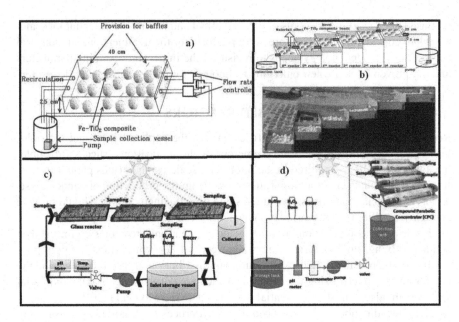

FIGURE 3.2 Line diagram of (a) pilot scale single non-concentrating reactor, (b) cascade reactor system with six packed-bed reactors connected in series, (c) pilot-scale three reactors connected in series, and (d) CPC reactor with the complete setup of the process.

was evaluated by analyzing the various mineralization as well as toxicity studies. As compared with other AOPs, the cost of the dual process was far less.

Talwar, Verma, and Sangal (2020) studied the PFR approaching fixed bed reactor utilizing in-situ dual effect. The exit age distribution was evaluated for the axial dispersion model to be employed and validated. Novel composite material made up of foundry sand and fuller's earth were used for execution of in-situ dual effect. The three reactors in series (Figure 3.2c) were used for the optimization of various parameters. The tentative mechanism for the degradation of metronidazole was also proposed. The industry viability of the reported process was demonstrated by recycling the composite beads up to 80 cycles and evaluating the approximate cost for the removal of the drug.

Talwar et al. (2020) evaluated the feasibility of dual effect utilizing the compound parabolic concentrator (CPC). The CPC reactor behaved as a plug flow approaching model with a hydraulic retention time of 15 min, 6 parabolic reactors connected in series (Figure 3.2d). For the real time analysis of energy intensive parameters, scale-up cost trials have also been studied. The synergy was also evaluated for the dual effect and 2.19 folds were observed as compared to the individual processes.

3.5 SYNERGISTIC CALCULATIONS

The improvement in the degradation efficiency using the dual effect could also be depicted in terms of the synergy obtained. The synergy of the dual process over the individual processes was calculated using kinetic rate constant of individual

TABLE 3.1
Synergistic Studies for Various Hybrid Processes

S. No.	% of Synergy/ Times Synergy	References
1.	32.4	Bansal and Verma (2018a)
2.	85	Talwar, Verma, and Sangal (2019)
3.	2.5	Mazille, Schoettl, and Pulgarin (2009)
4.	1.86	Mecha et al. (2017)
5.	2.19	Talwar et al. (2020)

processes of photocatalysis, photo-Fenton, and dual process. The rate constant was evaluated using the data from the batch reactor. The synergy of the integrated system over the individual process (Madhavan, Grieser, and Ashokkumar 2010; Rajeswari and Kanmani 2009) was studied using Equations 3.10–3.12.

Synergy of dual over photocatalysis process:

$$\% \ Synergy = 100 \times \{(k_{dual} - k_{photocatalysis})\}/k_{dual} \tag{3.10}$$

Synergy of dual over photo-Fenton process:

$$\% \ Synergy = 100 \times \left\{\left(k_{dual} - k_{photoFenton}\right)\right\} / k_{dual} \tag{3.11}$$

Synergy of dual over both photocatalysis and photo-Fenton process, i.e. overall synergy:

$$\% \ Synergy = 100 \times \{(k_{dual} - (k_{photocatalysis} + k_{photoFenton}))\} / k_{dual} \tag{3.12}$$

Using the above equations, Bansal and Verma studied the synergy for the dual process and 28.5% was observed (Table 3.1). Bansal and Verma (2018a) confirmed the overall synergy to be 32.4% for dual process using the single pilot scale set up. Talwar, Verma, and Sangal (2019) showed 85% of the synergy for the dual effect as compared to individual processes for the degradation of metronidazole.

3.6 DURABILITY/RECYCLABILITY OF COMPOSITE

The main challenging task in using composite beads is to establish the dual effect for the continuous number of cycles, i.e. continuous leaching of iron from the support besides maintaining the surface activity of TiO_2. The immobilized catalyst showed the recyclability up to 10 recycles without much loss of efficiency (Sabater, Mata, and Peris 2014). With the Pd-loaded catalyst, the graphene surface showed recyclability up to 10 recycles (Sabater, Mata, and Peris 2014). Most of the studies have shown recyclability only up to 30 recycles, but the composite supports prepared for the dual effect have shown excellent recyclability. The composite beads were recycled for over 200 runs, devoid of much decrease in the efficiency of degradation (Bansal, Verma, and Talwar 2018; Talwar, Verma, and Sangal 2020 2019). The studies confirmed the viability of process at large scale.

3.7 CONCLUSION

The in-situ dual effect has proven to be effective among other AOPs. The $Fe–TiO_2$ nanoparticles for the photocatalytic composite construction have proven to be advantageous. Subsequent structural modifications to the nanocatalyst TiO_2 have prompted the wider applications of $Fe–TiO_2$ composite. Dual-effect technology has been effective and has overcome hurdles faced by the individual AOPs (photo-Fenton and photocatalysis). Dual effect producing more hydroxyl radicals has proven to be effective in terms of degrading various pollutants and producing less harmful by-products. The dual effect at the large scale was also implemented using various reactors including reactors in series, CPC, and step reactor. The optimization of various parameters confirmed the effective degradation of pollutants. The durability of the support also showed the effectiveness of the process and confirms the commercial scale viability of the process. This research unlocks the new horizon toward the degradation of wastewater employing large scale dual effect.

REFERENCES

Adish Kumar, S., G. S. Sree Lekshmi, J. Rajesh Banu, and I. Tae Yeom. 2014. "Synergistic Degradation of Hospital Wastewater by Solar/TiO_2/Fe^{2+}/H_2O_2 Process." *Water Quality Research Journal of Canada* 49 (3): 223–233. https://doi.org/10.2166/wqrjc.2014.026.

Alvarez, Pedro J. J. 2013. "Nanotechnology for a Safe and Sustainable Water Supply : Enabling Integrated Water Treatment and Reuse" *Accounts of Chemical Research* 46 (3). https://doi.org/10.1021/ar300029v.

Antonio, Juan, A. Juan, M. Isabel Pariente, Fernando Martı, Theodora Velegraki, Nikolaos P. Xekoukoulotakis, and Dionissios Mantzavinos. 2008. "Heterogeneous Photo-Fenton Oxidation of Benzoic Acid in Water : Effect of Operating Conditions, Reaction by-Products and Coupling with Biological Treatment." *Applied Catalysis B : Environmental* 85: 24–32. https://doi.org/10.1016/j.apcatb.2008.06.019.

Ayati, Ali, Ali Ahmadpour, Fatemeh F. Bamoharram, Bahareh Tanhaei, Mika Mänttäri, and Mika Sillanpää. 2014. "A Review on Catalytic Applications of Au/TiO_2 Nanoparticles in the Removal of Water Pollutant." *Chemosphere* 107: 163–174. https://doi.org/10.1016/j.chemosphere.2014.01.040.

Bansal, Palak, and Anoop Verma. 2017a. "Novel Fe-TiO_2 Composite Driven Dual Effect for Reduction in Treatment Time of Pentoxifylline: Slurry to Immobilized Approach." *Materials and Design* 125: 135–145. https://doi.org/10.1016/j.matdes.2017.03.083.

Bansal, Palak, and Anoop Verma. 2017b. "Synergistic Effect of Dual Process (Photocatalysis and Photo-Fenton) for the Degradation of Cephalexin Using TiO_2 Immobilized Novel Clay Beads with Waste Fly Ash/Foundry Sand." *Journal of Photochemistry and Photobiology A: Chemistry.* 342: 131–142. https://doi.org/10.1016/j.jphotochem.2017.04.010.

Bansal, Palak, and Anoop Verma. 2018a. "In-Situ Dual Effect Studies Using Novel Fe-TiO_2 Composite for the Pilot-Plant Degradation of Pentoxifylline." *Chemical Engineering Journal* 332:682–694. https://doi.org/10.1016/j.cej.2017.09.121.

Bansal, Palak, and Anoop Verma. 2018b. "Pilot-Scale Single-Step Reactor Combining Photocatalysis and Photo-Fenton Aiming at Faster Removal of Cephalexin." *Journal of Cleaner Production* 195: 540–551. https://doi.org/10.1016/j.jclepro.2018.05.219.

Bansal, Palak, and Anoop Verma. 2018c. "N, Ag co-doped TiO_2 mediated modified in-situ dual process (modified photocatalysis and photo-Fenton) in fixed-mode for the degradation of Cephalexin under solar irradiations." *Chemosphere* 212: 611–619. https://doi.org/10.1016/j.chemosphere.2018.08.120.

Bejarano-Pérez, Néstor Javier, and Marco Fidel Suárez-Herrera. 2007. "Sonophotocatalytic Degradation of Congo Red and Methyl Orange in the Presence of TiO_2 as a Catalyst." *Ultrasonics Sonochemistry* 14 (5): 589–595. https://doi.org/10.1016/j.ultsonch.2006.09.011.

Belver, Carolina, Jorge Bedia, Almudena Go, and Manuel Pen. 2019. "Semiconductor Photocatalysis for Water Purification." *Nanoscale Materials in water Purificationt.* https://doi.org/10.1016/B978-0-12-813926-4.00028-8.

Bokare, Alok D., and Wonyong Choi. 2014. "Review of Iron-Free Fenton-like Systems for Activating H_2O_2 in Advanced Oxidation Processes." *Journal of Hazardous Materials* 275: 121–135. https://doi.org/10.1016/j.jhazmat.2014.04.054.

Buthiyappan, Archina, Abdul Raman, Abdul Aziz, Wan Mohd, and Ashri Wan. 2015. "Recent Advances and Prospects of Catalytic Advanced Oxidation Process in Treating Textile Effluents." *Reviews in chemical engineering* 32: 1–10. https://doi.org/10.1515/revce-2015-0034.

Byrne, Ciara, Gokulakrishnan Subramanian, and Suresh C. Pillai. 2018. Recent Advances in Photocatalysis for Environmental Applications." *Journal of Environmental Chemical Engineering* 6 (3): 3531–3555. https://doi.org/10.1016/j.jece.2017.07.080.

Choi, Jina, Hyunwoong Park, and Michael R. Hoffmann. 2010. "Effects of Single Metal-Ion Doping on the Visible-Light Photoreactivity of TiO_2." *Journal of Physical Chemistry C* 114: 783–792. https://doi.org/10.1021/jp908088x.

Choudhury, Biswajit, Bikash Borah, and Amarjyoti Choudhury. 2012. "Extending Photocatalytic Activity of TiO_2 Nanoparticles to Visible Region of Illumination by Doping of Cerium," *Photochemistry and Photobiology* 257–264. https://doi.org/10.1111/j.1751-1097.2011.01064.x.

Clarizia, L., D. Russo, I. Di Somma, R. Marotta, and R. Andreozzi. 2017. "Homogeneous Photo-Fenton Processes at Near Neutral PH: A Review." *Applied Catalysis B: Environmental* 209: 358–371. https://doi.org/10.1016/j.apcatb.2017.03.011.

Dalrymple, Omatoyo K., Daniel H. Yeh, and Maya A. Trotz. 2007. "Removing Pharmaceuticals and Endocrine-Disrupting Compounds from Wastewater by Photocatalysis." *Journal of Chemical Technology and Biotechnology* 82: 121–134. https://doi.org/10.1002/jctb.1657.

Destaillats, Hugo, A. J. Colussi, Jiju M. Joseph, and Michael R. Hoffmann. 2000. "Synergistic Effects of Sonolysis Combined with Ozonolysis for the Oxidation of Azobenzene and Methyl Orange." *Journal of Physical Chemistry A* 104: 8930–8935. https://doi.org/10.1021/jp001415+.

Elmehasseb, Ibrahim, Saleh Kandil, and Khaled Elgendy. 2020. "Advanced Visible-Light Applications Utilizing Modified Zn-doped TiO_2 Nanoparticles via Non-Metal In Situ Dual Doping for Wastewater Detoxification" *Optik – International Journal for Light and Electron Optics*, 164654. https://doi.org/10.1016/j.ijleo.2020.164654.

Essawy, Amr A. 2018. "Silver Imprinted Zinc Oxide Nanoparticles: Green Synthetic Approach, Characterization and Efficient Sunlight-Induced Photocatalytic Water Detoxification." *Journal of Cleaner Production* 10110–10120. https://doi.org/10.1016/j.jclepro.2018.02.214.

Esteves, Bruno M., Carmen S. D. Rodrigues, Rui A. R. Boaventura, Luís M. Madeira, and F. J. Maldonado-h. 2016. "Coupling of Acrylic Dyeing Wastewater Treatment by Heterogeneous Fenton Oxidation in a Continuous Stirred Tank Reactor with

Biological Degradation in a Sequential Batch Reactor" 166. https://doi.org/10.1016/j.jenvman.2015.10.008.

Fatimah, Siti, Abdul Halim, Azlina Harun Kamaruddin, and W. J. N. Fernando. 2009. "Bioresource Technology Continuous Biosynthesis of Biodiesel from Waste Cooking Palm Oil in a Packed Bed Reactor : Optimization Using Response Surface Methodology (RSM) and Mass Transfer Studies." *Bioresource Technology* 100 (2): 710–16. https://doi.org/10.1016/j.biortech.2008.07.031.

Fu, Min, Yalin Li, Siwei Wu, Peng Lu, Jing Liu, and Fan Dong. 2011. "Sol-Gel Preparation and Enhanced Photocatalytic Performance of Cu-Doped ZnO Nanoparticles." *Applied Surface Science* 258: 1587–1591. https://doi.org/10.1016/j.apsusc.2011.10.003.

Guz, A. N., and Ya Ya Rushchitskii. 2003. "Nanomaterials : On the Mechanics of Nanomaterials." *International Applied Mechanics* 39 (11): 1271–93.

Hodaifa, Gassan, and Gassan Hodaifa. 2017. "Real Olive Oil Mill Wastewater Treatment by Photo-Fenton System Using Artificial Eltraviolet Light Lamps." *Journal of Cleaner Production* 162: 743–753. https://doi.org/10.1016/j.jclepro.2017.06.088.

Huang, Miaoliang, Chunfang Xu, Zibao Wu, Yunfang Huang, Jianming Lin, and Jihuai Wu. 2008. "Photocatalytic Discolorization of Methyl Orange Solution by Pt Modified TiO_2 Loaded on Natural Zeolite." *Dyes and Pigments* 77: 327–334. https://doi.org/10.1016/j.dyepig.2007.01.026.

Ismail, A, and E. F. Eladl. 2019. "Photocatalytic Degradation of Reactive Blue 21 Using Ag Doped ZnO Nanoparticles" *Journal of Material and Environmental Sciences* 10 (12): 1258–1271.

Jamila, Ghulam Sughra, Sajjad Shamaila, Sajjad Ahmed, Khan Leghari, and Tariq Mahmood. 2019. "Graphical Abstract." *Journal of Physics and Chemistry of Solids*, 109233. https://doi.org/10.1016/j.jpcs.2019.109233.

Kanakaraju, Devagi, Beverley D. Glass, and Michael Oelgemöller. 2018. "Advanced Oxidation Process-Mediated Removal of Pharmaceuticals from Water: A Review." *Journal of Environmental Management* 219: 189–207. https://doi.org/10.1016/j.jenvman.2018.04.103.

Kerrami, A., L. Mahtout, F. Bensouici, M. Bououdina, S. Rabhi, E. Sakher, and H. Belkacemi. 2019. "Synergistic Effect of Rutile-Anatase Fe-Doped TiO_2 as Efficient Nanocatalyst for the Degradation of Azucryl Red." *Materials Research Express* 6: 8. https://doi.org/10.1088/2053-1591/ab2677.

Kim, Hyung Eun, Jaesang Lee, Hongshin Lee, and Changha Lee. 2012. "Synergistic Effects of TiO_2 Photocatalysis in Combination with Fenton-like Reactions on Oxidation of Organic Compounds at Circumneutral PH." *Applied Catalysis B: Environmental* 115–116: 219–224. https://doi.org/10.1016/j.apcatb.2011.12.027.

Kočí, K., K. Matějů, L. Obalová, S. Krejčíková, Z. Lacný, D. Plachá, L. Čapek, A. Hospodková, and O. Šolcová. 2010. "Effect of Silver Doping on the TiO_2 for Photocatalytic Reduction of CO_2." *Applied Catalysis B: Environmental* 3: 239–244. https://doi.org/10.1016/j.apcatb.2010.02.030.

Koppala, Sivasankar, Yi Xia, Libo Zhang, Jinhui Peng, Zhen Chen, and Lei Xu. 2019. "Hierarchical ZnO/Ag Nanocomposites for Plasmon-Enhanced Visible-Light Photocatalytic Performance." *Ceramics International* 45 (12): 15116–15121. https://doi.org/10.1016/j.ceramint.2019.04.252.

Laxma, P. Venkata, Beluri Kavitha, Police Anil, Kumar Reddy, and Ki-hyun Kim. 2017. "TiO_2 -Based Photocatalytic Disinfection of Microbes in Aqueous Media : A Review." *Environmental Research* 154: 296–303. https://doi.org/10.1016/j.envres.2017.01.018.

Mackenzie, John D., and Eric P. Bescher. 2007. "Chemical Routes in the Synthesis of Nanomaterials Using the Sol – Gel Process." *Accounts of Chemical Research* 40 (9): 810–818.

Madhavan, Jagannathan, Franz Grieser, and Muthupandian Ashokkumar. 2010. "Combined Advanced Oxidation Processes for the Synergistic Degradation of Ibuprofen in Aqueous Environments." *Journal of Hazardous Materials* 178 (1–3): 202–208. https://doi.org/10.1016/j.jhazmat.2010.01.064.

Mahlambi, Mphilisi M., Catherine J. Ngila, and Bhekie B. Mamba. 2015. "Recent Developments in Environmental Photocatalytic Degradation of Organic Pollutants : The Case of Titanium Dioxide Nanoparticles — A Review." *Journal of Nanomaterials* 2015: 1–29.

Malato, Sixto, Julián Blanco, Alfonso Vidal, and Christoph Richter. 2002. "Photocatalysis with Solar Energy at a Pilot-Plant Scale: An Overview." *Applied Catalysis B: Environmental* 37: 1–15. https://doi.org/10.1016/S0926-3373(01)00315-0.

Márquez, Juan José Rueda, Irina Levchuk, and Mika Sillanpää. 2018. "Application of Catalytic Wet Peroxide Oxidation for Industrial and Urban Wastewater Treatment: A Review." *Catalysts* 8: 673. https://doi.org/10.3390/catal8120673.

Mazille, F., T. Schoettl, and C. Pulgarin. 2009. "Synergistic Effect of TiO_2 and Iron Oxide Supported on Fluorocarbon Films. Part 1: Effect of Preparation Parameters on Photocatalytic Degradation of Organic Pollutant at Neutral PH." *Applied Catalysis B: Environmental* 89 (3–4): 635–644. https://doi.org/10.1016/j.apcatb.2009.01.027.

Mecha, Achisa C., Maurice S. Onyango, Aoyi Ochieng, and Maggy N. B. Momba. 2017. "Evaluation of Synergy and Bacterial Regrowth in Photocatalytic Ozonation Disinfection of Municipal Wastewater." *Science of the Total Environment* 601–602: 626–635. https://doi.org/10.1016/j.scitotenv.2017.05.204.

Miyawaki, Atsuko, Shunya Taira, and Fumihide Shiraishi. 2016. "Performance of Continuous Stirred-Tank Reactors Connected in Series as a Photocatalytic Reactor System." *Chemical Engineering Journal* 286: 594–601. https://doi.org/10.1016/j.cej.2015.11.007.

Nanda, Binita, and Dharitri Rath. 2020. "Materials Today : Proceedings Surface-Plasmon-Resonance Induced Photocatalysis by Cu (0)/Cu(II)@ g-C_3N_4/MCM-41 Nanosphere towards Phenol Oxidation under Solar Light." *Materials Today: Proceedings* 1–5. https://doi.org/10.1016/j.matpr.2020.01.537.

Ni, Meng, Michael K. H. Leung Ã, Dennis Y. C. Leung, and K. Sumathy. 2007. "A Review and Recent Developments in Photocatalytic Water-Splitting Using TiO_2 for Hydrogen Production" *ACS Applied Nano Materials* 11: 401–425. https://doi.org/10.1016/j.rser.2005.01.009.

Nikhila, M. P., Deepthi John, Mrinal R. Pai, and N. K. Renuka. 2020. "Nano-Structures & Nano-Objects Cu and Ag Modified Mesoporous TiO_2 Nanocuboids for Visible Light Driven Photocatalysis." *Nano-Structures & Nano-Objects* 21: 100420. https://doi.org/10.1016/j.nanoso.2019.100420.

Plodinec, Milivoj, Ivana Grčić, Marc G. Willinger, Adnan Hammud, Xing Huang, Ivana Panžić, and Andreja Gajović. 2018. "Facile Method to Enhance the Adhesion of TiO_2 Nanotube Arrays to Ti Substrate." *Journal of Alloys and Compounds* 6: 8001–8005. https://doi.org/10.1016/j.jallcom.2018.10.248.

Pradeep, T. 2009. "Noble Metal Nanoparticles for Water Purification : A Critical Review." *Thin Solid Films* 517 (24): 6441–6478. https://doi.org/10.1016/j.tsf.2009.03.195.

Rajesh Banu, J., R. Yukesh Kannah, S. Kavitha, M. Gunasekaran, and Gopalakrishnan Kumar. 2018. "Novel Insights into Scalability of Biosurfactant Combined Microwave Disintegration of Sludge at Alkali pH for Achieving Profitable Bioenergy Recovery and Net Profit." *Bioresource Technology* 267: 281–290. https://doi.org/10.1016/j.biortech.2018.07.046.

Rajeswari, R., and S. Kanmani. 2009. "Degradation of Pesticide by Photocatalytic Ozonation Process and Study of Synergistic Effect by Comparison with Photocatalysis and UV/Ozonation Processes." *Journal of Advanced Oxidation Technologies* 12 (2): 208–214.

Sabater, Sara, Jose A. Mata, and Eduardo Peris. 2014. "Catalyst Enhancement and Recyclability by Immobilization of Metal Complexes onto Graphene Surface by Noncovalent Interactions." *ACS Catalysis* 6: 2038–2047. https://doi.org/10.1021/cs5003959.

Savage, Nora, and Mamadou S. Diallo. 2005. "Nanomaterials and Water Purification : Opportunities and Challenges," *Journal of Nanoparticle Research* 7: 331–342. https://doi.org/10.1007/s11051-005-7523-5.

Sharma, Kritika, Steffi Talwar, Anoop Kumar Verma, Diptiman Choudhury, and Borhan Mansouri. 2020. "Innovative Approach of In-Situ Fixed Mode Dual Effect (Photo-Fenton and Photocatalysis) for Ofloxacin Degradation." *Korean Journal of Chemical Engineering* 37, 350–357. https://doi.org/10.1007/s11814-019-0427-3.

Shekofteh-gohari, Maryam, Aziz Habibi-yangjeh, and Masoud Abitorabi. 2018. "Magnetically Separable Nanocomposites Based on ZnO and Their Applications in Photocatalytic Processes: A Review" *Critical Reviews in Environmental Science and Technology* 1: 1–52. https://doi.org/10.1080/10643389.2018.1487227.

Singh, Lovjeet, Pawan Rekha, and Shri Chand. 2018. "Comparative Evaluation of Synthesis Routes of Cu/Zeolite Y Catalysts for Catalytic Wet Peroxide Oxidation of Quinoline in Fi Xed-Bed Reactor." *Journal of Environmental Management* 215: 1–12. https://doi.org/10.1016/j.jenvman.2018.03.021.

Sirés, Ignasi, Conchita Arias, Pere Lluís Cabot, Francesc Centellas, José Antonio Garrido, Rosa María Rodríguez, and Enric Brillas. 2007. "Degradation of Clofibric Acid in Acidic Aqueous Medium by Electro-Fenton and Photoelectro-Fenton." *Chemosphere* 66: 1660–1669. https://doi.org/10.1016/j.chemosphere.2006.07.039.

Song, Shuang, Haiping Ying, Zhiqiao He, and Jianmeng Chen. 2007. "Mechanism of Decolorization and Degradation of CI Direct Red 23 by Ozonation Combined with Sonolysis." *Chemosphere* 66: 1782–1788. https://doi.org/10.1016/j.chemosphere.2006.07.090.

Spasiano, Danilo, Raffaele Marotta, Sixto Malato, Pilar Fernandez-Ibañez, and Ilaria Di Somma. 2015. "Solar Photocatalysis: Materials, Reactors, Some Commercial, and Pre-Industrialized Applications. A Comprehensive Approach." *Applied Catalysis B: Environmental* 170–171: 90–123. https://doi.org/10.1016/j.apcatb.2014.12.050.

Steter, J. R., D. Dionisio, M. R.V. Lanza, and A. J. Motheo. 2014. "Electrochemical and Sonoelectrochemical Processes Applied to the Degradation of the Endocrine Disruptor Methyl Paraben." *Journal of Applied Electrochemistry* 50: 9601–9608. https://doi.org/10.1007/s10800-014-0742-7.

Tahir, Muhammad Bilal. 2019. "The Detoxification of Heavy Metals from Aqueous Environment Using Nano-Photocatalysis Approach: A Review." *Environmental Science and Pollution Research* 26(11): 10515–10528.

Talwar, Steffi, Vikas Kumar Sangal, and Anoop Kumar Verma. 2019. "In-Situ Dual Effect of Novel Fe-TiO$_2$ Composite for the Degradation of Phenazone." *Separation and Purification Technology* 211: 391–400. https://doi.org/10.1016/j.seppur.2018.10.007.

Talwar, Steffi, Anoop Kumar Verma, and Vikas Kumar Sangal. 2019. "Modeling and Optimization of Fixed Mode Dual Effect (Photocatalysis and Photo-Fenton) Assisted Metronidazole Degradation Using ANN Coupled with Genetic Algorithm." *Journal of Environmental Management* 250: 109428. https://doi.org/10.1016/j.jenvman.2019.109428.

Talwar, Steffi, Anoop Kumar Verma, and Vikas Kumar Sangal. 2020. "Plug Flow Approaching Novel Reactor Employing In-Situ Dual Effect of Photocatalysis and Photo-Fenton for the Degradation of Metronidazole." *Chemical Engineering Journal* 382: 122772. https://doi.org/10.1016/j.cej.2019.122772.

Talwar, Steffi, Anoop Kumar Verma, Vikas Kumar Sangal, and Urška Lavrenčič Štangar. 2020. "Once through Continuous Flow Removal of Metronidazole by Dual Effect of Photo-Fenton and Photocatalysis in a Compound Parabolic Concentrator at Pilot Plant Scale." *Chemical Engineering Journal* 388: 124184.https://doi.org/10.1016/j. cej.2020.124184.

Tisa, Farhana, Abdul Aziz, Abdul Raman, Wan Mohd, and Ashri Wan. 2014. "Applicability of Fl Uidized Bed Reactor in Recalcitrant Compound Degradation Through Advanced Oxidation Processes: A Review." *Journal of Environmental Management* 146: 260–275. https://doi.org/10.1016/j.jenvman.2014.07.032.

Tran, Nam, Patrick Drogui, Laurent Nguyen, and Satinder K. Brar. 2015. "Optimization of Sono-Electrochemical Oxidation of Ibuprofen in Wastewater." *Journal of Environmental Chemical Engineering* 3: 2637–2646. https://doi.org/10.1016/j.jece.2015.05.001.

Varatharajan, Bhaskaran, and S. Kanmani. 2007. "Treatability Study of Pharmaceutical Wastewater by Combined Solar Photo Fenton and Activated Sludge Process." *Journal of Industrial Pollution Control* 23 (1): 157–164.

Vieira Dos Santos, E., C. Sáez, P. Cañizares, C. A. Martínez-Huitle, and M. A. Rodrigo. 2017. "Treating Soil-Washing Fluids Polluted with Oxyfluorfen by Sono-Electrolysis with Diamond Anodes." *Ultrasonics Sonochemistry* 34: 115–122. https://doi.org/10.1016/j. ultsonch.2016.05.029.

Yan, Ying, Songshan Jiang, and Huiping Zhang. 2014. "Efficient Catalytic Wet Peroxide Oxidation of Phenol over Fe-ZSM-5 Catalyst in a Fixed Bed Reactor." *Separation and Purification Technology* 133: 365–374. https://doi.org/10.1016/j.seppur.2014.07.014.

Yan, Ying, Songshan Jiang, Huiping Zhang, and Xinya Zhang. 2014. "Preparation of Novel Fe-ZSM-5 Zeolite Membrane Catalysts for Catalytic Wet Peroxide Oxidation of Phenol in a Membrane Reactor." *Chemical Engineering Journal* 259: 243–251. https://doi. org/10.1016/j.cej.2014.08.018.

Yang, Lixia, Yan Xiao, Shaohuan Liu, Yue Li, Qingyun Cai, Shenglian Luo, and Guangming Zeng. 2010. "Photocatalytic Reduction of Cr(VI) on WO_3 Doped Long TiO_2 Nanotube Arrays in the Presence of Citric Acid." *Applied Catalysis B: Environmental* 94: 142–149. https://doi.org/10.1016/j.apcatb.2009.11.002.

Yu, Jiaguo, Quanjun Xiang, and Minghua Zhou. 2009. "Preparation, Characterization and Visible-Light-Driven Photocatalytic Activity of Fe-Doped Titania Nanorods and First-Principles Study for Electronic Structures." *Applied Catalysis B: Environmental* 90: 595–602. https://doi.org/10.1016/j.apcatb.2009.04.021.

Zhang, Lijie, and Thomas J. Webster. 2009. "Nanotechnology and Nanomaterials : Promises for Improved Tissue Regeneration," *Nanotoday* 4: 66–80. https://doi.org/10.1016/j. nantod.2008.10.014.

4 Microbiological Degradation of Organic Pollutants from Industrial Wastewater

*Ayantika Banerjee**
University of Science and Technology

*Moharana Choudhury**
Voice of Environment (VoE)

Arghya Chakravorty and Vimala Raghavan*
Vellore Institute of Technology

*Bhabatush Biswas**
National Institute of Technology

Siva Sankar Sana
North University of China

Rehab A. Rayan
Alexandria University

Nalluri Abhishek
Huazhong University of Science and Technology

Neeta L. Lala and Seeram Ramakrishna**
National University of Singapore

CONTENTS

---

* Authors have equal contribution on this chapter.

4.1   INTRODUCTION

Overwhelming metal contamination is viewed as one of the enduring worldwide eco-logical issues. In spite of the fast progressions made over the most recent couple of decades in the field of designing sciences and clinical well-being, substantial metals present in water sources and surrounding air ceaselessly incite unfriendly impacts on general well-being and the earth. Instead of its normal event, overwhelming metals are generally utilized in various applications in industry, farming, and barrier tasks. Modern effluents are debased with overwhelming metals, for example, zinc (Zn), lead (Pb), copper (Cu), cadmium (Cd) and nickel (Ni), which are lethal in nature (Vimala and Das, 2009). Metals are non-biodegradable and can be bio-aggregated in the evolved way of life, prompting cancer-causing impacts on plants, creatures, and people. Altogether, to restrain the introduction of lethal overwhelming metals to nature and general well-being, norms and rules have been built up by various asso-ciations, for example, the United States Environmental Protection Agency (USEPA), the World Health Organization (WHO) and the European Union (EU).

Aside from the shocking ecological issues, substantial metals are non-sustainable normal assets. Regular stores of substantial metals are being drained due to their exorbitant use in modern applications. Accordingly, the expulsion of substantial metals ought to be engaged alongside their recuperation. Diverse physicochemical advances for the expulsion of overwhelming metals from wastewater, both at the lab and modern scale, have been proposed: layer filtration, compound precipitation, particle trade and turn around assimilation, coagulation–flocculation, buoyancy, and electrochemical strategies (Fu and Wang, 2011).Despite the fact that broad research has been done on the utilization of these advancements, there are still a few downsides identified with their applications. Particularly on account of the low groupings of overwhelming metals, the vast majority of the regular innovations are costly and wasteful (Al-Mossawi, 2020; Say et al., 2001; Vimala et al., 2011); as a result of worldwide worry toward natural insurance and the advancement of greener remediation strategies for contamination remediation, biosorption has become one of the promising strategies for expelling metal and metalloid particles from wastewater (Das et al., 2008; Gautam et al., 2014). The utilization of nanoparticles (NPs) for the expulsion of overwhelming metals from wastewater has extraordinary intrigue. Various examinations have exhibited the fruitful utilization of nanotechnology for the remediation of metal contamination. Research has been directed on various nanomaterials, for example, carbon nanotubes (CNTs), dendrimers, nanostructured reactant films, nanosorbents, nanocatalysts, bioactive NPs, biomimetic layers, and molecularly engraved polymers (MIP) for expelling toxins. Utilization of naturally integrated NPs for the remediation of overwhelming metals from wastewater is a developing examination theme in the field of nanotechnology for wastewater executives (Jain et al., 2016; Järup, 2003; Ranjan et al., 2014). Considering this line of dynamic research, this chapter manages the use of nanosorbents with an extraordinary spotlight on organism-based nanotechnology for the expulsion of overwhelming metals from wastewater.

4.2   GLOBAL STATUS OF WATER DEMAND
AND ECONOMIC IMPACT

Since the 1950s, the worldwide population has doubled to the current 7.8 billion people and with the estimated growth rate of 1.05% per year, the population is increasing at 81 million people per year (Ozcan et al., 2020). This growing population is depending on just 2.5% of fresh water sources of the total earth's surface water. Also, the demand for water is rising at a rate of 1% every year with a threefold increase in global water used in the past 50 years. Globally, this demand is expected to increase at a similar rate until 2050, accounting for an increase of 20–30% above the current level of water use (UNESCO, 2019). While climate change is one of the major reasons of extreme weather conditions including droughts, resulting in water-scarce or water-dry zones around the world, the ecological footprints (EFs) with overexploitation of the available resources has further taken this very commodity to a toll (Ozcan et al., 2020; Rezić, 2011).

It has been reported in a study, that, more than 50% of the lakes and rivers of the world are polluted with industrial waste and discharges. Due to such an unintended growing pollution level, large amount of waterbodies have become unsafe and untenable for drinking and other usable purposes.

Virtual water is a concept developed by Allan in 1993 and it refers to the volume of water required to produce goods and services. This in turn is linked to the concept of "water footprint." This is given by the following expression (Hossain and Mertig, 2020):

Water footprint = domestic water withdrawal + water imports − water exports

In a recent study, it has been observed that around 16% of global fresh water is traded across the countries in the form of virtual water through the exchange of goods and services. Middle East nations like Turkey, Syria, and Iraq face serious drought-related problems due to absence of proper international law for water distribution from the rivers that flow across these nations [4]. In addition to that a country like the U.S. invests millions to decrease pollution in rivers, lakes, and other surface waters. A public report cited that, since 1960, over 1.9 trillion USD (2014) have been spent to abate surface water pollution. It accounted an expenditure of over $140 per person per year, or over 35 billion USD total per year (Keiser et al., 2019).

Recognized human basic rights also include safe drinking water and sanitation, as they are important for sustaining healthy livelihoods and, above all, to maintain the dignity of human beings. To this spirit, the international forums and bodies have devised ambitious goals to meet the growing demands and mitigate the crises of water. The sustainable development goals (SDGs) of the UN have explicitly mentioned the objective to improve water quality by 2030. The targets in the Agenda 2030 of SDGs are as follows (Alcamo, 2019):

• Reducing pollution;
• Eliminating dumping and minimizing release of hazardous chemicals and materials;
• Halving the proportion of untreated wastewater; and
• Substantially increasing recycling and safe reuse globally.

4.3   GLOBAL EFFORT TO STOP WATER POLLUTION

Nowadays, preventing pollution is a great challenge with the climate change issue. Rapid growth of urbanization and industrialization is the main issue of global pollution. Land, air, soil pollution, and the quality of water are degrading day by day. Polluted water is becoming a concern for tropic levels of ecosystem. Water pollution is now a great threat to our environment. Some polluted substances degrade the water body as well as the living organisms that are affected. The conditions of a water body such as pH, temperature, and acidity are affected by those polluted substances. Water has a big role in the agricultural field. Insufficient clean water is affecting irrigation and soil quality, which is the main reason of bioaccumulation.

Sustainable development needs to be considered where a society can get a standard of living without getting any threat from the environment. It also needs to solve all economic and social problems to protect the natural world from any kind of destruction and damage (Inyinbor Adejumoke et al., 2018). To achieve this, science bodies, research foundations and governments of all countries need to set a goal and aim to achieve that.

Among all environmental components, water is an vital resource for all living organisms, so the demand of clean water is getting greater worldwide. To get clean and safe water, a global effort is needed to face all challenges and problems.

4.4 SOURCES OF WATER POLLUTION

There are many sources for water pollution and also substances, which are the main reason for degrading water quality, such as chemicals, heavy metals, dye, organic pollutants, pharmaceutical chemicals, etc.

- Heavy metals are the main inorganic pollutant substances which have negative effects on plants, aquatic organisms, and human beings. Heavy metals can be present in soil and may be absorbed by plants, creating a reason of plant metabolism dysfunctioning (Lajayer et al., 2017). These substances release to the environment mainly from mining industries. Heavy metals in crops and aquatic organisms enter into human or animal bodies through the food chain and food web. Accumulating heavy metals into the body can cause skin irritation, headache, stomach-ache, diarrhea, organ dysfunctioning, hypertension, gastrointestinal problem, etc. (Dada et al., 2016). Besides this, exposure of lead and mercury is highly toxic to animals and humans. Recently, Arsenic (As) contained in groundwater is a huge problem. Groundwater in almost all Asiatic countries like India, China, the Lao People's Democratic Republic, Myanmar, Cambodia, and Pakistan are affected by arsenic (UNESCAP-UNICEF_WHO, May 2–4, 2001, Expert Group Meeting, Bangkok)
- Twenty-three countries have reported about the groundwater contamination via arsenic and its' impact on human beings and aquatic lives (Purkayastha et al., 2015). In India, the main river, Ganga, has been contaminated with arsenic for a long period of time. Taking arsenic-contaminated water is a cause of lung cancer, liver cancer, skin diseases etc.
- Organic pollutants are the second most important substance related to water pollution and it has a wide variety of toxicity. Organic pollutants include dyes, which have a good solubility in water and are used in many industries like textile, leather and tanning, and food. Endocrine disruptive chemicals (EDCs) refers to a group of chemicals that are classified as emerging contaminants, and an accumulation of this group of chemicals can cause hormonal-imbalance and affect the reproduction system of humans (Gillan, 2016; Jung et al., 2015). Pharmaceutical samples are considered as a great threat to human and aquatic life (Archer et al., 2017).
- Pathogens are also a cause of unhealthy water. Pathogens like *Cryptosporidium*, *Salmonella*, *Shigella*, etc. can be present in highly contaminated water and make the water unsafe for drinking (Inyinbor Adejumoke et al., 2018).
- Personal care products like shampoo, perfume, etc., found in groundwater and other aquatic systems, can pollute the water (Noguera-Oviedo and Aga, 2016).

4.5 MITIGATION AND REMEDIATION

Water is an essential element for our human body. Water helps to digest foods, blood circulation, dissolve protein, and more in our bodies. So water is a much-needed element in this world, no doubt. Recently some studies stated that a very high percentage of world population lacks safe and clean water for their daily life (Costa et al., 2016; Purkayastha et al., 2015). Usually, we collect water from rivers, springs, ponds, or underground for our daily lives. Sometimes in many developing countries, one source of water is used for various needs of daily life like, drinking, washing, bathing, etc. So water sources need to be clean and uncontaminated. There are many ways to treat the wastewater.

- Anaerobic wastewater treatment is one of the best methods to remove the organic pollutants from waterbodies. There are several steps of anaerobic digestion methods to treat the water such as, hydrolysis, acid fermentation, methane, or mentation (Lettinga et al., 1984).
- A flotation process which is successfully applied as a cleaning method of the water body by using air bubbles (Sarbu and Lorand, 2006).
- Water contamination could be monitored by microbial activities.
- Wastewater treatment plants follow several technologies to treat water and make it as clean as possible.
- Others like septic tanks, denitrification, wetlands, etc. can prevent the waterbodies from pollution.

4.6 GLOBAL EFFORTS: POLICIES AND AWARENESS

Once water is contaminated and infected, it is costly and sometimes impossible to treat the water and reuse for drinking purposes. Polluted water body is harmful for the human body as well as aquatic life and ecosystem. So to keep the waterbody clean, we need to follow some rules and regulations to make some policies that can be implemented by governments, hold awareness programs to teach people how to keep water safe and clean. To get a sustainable environment and society's global effort is much needed.

- Policies should be made by each government and need to be implemented in all social stages of a society. A framework should be made with clear documents and laws and ensure that people will follow it. Agenda 21 has made two statements to arise the concerns about water pollution, "Fresh water should be seen as finite and vulnerable resource, essential to sustain a life, development and environment" and "Water should be considered as a social and economic good with a value reflecting its most valuable potential use." Apply some strategies like "polluter-pays principle," which would encourage people to use less stress on environmental components. Water-quality criteria should be set and regulated with conditions. Water pollution control boards should examine the quality of waterbodies in short time intervals. Legal actions need to be taken if a waterbody starts degrading its quality. Except for national policies, state policies should be made. Ensure the conservation, restoration, and sustainable use of freshwater. Appropriate management systems should be made to control and prevent pollution.
- Policies and strategies cannot come into effect if normal people don't take part in it. Protecting a waterbody from pollution means every household

waste should be treated well. People need to be concerned about the quality of the water they are taking to use for their daily lives. Sewages, medical wastes, and agricultural field wastes need to be treated before mixing with rivers or any waterbodies. Awareness programmed at state and national levels should be made to educate people about controlling and mitigating the pollution issues. Policymakers should encourage common people to raise their voice about pollution control ideas. Decision makers should create a small group with common people and should consult with them. Scientific experts can treat the wastewater, but the general public can help to keep the waterbody safe and clean. Public participation is a necessity to stop pollution. Some small programs should be held to increase the interest among the general public. Give methods to contact the public for public participation in this issue.

There are many more strategies to follow for getting clean and safe water. Clean water and salinity are a main goal to sustainable development. To achieve sustainable development and keep the environment clean and safe for living, global effort is most needed. Policymakers, scientists, governments, and the general public all need to create and follow a strategy to get clean and safe water in our daily life.

4.7 WASTEWATER TREATMENT METHODS

Water is one of the most sought after resources in the modern world; continuous depletion and contaminants have led to this state. The huge demand for pure drinking water has led to many different methods of water treatment. The most conventional methods like adsorption or chemical treatment fall back. The most important problem persisting today is the presence of multiple contaminants in wastewater and also new emerging pollutants. These combined contaminants are microorganisms, heavy metals, toxic chemicals, and dyes (Quan et al., 2015). Treatment of such combined pollutants is not effective with a single treatment process but involves multiple processes.

One novel approach for the treatment of wastewater employed today is the use of nanomaterials. Nanotechnology provides promising techniques employing nanomaterials, because conventional methods used find it very difficult to treat the complex wastewater output and provide quality and sustainable pure drinking water. Owing to their very small size, nanomaterials showcase different physical, chemical, and biological properties, and, hence, are effective treatment methods as compared to conventional methods. Nanotechnology can provide a clean way for environmentally friendly means of water treatment processes.

There are two major categories of nanomaterials – organic and inorganic. NPs are further divided into two, based on their origin as natural and incidental (Monica and Cremonini, 2009). There are four ways in which nanomaterials are employed in wastewater treatment: dendrimers, metal-containing nanomaterials, zeolites, and carbonaceous nanomaterials. Polymerosomes are NPs that find immense use in water treatment methods.

4.8 SYNTHESIS OF NANOMATERIALS

Nanoparticles are synthetically prepared by larger structures by a bottom-up approach; two major processes involved are gas phase synthesis and sol-gel processing. Another approach for the synthesis of nanomaterials is biosynthesis using microorganisms.

Biosynthesis of nanomaterials is a synthetic process leading to the synthesis of nanomaterials using microorganisms which are cheaper and is also a green approach. Biosynthesis of nanomaterials is done with the help of plants, bacteria, yeast, and fungi, while the use of plants is much considered because of the simplicity of the process.

4.8.1 WASTEWATER TREATMENT USING NANOTECHNOLOGY

Nanotechnology implies the use of nanoscale filtration techniques, adsorption, and breakdown of contaminants by NP catalysts, in the cleanup of contaminated water, which has been summarized by Smith (2006). Adsorption, photocatalysis, membrane technology, disinfection with metal ions, and nanobioremediation are some techniques that are used alone or in combination for pollutant treatment.

- **Adsorption:** Adsorption is used mostly for the removal of heavy metal ions and organic compounds from water. It involves the use of CNTs, metal oxides, and dendrimers. Nanosized metal oxides like ferric oxides, manganese oxides, and aluminum oxides are currently used. CNTs and dendrimers prevail as the most effective treatment process. Adsorption of polyatomic ions like malate and arsenate has been successfully treated by malachite NPs by Saikia et al. and Wang et al. who described the use of magnetite NPs for the removal of heavy metal ions like lead and chromium wherein competitive adsorption was noticed. Chemical adsorption is also known as chemisorption, which is site-specific and chemisorbed. Here the molecules, when absorbed, reacts with the surface or interface to form a chemical or ionic bonds in-between molecules, following Coulomb's Law. On the other hand, physical adsorption doesn't involve sharing of electrons and is not site-specific like chemical adsorption. The most common commercial adsorbent used in wastewater treatment is activated carbon, zeolites, and synthetic polymers. Activated carbon (AC) has a wide range of capabilities to remove heavy metals from the water because of its' large microporous and mesoporous volume and high surface area. Extensive research is going on based on the usefulness of AC in wastewater treatment (Saikia et al., 2011; Wang et al., 2014a).
- **Photocatalysis:** Photocatalysis is one among the most appreciated technique for removal of microbial contaminants. In photocatalysis, a light active catalyst helps in the degradation of pollutants by the release of highly reactive radicals. Nanoparticles associated with advanced oxidative processes (AOPs) have been developed for wastewater treatment. Titanium oxide NPs are used for photocatalysis owing to its good photoactivity and non-toxicity highly active oxidants which help in the elimination of

microbial contaminants in wastewater. S.-H. Lee et al., used TiO_2 nano-composites with multiwalled CNTs showing complete inactivation of bacterial endospores (Lee et al., 2005; Tuzen et al., 2008).

- **Membrane technology: Membrane filtration and reverse osmosis:** Membrane filtration is also an important technology to remove organic solids, as well as inorganic compounds and suspended solids through making a barrier to transport metals between two phases. There are several membrane filtration technologies depending on the size of particulate matter needs to be removed, such as, ultrafiltration, microfiltration, nanofiltration, and reverse osmosis.

Ultrafiltration utilizes permeable membranes which can able to separate heavy metals, suspended solids and organic compounds of 5–20 nm pore size. This technique follows the Darcy's Law (Hu and Apblett, 2014). There are two types of ultra-filtration such as Miceller enhanced UF (MEUF) and Polymer Enhanced UF (PEUF) to make the membrane filtration more efficient. Samper et al., used MEUF to remove Cd(II), Cu(II), Pb(II), and Zn(II) by using two anionic surface. Cu(II) ion can be removed by using dendron enhanced UF (DEUF) and poly amidoamine dendrimers with ethylene diamine (EDA) core and NH_2 group (Diallo et al., 2005; Tabe, 2014).

Microfiltration (MF) follows the same technique as ultrafiltration (UF). MF membrane have large pore size which rejects large particulates and microorganisms, while UF membrane has smaller pore size which rejects nanoparticulates respectively.

Nanofiltration is a process which follows the techniques between UF and reverse osmosis. NF has a better capability to remove heavy metal ions like copper (Ahmad and Ooi, 2010), chromium (Muthukrishnan and Guha, 2008), arsenic (Figoli et al., 2010) etc.

Osmosis is a process of solvent flow through a semi-permeable membrane, by using the pressure it allows the purified fluid to pass through the membrane. The pressure usually applied on the side of membrane where salts are added, and as the pore size of the membrane is too small so salt is unable to flow through it. The driving force for creating the osmotic pressure occurs through concentration gradient. Shahalam et al. (2002) stated that, it has more than 20% of the world's desalination capacity. Nanoparticles are used in the manufacture of polymeric and inorganic membranes which are effectively used for wastewater treatment process. Nanomembranes are seen as pretreatment process but with the addition of NPs the efficacy of the system can be improved. This can emerge as the most sought-after treatment process because most nanomembranes can be made and are selectively permeable. Nanomembranes play a role in desalination process compared to reverse osmosis. CNTs have shown increased permeability and small pore size which in turn results in the removal of microbial contaminants. Single walled CNTs showed much higher adsorption rate because the mechanism used is the production of oxidative stress and cell membrane disturbance to contaminants. CNTs is radically arranged to form filtration membranes which can successfully eliminate bacteria and viruses.

- **Metal ions:** Biological contaminants in water can be treated with metal ions. Embedded silver NPs, titanium dioxide NPs as photocatalyst, iron oxide nanomaterials, chitosan nanomaterials are used in the treatment of biological contaminants. Zhang et al., has successfully depicted the use of NPs for the successful removal of bacteria and viruses without the production of harmful byproducts. Silver NPs have been used extensively in the production of antimicrobial nanofibers, low cost portable microfilters (Zhang et al., 2010).

• **Nanobioremediation:** The use of NPs biosynthesized from plants, microbes, enzymes in the treatment of wastewater is known as nanobioremediation. In this process the complex pollutants are first broken down into smaller particles with the help of encapsulated enzymes and then acted upon by these microbes into much simpler compounds. Fungal NPs and plant NPs are used in bioremediation of heavy metals and petroleum pollutants. The use of nanosized iron particles in remediation proves a better technique.

With the increase in multiple contaminants in water, where the presence of one pollutant affects the removal of another pollutant effectively, one single technique employed seems ineffective. Combinations of different techniques, like adsorption and photocatalysis, have been successfully involved in the treatment of wastewater containing pharmaceutical pollutants and toxic metals. Adsorption, disinfection, and microbial control is also used in combination for the removal of toxic metals, microbes, and organic pollutants by Cai et al., with the use of MgO NPs. Photocatalysis, microbial control, and disinfection are also combined wherein the nanocomposite core is enclosed in a mesoporous shell. Even though nanomaterials help in treatment of wastewater, certain factors like the adsorption efficiency of the pollutant to the nanomaterial, pH of the contaminated water, amount of nanomaterial and the contact time with pollutants, proper disposability of the pollutant after treatment and the ecotoxicity of the NPs have also been examined (Cai et al., 2017).

4.9 ADVANTAGES OF USING NANOPARTICLES FOR WASTE/ POLLUTED WATER TREATMENT

Water containing unwanted substances which can be the cause of the degradation of its quality and for this reason water can become unusable turning into wastewater. Polluted substances such as organic trace compounds can be mixed into the river from various sources such as residential areas, commercial properties, agricultural fields, municipal wastes, etc. Rapid industrialization and urbanization are the cause of generating municipal sewages, which is one of the main sources of water pollution. Day by day, water bodies, like rivers and lakes, are becoming polluted, so the demand for clean water is increasing at an alarming rate. To achieve the demand for clean water, wastewater should be treated well so that it can be reused for each and every purpose. There are lots of technologies for wastewater treatment, but current researchers are focusing on different types of nanomaterials by using advance nanotechnology.

Research on nanotechnology is a promising aspect in medicine, food industry, robotics, and textiles (Singh et al., 2012).In the desalination and purification of water, nanotechnology is becoming a promising application(Diallo et al., 2005). In this chapter, this topic will be discussed regarding various aspects of nanotechnology and their applications.

4.9.1 NANOTECHNOLOGY: WHAT? AND WHY?

Water usually contains inorganic heavy metals, organics, and microorganisms that can be treated and managed through various advanced nanotechnologies, mainly through adsorption and membrane separation. Nanomaterials have unique size-dependent properties with their specific surface area and high adsorption rate. Nanotechnology suggests that this area of research can provide us a wide aspect to remove unwanted polluted substances from water to make it usable for daily life. Membranes and membrane processes like nanofiltration, nanocomposite, and self-assembling, are the best nanotechnologies for removing organic materials, heavy metals, radionuclides, etc. Nanotechnology-derived water filtration processes are much more cost-effective and efficient than other traditional techniques (Diallo et al., 2005; Singh et al., 2012).

4.9.2 APPLICATION OF NANOTECHNOLOGY IN WASTEWATER TREATMENT

For wastewater treatment by using nanotechnology the nanoscopic materials like CNTs or alumina fibers for nanofiltration, nanocatalysts, magnetic NPs and even nanosensors like titanium oxide nanowires, etc. (Prajapati and Sen, 2013). Table 4.1 provides the nanomaterials used for water filtration and purification and their application (Gehrke et al., 2015).

There are many effective methods of removing heavy metals and other organic materials by using different processes related nanotechnology, which has been discussed earlier.

TABLE 4.1

Illustration of the Nanomaterials with Their Respective Properties and Applications Used for Water Filtration

Nanomaterials	Properties	Applications
Nanoadsorbents	High surface energy and adsorption rate, but high production cost	Able to remove heavy metals, organic materials, bacteria
Nanometals and nanometal oxides (nanosilver, TiO_2, magnetic nanoparticles)	Supermagnetic, abrasion-resistant, short intraparticle diffusion distance compressible, photocatalytic and low production cost	Able to remove heavy metals, radionuclides, powders, pellets
Membrane and membrane processes (nanofiltration, nanocomposite, self-assembling, nanofiber-membranes)	Reliable and automated process, can provide a physical barrier for substances depending on their pore and molecule size, but required high amount of energy	Every field of water and wastewater treatment processes

4.10 DEVELOPMENT AND RECENT APPLICATION OF NANOTECHNOLOGY IN WASTEWATER TREATMENT

Some advanced materials used in nanotechnology are discussed below:

- **CNTs:** CNTs have been reported by Iijima (1991) and Bethune et al. (1993),which have become a revolutionary discovery in a wide aspect. CNTs are like cylinder-shaped macromolecules with a few nanometer radii. CNTs are used in the adsorption process for treating the wastewater and as CNTs are hydrophobic in nature, it has the ability to be the most powerful adsorbent. CNTs are able to remove heavy metals like polynuclear aromatic hydrocarbons (PAHs) (Yang et al., 2007), DDT and metabolites (Zhou et al., 2006), dyes, pesticides, and herbicides. CNTs are not considered to be a strong oxidant. Multi-walled carbon nanotubes (MWCNT), nanosilica, serve as the potentiometric membrane for removing Yb^{3+}, Er^{3+}, Pb^{2+}, and Cu^{2+} (Norouzi et al., 2010; Soltani et al., 2018). CNT-modified electrodes showed higher efficiency for stripping analysis than any other modified electrodes. With recent developments in wastewater treatment, CNTs have become an important factor for removing a variety of elements.
- **Metal oxide nanomaterials:** Metal oxides like TiO_2, ZnO, ZrO_2, and NiO have high reactivity and photolytic property and are considered as high-quality adsorbent because of their large surface area (Klaine et al., 2008). Singh et al. (2011) stated that, toxic ions such as Co^{2+}, Ni^{2+}, Cu^{2+}, Cd^{2+}, Pb^{2+}, and Hg^{2+} can be removed by ZnO nano-assemblies. Pharmaceuticals and personal care products (PPCPs) are a group of pollutants that are being released to the environment by industrial and urban wastes affecting the environment in a large way (Hu et al., 2012). PPCPs needs to be removed from water, too, but using the traditional techniques like coagulation/flocculation, filtration, and sedimentation does not give us satisfactory results on removing these compounds. While the photocatalytic degradation by TiO_2 nanowires used as adsorbent to remove PPCPs is suggested as an effective process. Effectiveness of MgO NPs and Mg NPs act as biocides against Gram-positive and Gram-negative bacteria (Stoimenov et al., 2002).
- **Metal NPs:** A metal NP like gold, palladium, and nanosilver have high surface reactivity and antimicrobial properties. These materials can be used as antimicrobials for coliform found in wastewater. Gold coated with palladium is very useful to remove tri-chloroethane from groundwater. Silver nanometals can show the antimicrobial activity even if the concentration is low.

Water is the essential part of all life on this earth. Due to increase in population, climate change, the water quality is degrading. Heavy metals, organic and inorganic substances are mixed with the water body through industrial waste, biological waste, and municipal wastes. The need for safe and clean water is essentially high and nanotechnology is becoming a promising option in the field of wastewater treatment. Some risk factors also related to nanotechnology has been discovered. The toxicity of nanometal ions are still in concern. Treating water by using nanometals can cause the bioaccumulation or bioremediation within the environment and can reach to higher tropic levels. Nanotechnology used for mine-water treatment is limited because of their surface-active property. Before using NPs or metals related to nanotechnology the characteristics and properties of all metals should be known and defined properly. From a green technology view point, the toxicity and environmental fate of nanosorbents are areas of great concern during material selection and design for wastewater treatment. In continuous systems, parameters such as inoculums size, pH, temperature, ionic strength and heavy metal concentrations play a major role in determining the kinetics of the heavy metal removal (Mukherjee et al., 2002; Tahir et al., 2019). To use nanotechnology in a large scale, research should be conducted on further process.

4.11   DIFFERENT NANOPARTICLES AND APPLICATIONS

There are various usages of NPs in wastewater treatment. Some of those have been discussed in Table 4.2.

4.11.1   Nanosorbents as Heavy Metals Removing Agent

Organization of new water treatment advancements using nanomaterials has gotten much consideration because of their exceptional physico-concoction and organic properties, such as particularly small size, high surface zone to volume proportion, surface modifiability, attractive properties, short intra-particle dissemination separation, alterable surface science, and biocompatibility (Amin et al., 2014; Meerburg et al., 2012). For the most part, adsorption is considered a cleaning venture to expel the natural and inorganic contaminants present in wastewater. The use of ordinary adsorbents is constrained due to restricted dynamic locales, surface zone, absence of selectivity, and adsorption energy. In addition, ordinary adsorbents require increasingly crude materials for handling and readiness when viewed as vitality concentrates where adsorbents are to be altered for their surface properties. Squander decrease can be accomplished by utilizing green nanotechnology so as to combine nanomaterials and use them as adsorbents. Progressive quantitative research ought to be done all together to approve these cases. Near investigations on vitality and material utilization are missing in the writing to supplant customary materials with NPs. Besides, investigating holes despite everything that exists on the points, for example, the well-being of utilizing nanosorbents and their latent capacity sway on the earth. Often these issues are appropriately tended to nanosorbents which are modest, environmentally friendly, and a reasonable option for "training nations" to treat modern wastewater as opposed to utilize the costly regular strategies (Liu et al., 2008; Sheet et al., 2014).

4.11.2   Mechanism of Heavy Metals Removal

Nanosorbents are nanoscale particles produced from natural or inorganic materials that have a high liking to retain substances. They have diverse physical and organic

TABLE 4.2
Types of Nanoparticles used for Various Applications

S. No.	Type of Nanoparticle	Shape, Size (nm)	Applications	References
1	NiO		Removal of Pb	Amira et al.(2015)
2	CdWO$_4$	Rod, 54	Removal of Bismarck brown R	Bushra et al. (2019)
3	FeNPs		Wastewater treatment	Bhavika et al.(2018)
4	FeS	200	Absorption of naphthol green B	Shuhua et al. (2019)
5	TiO$_2$	Spherical, 10–30	Photocatalytic activity	Perumal et al. (2012)
6	ZnO	Spherical, 21.5	Catalytic degradation of dyes (methyl orange, methyl red, and methylene blue)	Fatemeh et al.(2015)
7	AgNPs	Spherical, 25	Catalytic degradation of 4-Nitrophenol	Kannan Badri et al. (2013)
8	Fe–TiO$_2$	Quasi-nanospheres 9.37–10.33	Wastewater treatment	Ricardo et al.(2019)
9	Magnetite		Degradation of Ar-NO$_2$	Mathew et al.(2017)
10	Silica	Spherical	Corrosion resistance	Vijayalakshmi et al. (2017)
11	Ag@AgCl	Spherical, 15	Determination of mercury	Sadegh and Tayebeh (2019)
12	Pd	Spherical, 4-20	Dehalogenation of polychlorinated dioxins	Michael et al. (2014)
13	AgNPs	Spherical, 35	Degradation of azo-dyes	Chinnashanmugam et al. (2017)
14	CuNPs	Spherical, 22–33	Azo dye degradation and treatment of textile effluents	Muhammad et al. (2020)
15	AuNPs	Spherical, 72.32 ±21.80	Degradation of rhodamine B	Arpit et al. (2016)

properties than their typical size counterparts. For instance, nanomaterials of metals and metal oxides convey a higher proportion of surface region to molecule size that prompts the display of various optically active and electrical properties (El Saliby et al., 2008). They contain unsaturated surface particles which help them to shape solid synthetic bonds with the metal particles (Gao et al., 2011; Lemos et al., 2008). They can likewise be functionalized with different concocted gatherings for the increment of their fondness toward a specific compound, for example improved specific properties (Bhattacharya et al., 2013). These physio-concoction attributes demonstrate the adequacy of nanomaterials in water-cleaning frameworks. Application of these particles as adsorbent leads to the accessibility of a high number of molecules or atoms on the outside of contaminants; in this way, upgrading the adsorption limits of sorbent materials. Various investigations have shown the effective use of nanosorbents, for example, CNTs, nanoscale metal oxides, and nanofibers for the expulsion of substantial metals from wastewater (Sheet et al., 2014).

Among the distinctive nanosorbents announced in the literature, the carbon-based nanosorbents have gotten incredible consideration in the field of substantial metal remediation. Lee et al., have reported carbon-based nanosorbents by ethylene decay on tempered steelwork without the utilization of outside impetuses for the treatment of water containing Ni(II) particles. Various group sorption tests were performed by the creators to decide the impacts of beginning pH, starting metal fixation, and contact time on Ni(II) expulsion by the nanosorbents. The motor information fitted well to a pseudo-second-request model demonstrating that the procedure was the chemisorption type through the balance information fitted to the Freundlich isotherm. Further investigation by the Boyd dynamic model uncovered that the primary system of adsorption was constrained by limited layer dispersion. This investigation proposes that prepared carbon-based nanosorbent is a promising sorbent for the sequestration of Ni(II) from fluid arrangements (Lee et al., 2012).

Di et al. have reported that Cr adsorption on Cerium NPs is upheld on adjusting carbon nanotubes (CeO$_2$/ACNTs) when examined. This epic adsorbent was set up by the synthetic response of CeCl$_3$ with NaOH in a CNT arrangement followed by heat treatment. The most extreme adsorption limit of this adsorbent at pH 7.0 has shown 30.2 mg g^{-1} at a balance of Cr(VI) convergence of 35.3 mg l^{-1} (Di et al., 2006). Li et al. and Lu and Chiu have reported that contemplated the adsorption limit of Zn and Pb onto decontaminated CNTs. The most extreme adsorption limit was seen to be 43.7 mg g^{-1} for Zn and 30.3 mg g^{-1} for Pb, individually. As per the creators, the adsorption information for Zn best fitted the Langmuir isotherm, though Pb adsorption was seen as pH-ward and it followed the Freundlich isotherm model (Li et al., 2005; Lu and Chiu, 2006).

4.12 HEAVY METALS REMOVAL BY FUNGI-MEDIATED NANOSORBENTS

For modern applications, the immobilization of the parasitic biomass is basic to keep up microbial cell movement in a dangerous domain, the re-utilization of biomass, the utilization of traditional response frameworks, and for the absence of biomass-fluid detachment necessities. To accomplish the correct size, mechanical quality, non-bending nature, and porosity of the biosorbent, immobilization of contagious biomass on various materials is performed. Various systems have been drilled for the immobilization of contagious biomass. For instance, reticulated froths, actuated carbon and glass raschig rings have been utilized as idle backings for the immobilization of biomass. For one case, polyurethane froths were utilized as supporting materials to immobilize the biomass of *Aspergillus terreus* so as to treat substantial metals, for example, iron, chromium, and nickel (Dias et al., 2002; Kitching et al., 2015). Polymeric frameworks, for example, calcium alginate, polyacrylamide, polysulfone, polyethylenimine, and polyhydroxoethylmethacrylate (polyHEMA) are likewise utilized for the immobilization of biomass. Immobilization of contagious biomass with NPs is another examination line in the field of biosorption of substantial metals. Iron oxide nanomaterials have additionally demonstrated great potential for the immobilization of biomass because of their synthetic idleness and ideal biocompatibility (Xu et al., 2012). All in all, nanomaterials should be steady to maintain a strategic distance from accumulation and accomplish low statement rates. Despite the fact that NPs are powerful for the adsorption of overwhelming metals, they have a few downsides, for example, low mechanical quality and stopping up of channels in consistent course through reactors. It is accounted for that nanomaterials, will in general, total its arrangement. All the more generally, the electrostatic what's more, van der Waals collaborations influence the strength of colloidal NPs. Various arrangements, for example, the utilization of stabilizer, electrostatic surfactant, and steric polymers have been generally proposed. For example, TiO_2 and SiO_2 NPs immobilized on various supporting materials have been utilized to treat Cd(II) whereby the supporting material is assisted on expanding the penetrability of the arrangement and the total does not go through the channel paper (Kalfa et al., 2009). Fungal biomass has stood out in wilderness explore zones so as to get compelling nanobiocomposites for the expulsion of metal contaminations since they offer points of interest, for example, great physical, synthetic, and natural steadiness (Perullini et al., 2010).

Filamentous contagious biomass stacked TiO_2 NPs utilized this blend as a sorbent for Pb evacuation. The adsorbent had a more drawn out life expectancy contrasted with the unadulterated NPs of TiO_2 and the adjusted biosorbent was seen as modest in light of the fact that the contagious biomass utilized was gotten as a side effect of an oil plant. Alongside this, the pre-concentration level for this nanobiocomposite is seen to be much higher (multiple times) in contrast with other biocomposites of organisms, for example, *Aspergillus fumigatus* immobilized Diaion HP-2MG (Soylak et al., 2006), *Penicillium italicum* stacked on Sepabeads SP(Tuzen et al., 2006), *Pseudomonas aeruginosa* immobilized onto multiwall carbon and *Aspergillus niger* stacked on silica gel (Baytak et al., 2007).

Appearing in a few late writing reports, these sorts of novel nanobiosorbents will have extensive applications in the field of substantial metal expulsion from wastewater. Such immobilization procedures help to pre-condition the parasitic biosorbent for an enormous scope of applications. Albeit, a few immobilization media and methods have been tried, little data is accessible on joining nanotechnology with existing biotechniques for reasonable applications. In expansion, screening studies ought to be performed to choose the best microorganism that can be utilized for immobilization, and the likelihood to scale up these nanobiocomposites must be evaluated (Tabi and Verdon, 2014).

4.13   METAL IONS DESORPTION WITH POST WATER
TREATMENT SCENARIO

Studies directed by Lu and Chiu and Li et al. have uncovered that metal particles adsorbed onto nanotubes can be handily expelled by modifying the pH estimations of the arrangement utilizing both HCl and HNO_3. In any case, the uses of nanosorbents in wastewater treatment will perpetually prompt the arrival of NPs into the earth. Following their versatility, bioavailability and perseverance in nature are the basics to evaluate their latent capacity chance in the given condition (Li et al., 2005; Lu and Chiu, 2006). Conventional techniques for the evacuation of particulate issues during wastewater treatment like flocculation, sedimentation, and filtration will not be compelling to evacuate these NPs because of their small size (Bhattacharya et al., 2013). Thus, nitty-gritty basic research is used to evaluate how nanomaterials are discharged into nature and economical eco-innovations ought to be created to treat those discharged NPs in water.

4.14 WORKING CHALLENGES

The utilization of nanotechnology-based biosorbents has demonstrated promising outcomes to treat wastewater contaminated with substantial metals. Albeit, a large portion of the nanobiosorbents have so far been explored distinctly at the research center scale. For full-scale applications, the determination of an appropriate contagious biomass, financial and vitality utilization parts of immobilization, pilot-scale testing to affirm the reuse and recovery limit of the nanosorbent, ought to be researched. From a green innovation perspective, the poisonous quality and ecological destiny of nanosorbents are regions of worry during material choice and structure for wastewater treatment. In consistent frameworks, parameters for example, inoculum size, pH, temperature, ionic quality, and substantial metal focuses, assume a significant job in deciding the energy of substantial metal expulsion (Kim et al., 2020), as mentioned in Table 4.3.

4.15 BIOGENIC NANOPARTICLES AND ITS ADVANTAGES

Lately, biogenic nanoparticles (BNPs) have been receiving substantial research concern after the evolution of the paradigm of microorganisms interacting with minerals (Sarı and Tuzen, 2009). For their distinctive characteristics, BNPs have electronic, biomedical, water and wastewater treatment technology (W/WWT) applications (Falih, 1997). Nowadays, the technology of producing BNPs through microorganisms' interaction is considerably recognized (Dias et al., 2002). Meanwhile, producing nanoparticles (NPs) by traditional techniques is usually dangerous and costly because of using toxic stabilizing and reducing agents (Dwivedi et al., 2013). Nevertheless, BNPs could be fabricated safely by employing microorganisms as a reductant (Ceribasi and Yetis, 2004). Additionally, water contaminated with heavy metals could likewise be utilized rather than salts for manufacturing BNPs, which is a technique for safely using industrial and domestic wastewater and restoring resources (Aftab et al., 2013). The dramatic rise in minerals' cost that are being employed as raw materials in most industries involving manufacturing cement, tanning leather, metallurgy, electronics, and electroplating, has driven research attention to restoring these minerals from contaminated soils and water (Luef et al., 1991).

4.16   ALL BIOGENIC NANOPARTICLES

TABLE 4.3
The Biosorption of Different Fungus

Name of Fungus	Type of Treatment	Heavy Metal and Adsorption Capacity (mg g^{-1})	pH	References
Phanerochaete chysosporium	FeO nanoparticles immobilized pellets	a) Pb(II) – 176.33	5.0	Xu et al.(2012)
Phanerochaete chysosporium	Dry biomass	b) Pb(II) – 45.25 c) Cd(II) – 13.24 d) Cu(II) – 10.72	6.0	Say et al.(2001)
Phanerochaete chysosporium	Inactivated pellet	a) Cd(II) – 15.5 b) Pb(II) – 12.34	4.5	Li et al.(2004)
Phanerochaete chysosporium	Washed cells	a) Pb(II) – 90.0 b) Cu(II) – 43.0 c) Cd(II) – 17.0	5.0	Gopal et al.(2002)
Phanerochaete chysosporium	Pellet with modified surface	a) Cr(VI) – 279.9	3.0	Chen et al.(2011)
Phanerochaete chysosporium	Resting cells	b) Pb(II) – 73.56 c) Ni(II) – 77.96	4.0	Ceribasi and Yetis (2004)
Lentinus edodes	Inactivated live pellets	a) Cd(II) – 78.6 b) Zn(II) – 33.7 c) Hg(II) – 336.3	6.0	Falih (1997)
Aspergillus flavus	Dry biomass	d) Zn(II) – 287.8	5.0	Aftab et al.(2013)
Aspergillus flavus	Polyurethane foam form	e) Ni(II) – 19.6 f) Fe(II) – 164.5 g) Cr(VI) – 96.5	4.5	Dias et al.(2002)
Aspergillus flavus	Dry biomass	h) Pb(II) – 12.44 i) Cr(II) – 0.05 j) Ni(II) – 0.53	NA	Dwivedi et al.(2013)
Mucor rouxii	NaOH treated biomass	a) Zn(II) – 7.75 b) Ni(II) – 11.09 c) Cd(II) – 8.46 d) Pb(II) – 35.69	5.0	Yan and Viraraghavan (2003)
Mucor rouxii	Non-living biomass	e) Zn(II) – 16.62 f) Ni(II) – 6.34 g) Cd(II) – 8.36 h) Pb(II) – 25.22	5.0	Yan and Viraraghavan (2003)
Amanita rubescents	Dry biomass	i) Cd(II) – 27.3 j) P(II) – 38.4	5.0	Sarı and Tuzen (2009)

4.17 ADVANTAGES OF BNPs

In addition to safe production, BNPs have been used to generate electricity, clean contaminated water and soils, and dispose or restore valuable and rare minerals (Qu et al., 2017), hence, there is a growing research interest in using BNPs for W/WWT. Because of the distinct bacterial carrier matrix, BNPs have been applied as adsorbents, biocatalysts, oxidants, and reductants in disposing toxic and emerging pollutants (T&EPs) from wastewater and drinking water (Gusseme et al., 2012). Most investigations declared that the diversity of reducing substance created via cells of the bacteria (charged opposite to targeted contaminants) and the unique functioning groups such as carboxylate, hydroxyl, amide I, II and II or methyl are mainly accounted for disposing pollutants through ion-exchange procedure and electrostatic interactions (Ahluwalia et al., 2016). Additionally, supplying an electron donor such as formate or hydrogen gas stimulates BNPs to degrade T&EPs (Suja et al., 2014). Some researchers have shown that the molecular oxygen could be fabricated by Mn-oxidizing bacteria to bio-degrade recalcitrant contaminants through integrating co-precipitation, electrostatic interaction, and co-metabolism of the bacteria (Su et al., 2014). Despite BNPs have demonstrated outstanding results in disposing T&EPs, yet, some issues are primarily challenging in commercializing BNP-derived W/WWT techniques (Ali et al., 2019).

4.18 METHODS TO DEVELOP BIOGENIC NANOPARTICLES

Guided by the scientific advances in BNPs manufacturing and significant applications in several disciplines, many pieces of research have examined their various characteristics (Shamaila et al., 2016). Diverse types of BNPs such as semiconductor, organic, inorganic, chalcogenide, or quantum dots, could be developed utilizing variants of microorganisms coupled with different applications (Mal et al., 2016). Generally, the minerals' salts such as silver, selenium, palladium, and platinum, have been utilized to develop BNPs, and thus their applications have been examined for various technologies (Holmes and Gu, 2016; Siddiqi and Husen, 2016). However, some investigations have underlined the scope of BNPs in W/WWT (Nancharaiah et al., 2016).

Nonetheless, the information level was limited to only few BNPs such as bio-Mn oxides NPs, bio-palladium, and biogenic Fe(II) (Ali et al., 2019). Consequently, this section presents an extensive update on using BNPs in disposing of T&EPs including protocols for manufacturing and methods for removing contaminants.

4.18.1 Production Process

Bio-reduction is primarily a biologic method to produce BNPs employing varieties of plants, microorganisms, and templates (Sun et al., 2016). A wide spectrum of NPs could be manufactured utilizing many biological substances, for example, fungi, bacteria, algae, yeast, viruses, and plants. Every substance possesses its distinct means of biochemical mechanism for generating NPs, for instance, oxidation/

reduction of mineral ions by sugars, enzymes, proteins, aldehyde, polyphenol carboxyl groups with plant varieties and microorganisms (Lu et al., 2017). Varied sets
of BNPs could be fabricated applying several mineral ions, for example, silver, gold,
alloy, titanium, silica, palladium, antimony sulfide, selenium, and others by various biologic substances (Qu et al., 2017). Though microorganisms, especially those
that develop NPs through environmental bio-mineralization or biologic, are more
interesting. Biologic means, for instance, biologic-controlled mineralization (BCM)
and biologically induced mineralization (BIM) are famous for manufacturing BNPs
(Martins et al., 2017). During BCM procedures, BNPs are made inside the cells
within definitive circumstances whereby the nucleation and growth of the NPs are
totally dominated by the microbes. The known magnetotactic bacteria (MTB) are
created magnetosomes (MS) by the BCM procedure that has been used in eliminating and retrieving mineral ions from industrial wastewater (Ali et al., 2017).

During BIM, the cell wall/membrane filtrate of the bacteria is manipulated using a
solution of mineral ions and BNPs developed by precipitation and/or reduction procedure (Yue et al., 2016). Additionally, microorganisms fabricate BNPs through intracellular or extracellular means of complexation, bio-mineralization/bio-accumulation, and
biosorption. Basically, the microorganisms mediate the interaction between an electron donor organic substance and an electron acceptor mineral ions' solution, which
contributes significantly to the aerobic and the anaerobic BNP synthesis process (Raja
et al., 2016). Moreover, mineral ions will reduce on the walls of the cell for the existing
various metabolites such as carboxyl, amino, thiol, imidazole, guanidine, and others
(Furgal et al., 2015). Later, BNPs would evolve via surface growth and nucleation procedure. Yet the shape and the structure could be regulated through manipulating the
manufacturing conditions such as temperature, pH, oxygen, and others (Su et al., 2014).

4.19   USING WASTEWATER-RETRIEVED METALS
IN SYNTHESIZING BNPs

Usually, wastewater is deemed to be a bio-factory to restore nutrients for use in
agriculture and to produce energy (Zan and Wu, 2016). But, for the rising prices and
the minimal availability of minerals, water contaminated with heavy metals is being
used to restore minerals, which are now getting more attention (Mal et al., 2016).
Various techniques such as ion-exchange, flotation, adsorption, membrane filtration,
and electro-chemical treatment have been created and enhanced to use wastewater in
eliminating and restoring mineral ions. Yet such techniques are challenged by poor
rates of elimination and restoration, utilizing toxic chemicals, generating by-products,
complicated conditions for implementation, and rising expenses among others. In
this regard, sorption is surpassing other approaches due to the simple and adaptable
conditions for implementation, the humble price, and the yield of high-quality effluents (Puyol et al., 2017). Moreover, the different functional groups on the surface of
sorbents could cause mineral ions to be adsorbed preferentially. Still, the procedure
of producing and operating adsorbents for a certain use is expensive(Zhuang et al.,
2015). Therefore, the techniques of biotechnology are providing various economical
means to concentrate and transform the valuable supply of minerals recovered from
wastewater to worthy materials (Siddiqi and Husen, 2016).

Applying microorganisms to restore minerals demands fewer waste-generating chemicals and labor than the available remaining techniques (Huang et al., 2015). Additionally, various sorts of procedures such as precipitation, ion-exchange, complexation, and adsorption could be managed only by modifying chemical groups, for example, phosphoryl, hydroxyl, carboxyl, carbonyl, or thiol moieties upon BNPs' surface (Puyol et al., 2017). Several studies stated that mineral salt solution has been utilized in BNPs synthesis (Ali et al., 2019), while literal wastewater has seldom been used for such purpose (Sun et al., 2016). In this regard, a variety of bioreactors were created to use metal-contaminated water in restoring minerals. The researchers found that the performance of restoring minerals from wastewater relies largely on the effluent's content and the matrix since wastewater sometimes carries an undesirable blend of multi-contaminants that could impede or decrease the restoration efficiency. Moreover, the effluents mineral concentration and their separation could as well affect this process. Hence, restored biogenic minerals should be insoluble in their effluents (Ali et al., 2019).

4.20   FACTORS IMPACTING THE USE MICROORGANISMS TO DEVELOP BNPs

Several types of mineral ion solutions and microorganisms have been used in developing BNPs. Their characteristics and shape depend on the precursor, the microorganisms, and the implementing circumstances. Some factors could impact the development of BNPs, for example, the blending concentration of an electron donor and acceptor, the kind of microorganism, temperature of incubation, the pH, and the contacting time (Tuo et al., 2013). Moreover, such factors could alter the capacity and the shapes of BNPs to eliminate a certain contaminant (Narayanan et al., 2013), yet quite few studies are made. Generally, investigators are only emphasizing applying various microorganisms to develop BNPs (Ali et al., 2019).

4.21   BNP CHARACTERIZATION APPROACHES REGARDING ACTIVITY

Characterization is the key element of NPs and it demonstrates their efficiency according to its dimensions, surface area, electric conduction, surface chemistry, steric features, functioning groups, charge and quantum aspects, covering style, hydrophobicity, liquid solubility, morphology and topology characteristics, indicators, fluorescence, UV/visible absorption, physico-chemical contents, skin penetration, and crystal structure (Qu et al., 2017). Several characterization approaches have been applied for characterizing microorganism-produced BNPs such as x-ray diffraction (XRD) and powder x-ray diffraction (PXRD) to determine the crystal morphology; selected area electron diffraction (SAED) and transmission electron microscopy (TEM) to determine the precise micro-structural parameter; Fourier transformed infrared analysis (FTIR) to determine the possible BNP's stabilizing and reducing functioning groups; energy dispersive spectroscopy (EDS) and scanning electron microscopy (SEM) to inspect the shapes; vibrating sample magnetometer (VSM) to determine the magnetic extent and to outline M-H hysteresis loop and Braunauer–Emmett–Teller (BET) to outline the N_2- adsorption-desorption isotherm (Wang et al., 2014b).

Moreover, approaches for electro-microscopy such as energy dispersive x-ray fluorescence (EDXRF), x-ray photoelectron spectroscopy (XPS), atomic force microscope (AFM), atomic absorption spectrophotometer (AAS), energy dispersive x-ray spectroscopy (EDX), x-ray absorption spectroscopy (XAS) and ultra violet-visible spectrophotometer (UV–VIS) have also been used to determine the chemical contents, size, morphology, and topology of the synthesized BNPs (Ahluwalia et al., 2016). Such approaches are sometimes applied for characterizing the developed BNPs to examine the reduction of metals and whether the reduction is intracellular or extracellular. In contrary, few studies are available to address the relations between the morphology, dimensions, characterization, and efficiency of BNPs in eliminating contaminants (Ali et al., 2019).

4.22 CHEMISTRY OF REMEDIATION USING BIOGENIC NANOPARTICLES

The use of living systems and plant extracts to detoxify environmental pollution is known as "bioremediation." It involves various remediation strategies like bioaugmentation, biostimulation, GMOs, phytoremediation, and biomineralization. The major constraint of bioremediation is a time-consuming process. To accelerate the remediation rate, new improved technology integrating nanotechnology and bioremediation termed nanobioremediation has been developed. Nanobioremediation involves usage of biosynthesized nanomaterials usually in the form of NPs that are effectively used for treatment of water pollutants. Conventionally, NPs are synthesized by physical or chemical routes. Recently, biosynthesis of NPs is an emerging green technology. In this process, NPs are produced from microorganisms like bacteria, fungi, algae, and other biological sources. Biosynthesis of NPs proves an alternate eco-friendly, cost effective option compared to other processes for large scale synthesis of NPs. Apart from the above-mentioned advantages, bio-synthesis eliminates the release of effluent toxins created after the chemical synthesis of NPs. Biosynthesis is more advantageous than other processes owing to the fact that it is environmentally friendly, and has size and characteristic customization capabilities. TiO_2 biosynthesis has been carried out by using *Bacillus* sp. and *Lactobacillus* sp. (Jha et al., 2009; Kirthi et al., 2011). Biosynthesis from plant extracts perform dual action as reducing and capping agents, thus controlling the size and shape of the NPs. Nanoparticles are also produced from fungal strains, Bansal et al., produced silica and titania NPs from fungus species. Silver NPs are also produced with the help of fungi, since it helps with huge metabolite production in a short time span. Silver NPs finds much usage in various streams from electronics to pharmaceuticals. Biosynthesis from plant extracts proves a much sought after the process as incubation period and other factors do not apply as in microbes (Bansal et al., 2005; Vijayalakshmi et al., 2017).

Pollutant treatment by NPs involves three steps: detection, prevention, and remediation. Bioremediation is an emerging remediation technology, and it involves the breakdown of complex structures into simpler units with the help of biosynthesized

NPs. Various morphological forms in nanoregime such as nanotubes, nanoshells, nanocomposites, nanofibers, nanoclusters are used in nanobioremediation. CNTs and zero-valent iron particles are used extensively in the treatment process. Nanoshells, spherical particles having a dielectric core and a thin metallic shell are used in the treatment of industrial effluents containing organic dyes.

Reactive techniques are often applied ex-situ or in-situ, wherein ex-situ involves degradation of organic contaminants without harmful by-products, and in-situ involves removal of heavy metal contaminants by sequestration. In the ex-situ process, biogenic particles enhance the remediation of organic contaminants. In in-situ treatment, nanoscale, zerovalent irons or other NPs are introduced into contaminated sites, which nullify the contaminant concentrate. The colloidal nature of NPs helps in the deeper penetration into contaminated sites. Gold and silver NPs are widely used for the degradation of organic dyes; however, iron NPs are used widely in environmental clean-up (Tratnyek and Johnson, 2006). Biofilm formation is one threat posed in drinking water systems; as it also induces corrosion in the water. The use of nanocrystalline semiconductor thin films prevents such biofilm formations.

Photocatalysis: Photocatalysis proves to be an economic process with ease to handle in large scale applications. Higher photocatalytic activity of TiO_2 is used for environmental cleanup. TiO_2 in different shapes and sizes is employed in photocatalysis. The photocatalyst is located in the center of the substrate where there is microbial attachment. During irradiation of light, the bacterial cell walls get disrupted due to the generation of free radicals on the particle surface. TiO_2NPs were biosynthesized from *Bacillussubtilis*, made into slurry with distilled water and coated onto a glass slide. It was then allowed for biofilm formation. During irradiation hydroxy radical generated results in the production of H_2O_2 which destroys the microbes within the biofilm. The production of H_2O_2 can be augmented by enhancing the active surface area (Dhandapani et al., 2012; Perumal et al., 2012). It has been found that NPs with metal doping, non-metal doping, and co-doping can be used to increase light response. Fe-doped NPs prepared by green synthesis efficiently use solar radiation in photocatalytic activity in the breakdown of organic pollutants.

Nanoparticles from plant extract; fungi: Iron NPs produced from spinach leaves are used in treating eutrophication, a serious global concern (Turakhia et al., 2018). Iron NPs are employed successfully in the removal of BOD and COD from wastewater.

Silver NPs are also produced from Reishi mushroom extracts. This silver NP has shown high DNA cleavage activity and is found to act against a spectrum of gram-positive and gram-negative bacteria. Gold NPs are also isolated from metal resistant fungi and are successfully used in the treatment of rhodamine dye, which is carcinogenic and found to be the reason for developmental issues in humans (Bhargava et al., 2016). Fungal species are isolated and its gold tolerance was assayed, inoculated, and extracellular AuNPs were produced. The reduction of rhodamine B dye was done with biosynthesized NPs. This reaction was initiated by adding gold NPs with the dye in the presence of a reducing agent. This reduction mechanism is a two-step process (Bastus et al., 2014) where electrons accumulate from borohydride ions onto the surface of the NP, diffusion of rhodamine B toward the NP, and its reduction by excess electrons. The proteins in biosynthesized NPs induce adsorption of the dye.

Nanoparticles from microbes: Metal nanocatalysts synthesized from microbes exhibit useful catalytic and antimicrobial properties (Lala et al., 2007). Palladium nanocatalysts are produced from palladium tolerant microbes and used in wastewater treatment. Samples were collected, tolerant species isolated and cultured. Palladium nanocatalysts were produced, which showed catalytic activity in dehalogenation reaction. They attach to functional groups on cell envelopes and catalyze the reductive dechlorination of polychlorinated dibenzodioxins and detoxify the compound.

Biogenic nanomagnetite (BnM) is used as a catalyst to treat organic water pollutants Nitrobenzene and tetrachloroethylene. BnM is synthesized from *Geobacter sulfurreducens*. Nitrobenzene was reduced and a reduction product was observed. The addition of palladium as a catalyst further improved its reduction reaction. Palladium-BnM when treated with tetrachloroethylene, results in fast reaction and degradation of the contaminant.

Nanoparticles are also finding their way in air pollution control and for its treatment. Advancements in this nanotechnology-related application are still in the developmental phase. The major issue here is being the release of NPs into the environment and nanotoxicity effects in human while other living beings have not yet been monitored.

REFERENCES

Aftab K., Akhtar K., Jabbar A., Bukhari I. H., and Noreen R. (2013) Physico-chemical study for zinc removal and recovery onto native/chemically modified Aspergillus flavus NA9 from industrial effluent. *Water Research* 47, 4238–4246. https://doi.org/10.1016/j.watres.2013.04.051

Ahluwalia S., Prakash N. T., Prakash R., and Pal B. (2016). Improved degradation of methyl orange dye using bio-co-catalyst Se nanoparticles impregnated ZnS photocatalyst under UV irradiation. *Chemical Engineering Journal* 306, 1041–1048.

Ahmad, A. L., and Ooi, B. S. (2010). A study on acid reclamation and copper recovery using low pressure nanofiltration membrane. *Chemical Engineering Journal* 156, 257–263.

Alcamo J. (2019). Water quality and its interlinkages with the Sustainable Development Goals. *Current Opinion in Environmental Sustainability* 36, 126–140.

Ali I., Peng C., Khan Z. M., and Naz I. (2017). Yield cultivation of magnetotactic bacteria and magnetosomes: a review. *Journal of Basic Microbiology* 57(8), 643–652.

Ali I., Peng C., Khan Z. M., Naz I., Sultan M., Ali M., ... and Ye T. (2019). Overview of microbes based fabricated biogenic nanoparticles for water and wastewater treatment. *Journal of Environmental Management* 230, 128–150.

Al-Mossawi M. A. (2020). Biological approach for recycling waste water in Iraq. *Air, Soil and Water Research* 7(1). https://doi.org/10.4137/ASWR.S17611

Amin M., Alazba A., and Manzoor U. (2014) A review of removal of pollutants from water/wastewater using different types of nanomaterials. *Advances in Materials Science and Engineering* 2014, 1–24. https://doi.org/10.1155/2014/825910

Amira M. M., Fatma A. I., Seham A. S., and Nadia A. Y. (2015). Adsorption of heavy metal ion from aqueous solution by nickel oxide nano catalyst prepared by different methods. *Egyptian Journal of Petroleum* 24, 27–35.

Archer E., Petrie B., Kasprzyk-Hordern B., and Wolfaardt G. M. (2017). The fate of pharmaceuticals and personal care products (PPCPs), endocrine disrupting contaminants (EDCs), metabolites and illicit drugs in a WWTW and environmental waters. *Chemosphere* 174, 437–446.

Arpit B., Navin J., Mohd A. K., Vikram P., Venkataramana D. R., and Jitendra P. (2016). Utilizing metal tolerance potential of soil fungus for efficient synthesis of gold nanoparticles with superior catalytic activity for degradation of rhodamine B. *Journal of Environmental Management* 183,22–32.

Bansal V., Rautaray D., Bharde A. et al. (2005) Fungus-mediated biosynthesis of silica and titania particles. *Journal of Materials Chemistry* 15, 2583–2589.

Bastus N.G., Merkoci F., Piella J., and Puntes V. (2014) Synthesis of highly monodisperse citrate-stabilized silver nanoparticles of up to 200 nm: kinetic control and catalytic properties. *Chemical Materials* 26, 2836–2846.

Baytak S., Kocyigit A., and Türker A. R. (2007) Determination of lead, iron and nickel in water and vegetable samples after preconcentration with *Aspergillus niger* loaded on silica gel. *Clean-Soil Air Water* 35, 607–611. https://doi.org/10.1002/clen.200700124

Bethune D. S., Kiang C. H., De Vries M. S., Gorman G., Savoy R., Vazquez J., & Beyers R. (1993). Cobalt-catalysed growth of carbon nanotubes with single-atomic-layer walls. *Nature* 363(6430), 605–607.

Bhargava A., Jain N., Khan M. A., et al. (2016) Utilizing metal tolerance potential of soil fungus for efficient synthesis of gold nanoparticles with superior catalytic activity for degradation of rhodamine B. *Journal of Environmental Management* 183, 22–32.

Bhattacharya S., Saha I., Mukhopadhyay A., Chattopadhyay D., Ghosh U. C., and Chatterjee D. (2013) Role of nanotechnology in water treatment and purification: potential applications and implications. *International Journal of Chemical Science and Technology* 3, 59–64.

Bhavika T., Paras T., and Sejal S. (2018). Green synthesis of zero valent iron nanoparticles from *Spinacia oleracea* (spinach) and its application in waste water treatment. *IAETSD Journal for Advanced Research in Applied Sciences* 5, 46–50.

Bushra F., Sharf Ilahi S., Rabia A., and Saif Ali C. (2019). Green synthesis of f-CdWO4 for photocatalytic degradation and adsorptive removal of Bismarck Brown R dye from water. *Water Resources and Industry* 22, 100119.

Cai Y., Li C., Wu D., Wang W., Tan F., Wang X., and Qiao X. (2017). Highly active MgO nanoparticles for simultaneous bacterial inactivation and heavy metal removal from aqueous solution. *Chemical Engineering Journal* 312, 158–166.

Ceribasi I. H. and Yetis U. (2004) Biosorption of Ni (II) and Pb (II) by *Phanerochaete chrysosporium* from a binary metal system-kinetics. *Water SA* 27, 15–20. https://doi.org/10.4314/wsa.v27i1.5004

Chen G. Q., Zhang W. J., Zeng G. M., Huang J. H., Wang L., and Shen G. L. (2011) Surface-modified *Phanerochaete chrysosporium* as a biosorbent for Cr (VI)-contaminated wastewater. *Journal of Hazardous Materials* 186, 2138–2143. https://doi.org/10.1016/j.jhazmat.2010.12.123

Chinnashanmugam S., Rajendiran R., Thanamegam K., Krishnan M., Digambar K., and Prathapkumar H. S. (2017). Synthesis of silver nanoparticles using bacterial exopolysaccharide and its application for degradation of azo-dyes. *Biotechnology Reports* 15, 33–40.

Costa D., Burlando P., and Priadi C. (2016). The importance of integrated solutions to flooding and water quality problems in the tropical megacity of Jakarta. *Sustainable Cities and Society* 20, 199–209.

Dada O. A., Adekola F. A., and Odebunmi E. O. (2016). Kinetics and equilibrium models for sorption of Cu (II) onto a novel manganese nano-adsorbent. *Journal of Dispersion Science and Technology* 37(1), 119–133.

Das N., Vimala R., and Karthika P. (2008). Biosorption of heavy metals – an overview. *Indian Journal of Biotechnology* 7, 159–169.

Dhandapani P., Maruthamuthu S., and Rajagopal G. (2012) Bio-mediated synthesis of TiO_2 nanoparticles and its photocatalytic effect on aquatic biofilm. *Journal of Photochemistry and Photobiology B* 110(2), 43–49

Di Ze-Chao, Ding Jun, Peng Xian-Jia, Li Yan-Hui, Luan Zhao-Kun, and Liang Ji. (2006) Chromium adsorption by aligned carbon nanotubes supported ceria nanoparticles. *Chemosphere* 62, 861–865.

Diallo M. S., Christie S., Swaminathan P., Johnson J. H., and Goddard W. A. (2005). Dendrimer enhanced ultrafiltration. 1. Recovery of Cu (II) from aqueous solutions using PAMAM dendrimers with ethylene diamine core and terminal NH_2 groups. *Environmental Science &Technology* 39(5), 1366–1377.

Dias M., Lacerda I., Pimentel P., De Castro H., and Rosa C. (2002) Removal of heavy metals by an *Aspergillus terreus* strain immobilized in a polyurethane matrix. *Letters in Applied Microbiology* 34, 46–50. https://doi.org/10.1046/j.1472-765x.2002.01040.x

Dwivedi S., AlKhedhairy A. A., Ahamed M., and Musarrat J. (2013) Biomimetic synthesis of selenium nanospheres by bacterial strain JS-11 and its role as a biosensor for nanotoxicity assessment: a novel Se-bioassay. *PLoS One* 8, e57404. https://doi.org/10.1371/journal.pone.0057404

El Saliby I., Shon H., Kandasamy J., and Vigneswaran S. (2008) Nanotechnology for wastewater treatment: in brief. *Encyclopedia of Life Support System (EOLSS)*. Eolss Publishers Co., UK.

Falih A. M. (1997) Influence of heavy-metals toxicity on the growth of *Phanerochaete chrysosporium*. *Bioresource Technology* 60, 87–90. https://doi.org/10.1016/S0960-8524(96)00177-0

Fatemeh D., Ali M., and Alireza M. (2015). Green Synthesis of ZnO nanoparticles and its application in the degradation of some dyes. *Journal of American Ceramic Society* 98(6), 1739–1746.

Figoli A., Cassano A., Criscuoli A., Mozumder M. S. I., Uddin M. T., Islam M. A., and Drioli E. (2010). Influence of operating parameters on the arsenic removal by nanofiltration. *Water Research* 44(1), 97–104.

Fu F., and Wang Q. (2011) Removal of heavy metal ions from wastewaters: a review. *Journal of Environmental Management* 92, 407–418. https://doi.org/10.1016/j.jenvman.2010.11.011

Furgal K. M., Meyer R. L., and Bester K. (2015). Removing selected steroid hormones, biocides and pharmaceuticals from water by means of biogenic manganese oxide nanoparticles in situ at ppb levels. *Chemosphere* 136, 321–326.

Gao W., Majumder M., Alemany L. B., Narayanan T. N., Ibarra M. A., Pradhan B. K., and Ajayan P. M. (2011) Engineered graphite oxide materials for application in water purification. *ACS Applied Material Interfaces* 3, 1821–1826. https://doi.org/10.1021/am200300u

Gautam R. K., Mudhoo A., Lofrano G., and Chattopadhyaya M. C. (2014) Biomass-derived biosorbents for metal ions sequestration: adsorbent modification and activation methods and adsorbent regeneration. *Journal of Environmental Chemical Engineering* 2, 239–259. https://doi.org/10.1016/j.jece.2013.12.019

Gehrke I., Geiser A., and Somborn-Schulz A. (2015). Innovations in nanotechnology for water treatment. *Nanotechnology, Science and Applications* 8, 1.

Gillan, D. C. (2016). Metal resistance systems in cultivated bacteria: are they found in complex communities? *Current Opinion in Biotechnology* 38, 123–130.

Gopal M., Pakshirajan K., and Swaminathan T. (2002) Heavy metal removal by biosorption using *Phanerochaete chrysosporium*. *Applied Biochemical Biotechnology* 102, 227–237. https://doi.org/10.1385/ABAB:102-103:1-6:227

Gusseme B. D., Soetaert M., Hennebel T., Vanhaecke L., Boon N., and Verstraete W. (2012). Catalytic dechlorination of diclofenac by biogenic palladium in a microbial electrolysis cell. *Microbial Biotechnology* 5(3), 396–402.

Holmes A. B., and Gu F. X. (2016). Emerging nanomaterials for the application of selenium removal for wastewater treatment. *Environmental Science: Nano* 3(5), 982–996.

Hossain M. B., and Mertig A. G. (2020). Socio-structural forces predicting global water footprint: socio-hydrology and ecologically unequal exchange. *Hydrological Sciences Journal* (just-accepted). https://doi.org/10.1080/02626667.2020.1714052

Hu A., and Apblett A. (Eds.). (2014). *Nanotechnology for water treatment and purification*. Springer International Publishing, Switzerland.

Hu A., Zhang X., Luong D., Oakes K. D., Servos M. R., Liang R., ... and Zhou Y. (2012). Adsorption and photocatalytic degradation kinetics of pharmaceuticals by TiO_2 nanowires during water treatment. *Waste and Biomass Valorization* 3(4), 443–449.

Huang J., Lin L., Sun D., Chen H., Yang D., and Li Q. (2015). Bio-inspired synthesis of metal nanomaterials and applications. *Chemical Society Reviews* 44(17), 6330–6374.

Iijima, S. (1991). Helical microtubules of graphitic carbon. *Nature* 354(6348), 56–58.

Inyinbor Adejumoke A., Adebesin Babatunde O., Oluyori Abimbola P., Adelani Akande Tabitha A., Dada Adewumi O., and Oreofe Toyin A. (2018). Water pollution: effects, prevention, and climatic impact. *Water Challenges of an Urbanizing World* 33. DOI: 10.5772/intechopen.72018.

Jain A., Ranjan S., Dasgupta N., and Ramalingam C. (2016) Nanomaterials in food and agriculture: an overview on their safety concerns and regulatory issues. *Critical Reviews in Food Science and Nutrition* 58(2), 297–317.

Järup L. (2003) Hazards of heavy metal contamination. *British Medical Bulletin* 68, 167–182. https://doi.org/10.1093/bmb/ldg032

Jha A. K., Prasad K., and Kulkarni A. R. (2009). Synthesis of TiO_2 nanoparticles using microorganisms. *Colloids and Surfaces B: Biointerfaces* 71(2), 226–229.

Jung C., Son A., Her N., Zoh K. D., Cho J., and Yoon Y. (2015). Removal of endocrine disrupting compounds, pharmaceuticals, and personal care products in water using carbon nanotubes: a review. *Journal of Industrial and Engineering Chemistry* 27, 1–11.

Kalfa O. M., Yalçınkaya O., and Türker A. R. (2009) Synthesis of nano B_2O_3/TiO_2 composite material as a new solid phase extractor and its application to preconcentration and separation of cadmium. *Journal of Hazardous Materials* 166, 455–461. https://doi.org/10.1016/j.jhazmat.2008.11.112

Kannan Badri N., Hyun Ho P., and Natarajan S. (2013). Extracellular synthesis of mycogenic silver nanoparticles by *Cylindrocladium floridanum* and its homogeneous catalytic degradation of 4-nitrophenol. *Spectrochimica Acta Part A* 116, 485–490.

Keiser D. A., Kling C. L., and Shapiro J. S. (2019). The low but uncertain measured benefits of US water quality policy. *Proceedings of the National Academy of Sciences* 116(12), 5262–5269.

Kim K., Hur J. W., Kim S., Jung J. Y., and Han H. S. (2020). Biological wastewater treatment: Comparison of heterotrophs (BFT) with autotrophs (ABFT) in aquaculture systems. *Bioresource Technology* 296, 122293.

Kirthi A. V., Rahuman A. A., Rajakumar G. et al. (2011) Biosynthesis of titanium dioxide nanoparticles using bacterium *Bacillus subtilis*. *Material Letters* 65, 2745–2747.

Kitching M., Ramani M., and Marsili E. (2015). Fungal biosynthesis of gold nanoparticles: mechanism and scale up. *Microbial Biotechnology* 8(6), 904–917.

Klaine S. J., Alvarez P. J., Batley G. E., Fernandes T. F., Handy R. D., Lyon D. Y., and Lead J. R. (2008). Nanomaterials in the environment: behavior, fate, bioavailability, and effects. *Environmental Toxicology and Chemistry: An International Journal* 27(9), 1825–1851.

Lajayer B. A., Ghorbanpour M., and Nikabadi S. (2017). Heavy metals in contaminated environment: destiny of secondary metabolite biosynthesis, oxidative status and phytoextraction in medicinal plants. *Ecotoxicology and Environmental Safety* 145, 377–390.

Lala N. L., Ramaseshan R., Bojun L., Sundarrajan S., Barhate R. S., Ying-jun L., and Ramakrishna S. (2007). Fabrication of nanofibers with antimicrobial functionality used as filters: protection against bacterial contaminants. *Biotechnology and Bioengineering* 97(6), 1357–1365.

Lee S. H., Pumprueg S., Moudgil B., and Sigmund W. (2005). Inactivation of bacterial endospores by photocatalytic nanocomposites. *Colloids and Surfaces B: Biointerfaces* 40(2), 93–98.

Lee X. J., Lee L. Y., Foo L. P. Y., Tan K. W., and Hassell D. G. (2012). Evaluation of carbon-based nanosorbents synthesised by ethylene decomposition on stainless steel substrates as potential sequestrating materials for nickel ions in aqueous solution. *Journal of Environmental Sciences* 24(9), 1559–1568.

Lemos V. A., Teixeira L. S. G., Bezerra M. A., Costa A. C. S., Castro J. T., Cardoso L. A. M., de Jesus D. S., Santos E. S., Baliza P. X., and Santos L. N. (2008) New materials for solid-phase extraction of trace elements. *Applied Spectroscopy Reviews* 43, 303–334. https://doi.org/10.1080/05704920802031341

Lettinga G., Pol L. H., Koster I. W., Wiegant W. M., De Zeeuw W. J., Rinzema A., … and Hobma S. W. (1984). High-rate anaerobic waste-water treatment using the UASB reactor under a wide range of temperature conditions. *Biotechnology and Genetic Engineering Reviews* 2(1), 253–284.

Li Q., Wu S., Liu G., Liao X., Deng X., Sun D., Hu Y., and Huang Y. (2004) Simultaneous biosorption of cadmium (II) and lead (II) ions by pretreated biomass of *Phanerochaete chrysosporium*. *Separation and Purification Technology* 34, 135–142. https://doi.org/10.1007/BF00167924

Li Y.-H., Di Z., Ding J., Wu D., Luan Z., Zhu Y. (2005) Adsorption thermodynamic, kinetic and desorption studies of Pb2+ on carbon nanotubes. *Water Research* 39, 605–609. https://doi.org/10.1016/j.watres.2004.11.004

Liu J.-F., Z-s Z., and G-b J. (2008) Coating Fe_3O_4 magnetic nanoparticles with humic acid for high efficient removal of heavy metals in water. *Environmental Sciences Technology* 42, 6949–6954. https://doi.org/10.1021/es800924c

Lu C., and Chiu H. (2006) Adsorption of zinc (II) from water with purified carbon nanotubes. *Chemical Engineering Sciences* 61, 1138–1145. https://doi.org/10.1016/j.ces.2005.08.007

Lu Y., Xu L., Shu W., Zhou J., Chen X., Xu Y., and Qian G. (2017). Microbial mediated iron redox cycling in Fe (hydr) oxides for nitrite removal. *Bioresource Technology* 224, 34–40.

Luef E., Prey T., and Kubicek C. P. (1991) Biosorption of zinc by fungal mycelial wastes. *Applied Microbiology Biotechnology* 34, 688–692. https://doi.org/10.1007/BF00167924

Mal J., Nancharaiah Y. V., Van Hullebusch E. D., and Lens P. N. L. (2016). Metal chalcogenide quantum dots: biotechnological synthesis and applications. *RSC Advances* 6(47), 41477–41495.

Martins M., Mourato C., Sanches S., Noronha J. P., Crespo M. B., and Pereira I. A. (2017). Biogenic platinum and palladium nanoparticles as new catalysts for the removal of pharmaceutical compounds. *Water Research* 108, 160–168.

Mathew P. W., Richard S. C., Nimisha J., Victoria S. C., Apalona M., Boyuan Z., Catherine M. D., Bart E. V., Thomas H., and Jonathan R. L.(2017). Highly efficient degradation of organic pollutants using a microbially-synthesized nanocatalyst. *International Biodeterioration & Biodegradation* 119, 155–161.

Meerburg F., Hennebel T., Vanhaecke L., Verstraete W., and Boon N. (2012). Diclofenac and 2-anilinophenylacetate degradation by combined activity of biogenic manganese oxides and silver. *Microbial Biotechnology* 5(3), 388–395.

Michael S., Thomas H., Christian S., Mandy K., Wilhelm G. L., Leonard B., Rolf-Alexander D., Karin A. K., and Michael B. (2014). Synthesis of novel palladium (0) nanocatalysts by microorganisms from heavy-metal-influenced high-alpine sites for dehalogenation of polychlorinated dioxins. *Chemosphere* 117, 462–470.

Monica R. C. and Cremonini R. (2009). Nanoparticles and higher plants. *Caryologia* 62(2), 161–165.

Muhammad N., Muhammad S., Temoor A., Muhammad B. K. N., Sabir H., Fengming S., and Irfan M. (2020). Use of biogenic copper nanoparticles synthesized from a native Escherichia sp. as photocatalysts for azo dye degradation and treatment of textile effluents. *Environmental Pollution* 257, 113514.

Mukherjee P., Senapati S., Mandal D., Ahmad A., Khan M. I., Kumar R., and Sastry M. (2002). Extracellular synthesis of gold nanoparticles by the fungus *Fusarium oxysporum*. *ChemBioChem* 3(5), 461–463.

Muthukrishnan M., and Guha B. K. (2008). Effect of pH on rejection of hexavalent chromium by nanofiltration. *Desalination* 219, 171–178.

Nancharaiah Y. V., Mohan S. V., and Lens P. N. L. (2016). Biological and bioelectrochemical recovery of critical and scarce metals. *Trends in Biotechnology* 34(2), 137–155.

Narayanan K. B., Park H. H., and Sakthivel N. (2013). Extracellular synthesis of mycogenic silver nanoparticles by *Cylindrocladium floridanum* and its homogeneous catalytic degradation of 4-nitrophenol. *Spectrochimica Acta Part A: Molecular and Biomolecular Spectroscopy* 116, 485–490.

Noguera-Oviedo K., and Aga D. S. (2016). Lessons learned from more than two decades of research on emerging contaminants in the environment. *Journal of Hazardous Materials* 316, 242–251.

Norouzi P., Pirali-Hamedani M., Ganjali M. R., and Faridbod F. (2010). A novel acetylcholinesterase biosensor based on chitosan-gold nanoparticles film for determination of monocrotophos using FFT continuous cyclic voltammetry. *International Journal of Electrochemical Science* 5, 1434–1446.

Ozcan B., Tzeremes P. G., and Tzeremes N. G. (2020). Energy consumption, economic growth and environmental degradation in OECD countries. *Economic Modelling* 84, 203–213.

Perullini M., Jobbágy M., Mouso N., Forchiassin F., and Bilmes S. A. (2010). Silica-alginate-fungi biocomposites for remediation of polluted water. *Journal of Material Chemistry* 20, 6479–6483. https://doi.org/10.1039/C0JM01144D

Perumal D., Sundram M., and Gopalakrishnan R. (2012). Bio-mediated synthesis of TiO2 nanoparticles and its photocatalytic effect on aquatic biofilm. *Journal of Photochemistry and Photobiology B* 110, 43–49.

Prajapati P. M. and Sen D. J. (2013). Nanotechnology: a new technique in water purification. *International Journal of Innovative Ideas and Research and Development* 1(1), 28–32.

Purkayastha S. P., Choudhury M., and Chinmoy Paul D. D. (2015). Arsenic contamination in ground water is a serious threat in the North Karimganj block of Karimganj district, Southern part of Assam, India. *Journal of Chemical and Pharmaceutical Research* 7(8), 371–378.

Puyol D., Batstone D. J., Hülsen T., Astals S., Peces M., and Krömer J. O. (2017). Resource recovery from wastewater by biological technologies: opportunities, challenges, and prospects. *Frontiers in Microbiology* 7, 2106.

Qu Y., Pei X., Shen W., Zhang X., Wang J., Zhang Z. and Zhou J. (2017). Biosynthesis of gold nanoparticles by Aspergillum sp. WL-Au for degradation of aromatic pollutants. *Physica E: Low-Dimensional Systems and Nanostructures* 88, 133–141.

Quan X., Zhang X., and Xu H. (2015). In-situ formation and immobilization of biogenic nanopalladium into anaerobic granular sludge enhances azo dyes degradation. *Water Research* 78, 74–83.

Raja C. P., Jacob J. M., and Balakrishnan R. M. (2016). Selenium biosorption and recovery by marine *Aspergillus terreus* in an upflow bioreactor. *Journal of Environmental Engineering* 142(9), C4015008.

Ranjan S., Dasgupta N., Chakraborty A. R., Melvin Samuel S., Ramalingam C., Shanker R., Kumar A. (2014) Nanoscience and nanotechnologies in food industries: opportunities and research trends. *Journal of Nanoparticle Research* 16(6), 1–23.

Rezić I. (2011). Determination of engineered nanoparticles on textiles and in textile wastewaters. *TrAC Trends in Analytical Chemistry* 30(7), 1159–1167.

Ricardo A. S., Adriana P. H., David M., and Ana C. (2019). Fe-TiO$_2$ nanoparticles synthesized by green chemistry for potential application in waste water photocatalytic treatment. *Journal of Nanotechnology* 2019, 1–11.

Sadegh K., and Tayebeh S. (2019). Green and simple synthesis route of Ag@AgCl nanomaterial using green marine crude extract and its application for sensitive and selective determination of mercury. *Spectrochimica Acta Part A* 222, 117216.

Saikia J., Saha B., and Das G. (2011). Efficient removal of chromate and arsenate from individual and mixed system by malachite nanoparticles. *Journal of Hazardous Materials* 186(1), 575–582.

Samper E., Rodriguez M., De la Rubia M. A., and Prats D. (2009). Removal of metal ions at low concentration by micellar-enhanced ultrafiltration (MEUF) using sodium dodecyl sulfate (SDS) and linear alkylbenzene sulfonate (LAS). *Separation and purification technology* 65(3), 337–342.

Sarbu R., and Lorand T. (2006). Wastewater treatment by flotation. *Mining Engineering* 7, 155.

Sarı A. and Tuzen M. (2009). Kinetic and equilibrium studies of biosorption of Pb (II) and Cd (II) from aqueous solution by macrofungus (*Amanita rubescens*) biomass. *Journal of Hazardous Materials* 164, 1004–1011. https://doi.org/10.1016/j.jhazmat.2008.09.002

Say R., Denizli A., and Yakup Arıca M. (2001).Biosorption of cadmium (II), lead (II) and copper (II) with the filamentous fungus *Phanerochaete chrysosporium*. *Bioresoure Technology* 76, 67–70. https://doi.org/10.1016/S0960-8524(00)00071-7

Shahalam A. M., Al-Harthy A., and Al-Zawhry A. (2002). Feed water pretreatment in RO systems: unit processes in the Middle East. *Desalination* 150(3), 235–245.

Shamaila S., Sajjad A. K. L., Farooqi S. A., Jabeen N., Majeed S., and Farooq I. (2016). Advancements in nanoparticle fabrication by hazard free eco-friendly green routes. *Applied Materials Today* 5, 150–199.

Sheet I., Kabbani A., and Holail H. (2014). Removal of heavy metals using nanostructured graphite oxide, silica nanoparticles and silica/graphite oxide composite. *Energy Procedia* 50, 130–138. https://doi.org/10.1016/j.egypro.2014.06.016

Shuhua C., Na L., Li J., Yating L., Baiheng X., and Weizhi Z. (2019).Biodegradation of metal complex Naphthol Green B and formation of iron–sulfur nanoparticles by marine bacterium Pseudoalteromonas sp CF10-13. *Bioresource Technology* 273, 49–55.

Siddiqi K. S. and Husen A. (2016). Green synthesis, characterization and uses of palladium/platinum nanoparticles. *Nanoscale Research Letters* 11(1), 482.

Singh P. K. T. R., Vats S., Kumar D., and Tyagi S. (2012). Nanomaterials use in wastewater treatment. In *International Conference on Nanotechnology and Chemical Engineering*. Bangkuk, Thailand (Vol. 21). https://www.ugc.ac.in/mrp/paper/MRP-MAJOR-BIOT-2013-36837-PAPER.pdf

Singh S., Barick K. C., and Bahadur D. (2011). Surface engineered magnetic nanoparticles for removal of toxic metal ions and bacterial pathogens. *Journal of Hazardous Materials* 192(3), 1539–1547.

Smith A. (2006). Opinion: Nanotech–the way forward for clean water?. *Filtration & Separation* 43(8), 32–33.

Soltani N., Keshavarzi B., Sorooshian A., Moore F., Dunster C., Dominguez A. O., ... and Asadi S. (2018). Oxidative potential (OP) and mineralogy of iron ore particulate matter at the Gol-E-Gohar Mining and Industrial Facility (Iran). *Environmental Geochemistry and Health* 40(5), 1785–1802.

Soylak M., Tuzen M., Mendil D., and Turkekul I. (2006).Biosorption of heavy metals on Aspergillus fumigatus immobilized Diaion HP-2MG resin for their atomic absorption spectrometric determinations. *Talanta* 70, 1129–1135. https://doi.org/10.1016/j.talanta.2006.02.027

Stoimenov P. K., Klinger R. L., Marchin G. L., and Klabunde K. J. (2002). Metal oxide nanoparticles as bactericidal agents. *Langmuir* 18(17), 6679–6686.

Su J., Deng L., Huang L., Guo S., Liu F., and He J. (2014). Catalytic oxidation of manganese (II) by multicopper oxidase CueO and characterization of the biogenic Mn oxide. *Water Research* 56, 304–313.

Suja E., Nancharaiah Y. V., and Venugopalan V. P. (2014). Biogenic nanopalladium production by self-immobilized granular biomass: application for contaminant remediation. *Water Research* 65, 395–401.

Sun J., Liang P., Yan X., Zuo K., Xiao K., Xia J., ... and Qi M. (2016). Reducing aeration energy consumption in a large-scale membrane bioreactor: process simulation and engineering application. *Water Research* 93, 205–213.

Tabe S. (2014). Electrospun nanofiber membranes and their applications in water and wastewater treatment. In *Nanotechnology for Water Treatment and Purification* (pp. 111–143). Springer, Cham.

Tabi M. T., and Verdon D. (2014). New public service performance management tools and public water governance: the main lessons drawn from action research conducted in an urban environment. *International Review of Administrative Sciences* 80(1), 213–235.

Tahir M. B., Tufail S., Ahmad A., Rafique M., Iqbal T., Abrar M., ... and Ijaz M. (2019). Semiconductor nanomaterials for the detoxification of dyes in real wastewater under visible-light photocatalysis. *International Journal of Environmental Analytical Chemistry* 1–15. https://doi.org/10.1080/03067319.2019.1686494

Tratnyek P. G., and Johnson R. L. (2006) Nanotechnologies for environmental cleanup. *Nano Today* 1(2):44–48.

Tuo Y., Liu G., Zhou J., Wang A., Wang J., Jin R., and Lv H. (2013). Microbial formation of palladium nanoparticles by *Geobacter sulfurreducens* for chromate reduction. *Bioresource Technology* 133, 606–611.

Turakhia B., Turakhia P., and Shah S. (2018).Green synthesis of zero valent iron nanoparticles from *Spinacia oleracea* (spinach) and its application in waste water treatment. *Journal for Advanced Research in Applied Sciences* 5, 46–51.

Tuzen M., Melek E., and Soylak M. (2006). Celtek clay as sorbent for separation–preconcentration of metal ions from environmental samples. *Journal of Hazardous Materials* 136(3), 597–603.

Tuzen M., Saygi K. O., Usta C., and Soylak M.(2008).*Pseudomonas aeruginosa* immobilized multiwalled carbon nanotubes as biosorbent for heavy metal ions. *Bioresource Technology* 99, 1563–1570. https://doi.org/10.1016/j.biortech.2007.04.013

UNESCAP-UNICEF_WHO, (May 2–4, 2001) Expert Group Meeting, Bangkok.

UNESCO, (2019) The United Nations World Water Development Report 2019, United Nations Educational, Scientific and Cultural Organization, Paris.

Vijayalakshmi U., Vineet V., Chellappaand M., and Anjaneyulu U.(2017). Green synthesis of silica nanoparticles and its corrosion resistance behavior on mild steel. *Journal of Indian Chemical Society* 92, 675–678.

Vimala R. and Das N. (2009). Biosorption of cadmium (II) and lead (II) from aqueous solutions using mushrooms: a comparative study. *Journal of Hazardous Materials* 168(1), 376–382.

Vimala R., Charumathi D., and Das N. (2011). Packed bed column studies on Cd (II) removal from industrial wastewater by macrofungus *Pleurotus platypus*. *Desalination* 275(1–3), 291–296.

Wang J. F., Meng H. L., Xiong Z. Q., Zhang S. L., and Wang Y. (2014a). Identification of novel knockout and up-regulated targets for improving isoprenoid production in *E. coli*. *Biotechnology Letters* 36(5), 1021–1027.

Wang T., Jin X., Chen Z., Megharaj M., and Naidu R. (2014b). Simultaneous removal of Pb (II) and Cr (III) by magnetite nanoparticles using various synthesis conditions. *Journal of Industrial and Engineering Chemistry* 20(5), 3543–3549.

Xu P., Zeng G. M.,Huang D. L., Feng C. L., Hu S., Zhao M. H., Lai C., Wei Z., Huang C., and Xie G. X. (2012) Use of iron oxide nanomaterials in wastewater treatment: a review. *Science of the Total Environment* 424, 1–10. https://doi.org/10.1016/j.scitotenv.2012.02.023

Yan G., and Viraraghavan T. (2003). Heavy-metal removal from aqueous solution by fungus Mucor rouxii. *Water Research* 37, 4486–4496. https://doi.org/10.1016/S0043-1354(03)00409-3

Yang W., Thordarson P., Gooding J. J., Ringer S. P., and Braet F. (2007). Carbon nanotubes for biological and biomedical applications. *Nanotechnology* 18(41), 412001.

Yue L., Wang J., Zhang Y., Qi S., and Xin B. (2016). Controllable biosynthesis of high-purity lead-sulfide (PbS) nanocrystals by regulating the concentration of polyethylene glycol in microbial system. *Bioprocess and Biosystems Engineering* 39(12), 1839–1846.

Zan G., and Wu Q. (2016). Biomimetic and bioinspired synthesis of nanomaterials/nanostructures. *Advanced Materials* 28(11), 2099–2147.

Zhang D., Li G., and Jimmy C. Y. (2010). Inorganic materials for photocatalytic water disinfection. *Journal of Materials Chemistry* 20(22), 4529–4536.

Zhou Q., Xiao J., and Wang W. (2006). Using multi-walled carbon nanotubes as solid phase extraction adsorbents to determine dichlorodiphenyltrichloroethane and its metabolites at trace level in water samples by high performance liquid chromatography with UV detection. *Journal of Chromatography A* 1125(2), 152–158.

Zhuang W. Q., Fitts J. P., Ajo-Franklin C. M., Maes S., Alvarez-Cohen L., and Hennebel T. (2015). Recovery of critical metals using biometallurgy. *Current Opinion in Biotechnology* 33, 327–335.

5 Lignocellulosic Biomass Wastes to Bioenergy

Role of Microbial Enzymes for Second Generation Biofuels

Amit Kumar
Debre Markos University

Diwakar Aggarwal
Maharishi Markandeshwar (Deemed to be University)

Amit Kumar Bharti
Indian Institute of Technology Roorkee

Chhotu Ram
Adigrat University

CONTENTS

5.1 INTRODUCTION

The increasing world population, industrial development, and urbanization all over the world have increased energy demand. According to U.S. Energy Information Administration data, energy consumption in the world will rise by 56% between 2010 and 2040 (EIA, 2013). The huge increase in energy consumption from 154 to 240 quadrillion watt-hours will be very difficult to achieve from diminishing fossil fuel reserves. The development of second-generation biofuels is critically important because of the extensive use and depletion of fossil fuel resources, rising prices of fossil fuels, security issues regarding the fossil fuel supply, and environmental problems (Manavalan et al., 2015; Saini et al., 2018). Therefore, efforts have been made to search for sustainable alternatives for fossil fuels. Transformation of lignocellulosic materials to biofuels is a sustainable approach for fossil fuels (Garoma and Shackelford, 2014; Kumar et al., 2016c). Lignocellulosic materials are a renewable and abundant resource for second-generation biofuels. Lignocellulosic biomass from agricultural residues, wood waste, forestry waste, pulp and paper industry wastes, and municipal solid waste are available abundantly throughout the world and can be converted into several value-added products including biofuels. A disposal problem is created due to the accumulation of lignocellulosic wastes in large quantities. These lignocellulosic materials can be used for the manufacturing of paper, animal feed, composting, and generation of organic chemicals (Kumar et al., 2016c; Sanchez, 2009). Bioenergy products that can be produced from lignocellulosic materials include ethanol, butanol, bio-hydrogen, and bio-methane. Several other industrially important products such as acetone, acetic acid, citric acid, fumaric acid, lactic acid, xylitol, single-cell protein, lignocellulolytic enzymes, etc. can also be generated from lignocellulosic materials (Sanchez, 2009).

Enzymes play a vital role in the growth, repair, and maintenance of living beings. Enzymes are proteins that act as biocatalysts for several biochemical reactions. In a living system, enzymes are responsible for the breakdown of macromolecule for energy production and synthesis of new materials. Enzymes are derived from plants, animals, and microorganisms, but a major fraction of commercial enzymes are produced from microorganisms due to ease of growth, nutritional requirement, and downstream processing. Furthermore, microbes produce a higher quantity of enzymes in comparison to plants and animal sources. Most of the microbial enzymes are inducible and therefore, their production can be enhanced significantly with the addition of inducers in the production media (Ibrahim, 2008; Prakash et al., 2013).

5.2 LIGNOCELLULOLYTIC ENZYMES

Lignocellulolytic enzymes are synthesized by many microorganisms, like fungi, bacteria, and actinomycetes during their growth on lignocellulosic materials. Fungi are the major sources of cellulolytic, hemicellulolytic, and ligninolytic enzymes due to the higher yield of enzyme production. *Trichoderma*, *Aspergilli*, and *Penicillium* are the main fungal genera that are very well known for lignocellulolytic enzyme production. White-rot fungi are the most dominant group of fungi that grow on lignocellulosic biomass and produce the ligninolytic enzymes efficiently.

5.2.1 CELLULOLYTIC ENZYME SYSTEM

Cellulases are the cellulose-degrading enzymes that are a multi-component enzymatic system that is made of three groups of enzymes including endo-β-1, 4-glucanases, exo-β-1, 4-glucanase, and β-glucosidases.

1. Endo-β-1, 4-glucanases (EC 3.2.1.4) randomly cleave β-1, 4-glycosidic linkages on amorphous part of cellulose away from chain ends and releases the oligosaccharides of variable length.
2. Exo-β-1, 4-glucanase, or cellobiohydrolase (EC 3.2.1.91), that produces cellobiose by attacking cellulose from reducing and non-reducing chain ends. Exo-1, 4-β-glucanases that show the processive nature of catalysis is known as cellbiohydrolases.
3. β-glucosidases (EC 3.2.1.21), that converts cellobiose and soluble oligosaccharides into glucose (Bansal et al., 2009; de Castro and de Castro, 2012; Dutta et al., 2008; Horn et al., 2012a, 2012b). (Figure 5.1)

Hydrolysis of cellulose needs the synergistic action of different components of cellulases that collectively results in higher hydrolysis rates than the sum of individual rates caused by each cellulolytic enzyme (de Castro and de Castro, 2012). Cellulases are synthesized by a large variety of microorganisms including bacteria, fungi, and actinomycetes (Kuhad et al., 2011). The most privileged source for cellulase production is fungi because of their higher enzyme yields and the ability to produce

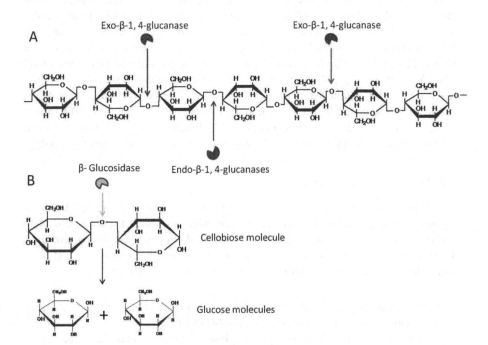

FIGURE 5.1 Action of cellulases (A) Action of endo-β-1, 4-glunanase and exo-β-1, 4-glunanase on cellulosic chain, (B) Action of β-glucosidase on cellobiose.

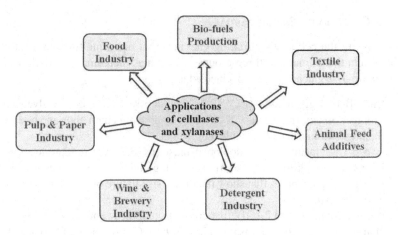

FIGURE 5.2   Industrial applications of cellulases and xylanases.

complete cellulase complex. Filamentous fungi such as *Trichoderma, Aspergillus, Penicillium* (Kuhad et al., 2011), *Fusarium* (Singh and Kumar, 1991; Soliman et al., 2013), *Huminicola* species (Gusakov, 2011) are well known as efficient producers of cellulases (Soliman et al., 2013). Cellulolytic fungi belonging to genera *Trichoderma* (*T. viride, T. longibrachiatum, T. reesei*) have been regarded as the most efficient producers of cellulase (Gusakov, 2011). *Trichoderma reesei* is the champion producer of cellulases but it lacks β-glucosidase production that results in the accumulation of cellobiose which causes end product inhibition during cellulase production and hydrolysis of lignocellulosic biomass. Several species of *Aspergillus* have been reported to produce a high level of β-glucosidase (Fang et al., 2010; Farinas et al., 2010; Mørkeberg et al., 2004). Currently, cellulases are the third largest industrial enzyme by dollar volume due to applications in several industries including biofuels, textile, pulp and paper, laundry, food, and feed (Figure 5.2) (Gautam and Sharma, 2014; Kuhad et al., 2011; Singhania et al., 2010). Major applications of cellulases in pulp and paper industry are in the deinking of recycled paper and refining of pulp.

5.2.2   XYLANOLYTIC SYSTEM

The enzymes that breakdown the hemicelluloses are collectively called hemicellulases; they are defined according to the substrate on which they act. Complete hydrolysis of hemicelluloses requires a spectrum of cooperatively acting enzymes. These enzymes are produced by a large number of microbes and mostly extracellular. These enzymes are (Chavez et al., 2006; Collins et al., 2005):

- Endoxylanases (E.C. 3.2.1.8); randomly cleave the backbone of xylan chain to release xylooligosaccharides of variable length.
- β-Xylosidases (E.C. 3.2.1.37); release xylose monomers from the non-reducing end of xylooligosaccharide and xylobiose.

- α-L-Arabinofuranosidases (E.C. 3.2.1.55); hydrolyze L-arabinofuranose side chains.
- α-D-Glucuronidases (E.C. 3.2.1.139); cleave methyl glucuronate residues.
- Acetyl xylan esterase (E.C. 3.1.1.72); hydrolyze acetate groups from the main chain.
- Feruloyl (E.C. 3.1.1.73) and coumaroyl esterase (E.C. 3.2.1.-): cleave the respective aromatic acids linked to the arabinofuranoside residues (Figure 5.3).

The presence of such multifunctional xylanolytic enzyme system is quite widespread among different microorganisms including fungi, actinomycetes and bacteria, and some of the most dominant xylanolytic enzyme producers such as *Aspergilli*, *Tricodermi*, *Penicillia*, *Streptomymycetes*, *Phanerochaetes*, *Chytridiomycetes*, *Ruminococci*, *Clostridia*, and *Bacilli* (Bastawde, 1992; Beg et al., 2001; Chavez et al., 2006; Sunna and Antranikian, 1997; Techapun et al., 2003). There are several biotechnological applications of xylanases such as animal feed processing and food additives, clarification of fruit juices and wines, biobleaching and processing of fabrics, bleaching of pulp, waste treatment, silage production, and bioconversion of lignocellulosic materials to simple sugars for the production of bioethanol and other industrial products (Bajaj and Manhas, 2012; Battan et al., 2012; Butt et al., 2008; Sharma et al., 2014) (Figure 5.2). Xylanases have also been used as an accessory enzyme along with cellulases in the de-inking of waste paper (Dutt et al., 2012; Maity et al., 2012).

5.2.3 LIGNIN DEGRADING ENZYMES

Lignin degradation is performed by white-rot fungi most effectively. Lignin peroxidases, manganese peroxidizes, and laccases are the major lignin-degrading enzymes that are produced by these fungi. The characteristics of these enzymes vary according to the microbial source. Lignin peroxidase is also known as ligninase. It is one of the most important enzymes involved in the degradation of lignin. Lignin peroxidase has high redox potential (700–1,400 mV) and it can degrade the compounds with high redox potential that are not oxidized by other enzymes. Lignin peroxide can oxidize both phenolic and non-phenolic compounds. It can cleave the recalcitrant non-phenolic units that comprise approximately 90% of lignin (Niladevi, 2009; Plácido and Capareda, 2015). Manganese peroxidase is another important enzyme produced by the lignin-degrading fungi. It is also a heme peroxidase and requires H_2O_2 for its activity. Manganese peroxidase has lower redox potential lignin peroxidase and generally, it does not oxidize non-phenolic lignin compounds. They are glycoproteins with a molecular weight between 38 and 62.5 kDa (Niladevi, 2009; Plácido and Capareda, 2015). Laccases are the multicopper oxidase enzyme, widely distributed in plants, fungi, and bacteria and have the ability to catalyze the oxidation of various phenolic and non-phenolic compounds as well as many environmental pollutants (Dwivedi et al., 2011).

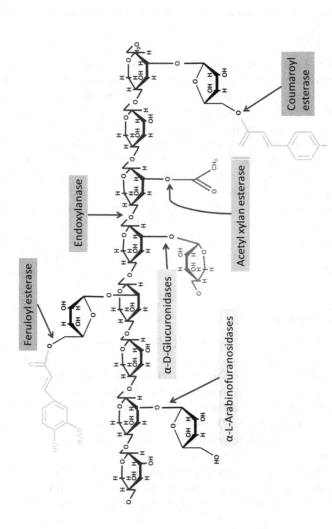

FIGURE 5.3 Action sites of different hemicellulases on xylan and its side. (Modified from Chavez et al., 2006; Chavez et al., 2006.)

5.2.4 COST-EFFECTIVE PRODUCTION OF LIGNOCELLULOLYTIC ENZYMES

The growth of microorganisms on a moist solid substrate is defined as solid-state fermentation SSF. Additionally, the solid substrate is also used as a carbon and energy source. It takes place in the absence or near absence of free water (Hölker et al., 2004; Pandey, 2003; Pandey et al., 2000). Submerged fermentation (SmF) is defined as the fermentation that is performed in the presence of an excess of free water (Singhania et al., 2010). Industrial enzymes are produced dominantly by genetically modified microorganisms under SmF (Delabona et al., 2012; Hölker et al., 2004). However, most of these enzymes can also be produced under SSF by wild type of microorganisms. Production of cellulolytic and xylanolytic enzymes under SSF is rapidly gaining interest as a cost-effective technology because microorganisms, especially fungi produce comparatively high titers of the enzyme due to similar fermentation conditions with the natural environment (Cen and Xia, 1999; Delabona et al., 2012; Singhania et al., 2010). SSF is also beneficial due to simple process operation, better contamination control, and a lesser amount of solvent for extraction of the enzyme, which considerably reduces the downstream processing cost (Pandey et al., 2000; Pathak et al., 2014). SSF is particularly suitable for filamentous fungi as they are better cellulase producers under SSF than SmF (Raghavarao et al., 2003; Shrestha et al., 2010; Yoon et al., 2014).

Another advantage of SSF is the utilization of crop residues as the substrate, acting as carbon and energy sources. Tengerdy (1996) showed a comparison of cellulase production between SmF and SSF systems and had concluded that there was a ten times reduction in the production cost when SSF was employed for production. SSF can, thus, be considered as a future technology for the production of commercial cellulases due to low-cost input and ability to utilize naturally available sources of cellulose (agro-industrial residues) as carbon and energy sources (Delabona et al., 2012; Pareek et al., 2014; Singhania et al., 2010). Several lignocellulosic materials including agro-residues, grasses, weed plants, and forest wastes have been utilized for the production of lignocellulolytic enzymes. Bharti et al(Bharti et al., 2018) utilized *Parthenium hysterophorous* biomass, a weed plant as cost-effective substrate for the production of cellulases and xylanases by *Talaromyces stipitatus* MTCC 12687 under SSF. Gautam et al. (2018) reported the high-level production of xylanase by *Schizophyllum commune* ARC-11 under SSF conditions by using rice straw as low-cost substrate. Kumar et al. (2016a) exploited several crop residues for cellulases and xylanases production and reported black gram (*Vigna mungo*) residue as a most effective substrate for cellulases and xylanases production by *Aspergillus nidulans* AKB-25 under SSF conditions. Gautam et al. (2015) reported xylanase production by *Aspergillus flavus* ARC-12 under SSF using pearl millet stover as low-cost substrate.

Furthermore, the production of cellulases and xylanases may be enhanced by the co-culture of fungi during SSF as they co-existed symbiotically on a solid substrate in nature. In co-cultivation or mixed-culture fermentation two or more organisms grow simultaneously. Co-culture enables production more efficient of enzyme mixtures for industrial processes in comparison to monocultures. During co-cultivation of different fungi complete cellulase system, balanced with all components synthesized and these components complement each other, which is favorable for lignocellulosic substrate hydrolysis (Gupte and Madamwar, 1997; Kumar et al., 2016d; Yoon et al., 2014). Kumar et al.(2016d) studied the co-cultivation of *A. nidulans* AKB-25 and *Penicillium* sp. AKB-24 and reported the improvement of FPase, endoglucanase, and exoglucanase production by 18%, 34%, and 11% respectively compared to monocultures.

5.3 PRODUCTION OF SECOND-GENERATION BIOFUELS

Biofuels are renewable fuels that are derived from biological materials including plants, microorganisms, animals, and wastes (Aro, 2016). Biofuels are the most favorable alternative for fossil fuels to tackle down the growing problem of global warming, environmental issues, and energy demand. Current research and developments are focused on biofuels such as bioethanol, biobutanol, biomethane, biohydrogen, and biodiesel. All of these biofuels except biodiesel are produced by the process of microbial fermentation of suitable carbohydrate feedstocks. Biofuels can be categorized in to first, second, and third-generation biofuels based on the feedstock utilized for conversion. First-generation biofuels are produced directly from edible food-based feedstocks which have several social, economical, and environmental issues such as increasing food price, and deforestation for farming, etc. Therefore, the production of first-generation biofuels in the future cannot be produced on a large scale without threatening food supplies (Barros-Rios et al., 2016; Ghosh et al., 2017; Tomei and Helliwell, 2016). Second-generation biofuels are produced from lignocellulosic feedstocks which do not compete with food supply. Second-generation biofuels are socially, economically, and environmentally sustainable due to the utilization of non-edible lignocellulosic biomass which are the most promising carbohydrate sources for the production of biofuels (Cheng et al., 2011; Ghosh et al., 2017).

Enzymatic hydrolysis of lignocellulosic biomass is an effective and environment-friendly method for the conversion of lignocellulosic materials into reducing sugars. The enzyme-based conversion of lignocellulosic biomass into bioenergy products such as bioethanol, biobutanol, and biohydrogen consists of four major steps which are:

1. Pretreatment of lignocellulosic feedstock to make it accessible to enzymes.
2. Enzymatic hydrolysis of cellulose and hemicelluloses to fermentable sugars.
3. Fermentation of released sugars (during pretreatment and fermentation) into bioenergy products such as ethanol, butanol, hydrogen, and methane. Simultaneous saccharification (hydrolysis of the pretreated substrate) and fermentation can also be combined in a single bioreactor.
4. Recovery of bioenergy products (Nguyen et al., 1999).

This review deals with recent developments in the enzymatic hydrolysis of pretreated lignocellulosic feedstocks (Table 5.1).

TABLE 5.1

Enzymatic Hydrolysis of Different Lignocellulosic Biomass

Source of Enzyme	Enzymes	Lignocellulosic Substrate	Hydrolysis (Total Reducing Sugars) Yield	References
Aspergillus nidulans AKB-25	Cellulase, xylanase, and β-glucosidase	Pearl millet stover	64.77%	Kumar et al. (2016b)
Talaromyces stipitatus MTCC 12687	Cellulase, xylanase, and β-glucosidase	*Parthenium hysterophorous* biomass	734 mg/g	Bharti et al. (2018)
MAPS Enzymes Private Ltd. *Aspergillus niger* ADH-11	Cellulase, hemicellulases, and β-glucosidase	Sugarcane bagasse	614 mg/g	Patel et al. (2017)
Aspergillus oryzae	Cellulase, xylanase, and β-glucosidase	*Saccharum spontaneum*	631.5 mg/g	Chandel et al. (2011)
Commercial enzyme	Cellulase, xylanase, and β-glucosidase	Rice hulls	428 mg/g (90%)	Saha and Cotta (2007)
Commercial enzyme	Cellulase-immobilized magnetic nanoparticles	Napier grass	42%	Ladole et al. (2017)
Aspergillus fumigates (CWSF-7)	Cellulase + commercial xylanase	*Pennisetum* grass–DG *Pennisetum* grass–HNG	478.7 mg/g 483.3 mg/g	Mohapatra et al. (2018)
Penicillium roqueforti ATCC10110	Cellulase, xylanase	Sugarcane bagasse	662.34 mg/g	de Almeida Antunes Ferraz et al. (2018)
Streptomyces flavogriseus AE64X *Streptomyces flavogriseus* AE63X	Cellulase, xylanase	*Arundo donax, Populus nigra,* and *Panicum virgatum*	In a range of 82–86%	Pennacchio et al. (2018)
Inonotus obliquus	Cellulase, xylanase, and β-glucosidase	Wheat straw Rice straw	130.24 mg/g 125.36 mg/g	Xu et al. (2018)
Thermomyces lanuginoosus VAPS-24	Xylanase	Rice bran	126.89 mg/g	Kumar et al. (2017)

5.3.1 ENZYMATIC HYDROLYSIS OF LIGNOCELLULOSIC BIOMASS FOR BIOFUELS PRODUCTION

Second-generation biofuels are derived from carbohydrates available in the cell wall of plants. The lignocellulosic plant cell wall consists mainly of three types of polymers that are cellulose, hemicelluloses, and lignin which are interlinked to each other in a hetero-matrix. Middle lamella is consists of pectic substances. The ligno-cellulosic plant cell wall is complex and resistant to degradation. The carbohydrates present in the lignocellulosic plant cell wall are broken down into reducing sugars either by acid hydrolysis of enzymes. Cellulose and hemicelluloses are hydrolyzed into fermentable sugars by cellulolytic and hemicellulolytic enzymes. Ligninolytic enzymes assist the release of reducing sugars by degrading the lignin content of the pretreated lignocellulosic feedstock. The conversion of lignocellulosic materials into reducing sugars is the main factor for the high cost of ethanol production; the step of enzyme production costs up to 40% of total production expenses during ethanol production from lignocellulosic biomass (Chandra et al., 2010; Kumar et al., 2016a). Cost-effective enzymatic hydrolysis of lignocellulosic feedstocks can lower the production cost of bioenergy products. To decrease the cost of saccharification different approaches such as high solid loadings of the substrate, recycling of enzymes for the reutilization, immobilization of enzymes, simultaneous saccharification, and fermentation have been utilized. Enzymatic hydrolysis of lignocellulosic biomass is greatly affected by various substrate and enzyme related factors.

5.3.2 RECYCLING OF ENZYMES FOR REUTILIZATION

The cost of enzymatic hydrolysis is a major part of the process of second-generation biofuels production. A large amount of cellulase and xylanases are required for the hydrolysis of lignocellulosic biomass. Therefore, a viable system of enzymes recycling is essential in order to reduce the overall production cost of biofuels. Several authors have reported the recycling of cellulases and xylanases after hydrolysis of pretreated lignocellulosic feedstocks (Table 5.2). The recycling of enzymes can make the process cost-effective. Recycling of enzymes can be carried out by re-adsorption onto fresh substrates or by desorption by an alkaline medium. During the process of hydrolysis, cellulases and xylanases adsorb on lignin and bind strongly. These enzyme molecules become unproductive during hydrolysis; this becomes a barrier for efficient hydrolysis. It has been shown that lignin does not inactivate these cellulolytic and hemicellulolytic enzymes. Several methods have been proposed to minimize the adsorption of enzymes on the lignin (Rodrigues et al., 2012). The addition of surfactants and additives during the hydrolysis can prevent the unproductive binding of cellulases and xylanases with lignin and enhances the yield of reducing sugars. Non-ionic surfactants such as Tween and Triton interact with lignin hydrophobically and prevent the binding of cellulases and xylanase with lignin which improves the saccharification yield. The addition of some proteins such as bovine serum albumin (BSA) during enzymatic hydrolysis has also improved the yield of reducing sugars (Van Dyk and Pletschke, 2012).

During enzymatic hydrolysis, enzymes are present in two different forms. Some molecules of enzymes remain free in solution (free enzymes) and the remaining enzyme molecules are bound to the residual substrate (cellulose and lignin). An effective method of enzyme recycling should be able to recover both free and adsorbed enzyme. It has been shown that free enzyme can be recovered by simple membrane filtration and affinity re-adsorption on fresh substrates. The enzyme-bound to the residual substrate is difficult to recover due to its non-productive binding with residual lignin. Several strategies

TABLE 5.2
Recycling of Enzymes for Reutilization

Enzymes	Strategy for Recovery of Enzyme	Recovery of Enzyme Activity	References
Cellulase	Alkaline elution	• 49% of cellulase recovered from wheat straw hydrolysate • 60% of activity was recovered on synthetic substrate, 4-methylumbelliferyl-b-D-cellobioside	Rodrigues et al. (2012)
β-glucosidase	Immobilization in calcium alginate	• β-glucosidase retained full activity during 20 times of re-use in cellulose hydrolysis of 48 h each	Tsai and Meyer (2014)
Cellulase, β-glucosidase	Adsorption of enzyme from supernatant to fresh substrate	• Using only recycled enzyme during second hydrolysis round, 10–71% cellulose conversion was achieved compared to first round of hydrolysis	Rosales-Calderon et al. (2017)
Cellulase	Recycling of enzyme associated with insoluble solids fraction after hydrolysis	• Significant amount of cellulase recovered by recycling of insoluble solids • Enzyme dosage decreased by 30% to achieve the same glucose yields in next hydrolysis round of pretreated corn stover	Weiss et al.(2013)
Cellulase, β-glucosidase, xylanase	Recycling of enzyme associated with insoluble solids fraction after hydrolysis	• The enzyme loading was reduced by approximately 50% to achieve the same glucose yields from pretreated corn stover	Xin et al.(2020)
Cellulase	Cellulase recycling by ultrafiltration and nanofiltration	• 74–89% of the protein content was recovered after hydrolysis of steam exploded wheat straw	Qi et al.(2012)

have been utilized for desorption of enzymes of bound enzyme which includes the use of surfactants, alkali, urea, and buffers of different pH. The detachment of cellulase from lignin is influenced by many factors including temperature, pH, ionic strength, and the amount of the surfactant (Tu et al., 2007, 2009; Weiss et al., 2013).

5.3.3 SYNERGY OF ENZYMES

The degree of synergy is defined as "the ratio of the rate or yield of product released by enzymes when used together to the sum of the rate or yield of these products when the enzymes are used separately in the same amounts as they were employed in the mixture" (Kumar and Wyman, 2009; Kumar et al., 2018; Van Dyk and Pletschke, 2012). Cellulases and xylanases hydrolyze of cellulose and hemicelluloses chains in order to produce fermentable sugars. Hydrolysis of cellulose is brought about by cellulase enzyme complex while hydrolysis of hemicelluloses requires enzymes such as endo-1, 4-β-xylanase, β-xylosidase, β-mannanase, β-mannosidase, α-glucuronidase, α- L-arabinofuranosidase, and acetyl xylan esterase (Saritha, 2014). Different types of synergies have been studied as follow next.

5.3.3.1 Synergy in Cellulase Systems

Synergy between cellulases has been identified between different cellobiohydrolases; between endo and exo-glucanases; between endo-glucanases; between cellobiohydrolases, endo-glucanases, and β-glucosidases. The highest degrees of synergy have generally been observed with highly crystalline cellulose substrates while a lower degree of synergy has been shown by cellulose having a higher content of amorphous reason. Synergism between cellulase components from the same species, acting on different substrates, has been well studied. Although, synergism between cellobiohydrolase and endo-glucanase components occurs but it is not necessary that cellobiohydrolase act synergistically with all endo-glucanase components equally. For example, during solubilization of cotton fiber, cellobiohydrolase from *Trichoderma koningii* acted synergistically with only two out of four isolated endoglucanase components from this species (Wood and McCrae, 1978). Generally, the degree of synergy depends on the nature of the substrate, nature of enzymes, and the assay conditions (Kumar et al., 2018; Lynd et al., 2002; Van Dyk and Pletschke, 2012; Woodward, 1991; Zhang et al., 2010). Kogo et al. (2017) studied the enzymatic hydrolysis of NH_4OH-treated rice straw by enzymes from *T. reesei* and *Humicola insolens* and found significant improvement in hydrolysis due to synergistic effect among combined enzyme preparations. Enzymes from *T. reesei*, *H. insolens*, and mixture (75%:25%, v/v) of *T. reesei* and *H. insolens* showed the hydrolysis yield 70.3%, 23.5%, and 79.8%, respectively.

5.3.3.2 Synergy between Different Components of Hemicellulases

Effective hydrolysis of hemicelluloses needs synergistic cooperation between main-chain and side-chain acting hemicellulases. Three types of synergies namely homo-synergy, hetero-synergy, and anti-synergy have been reported in hemicellulases. Homo-synergy is the synergy between two or more types of main-chain-cleaving hemicellulases or between two or more types of side-chain-cleaving hemicellulases. Hetero-synergy occurs between side-chain and main-chain hydrolyzing hemicellulases. When the action of one enzyme is inhibited by another enzyme is termed as anti-synergy (Brigham et al., 1996; Kovacs, 2009; Kumar et al., 2018). Endoxylanase and β-xylosidase are the main enzymes for saccharification of hemicelluloses and their conversion to biofuels. Zhuo et al. (2018) studied the synergistic effect between endoxylanase and β-xylosidase from *Pleurotus ostreatus* and *Irpex lacteus* that were expressed in *Escherichia coli* BL21 and *Pichia pastoris* GS115, respectively. The hydrolysis study of sodium hydroxide pretreated cornstalk by these heterologously expressed enzymes showed the degree of synergy by 2.26.

5.3.3.3 Synergy between Cellulases and Hemicellulases

The synergy between cellulases and xylanases has been extensively studied. It has been proved that the addition of xylanases to cellulases preparations significantly improves the hydrolysis of both cellulose and xylan available in pretreated lignocellulosic biomass. Furthermore, it also reduces the dosage of cellulases required for reasonable saccharification yield. This occurs due to the synergistic effect among cellulases and hemicellulases. Hemicelluloses are generally present not only on the outer surface of cellulosic fibers but are diffused into the interfibrillar space through fiber pores also. Here, hemicelluloses provide a physical barrier that limits the accessibility of the cellulases to the cellulose and effective hydrolysis of cellulose requires relatively high enzyme loadings. The cellulase accessibility of the substrate can be improved by the supplementation of hemicellulases to cellulase preparation due to the solubilization of xylan. The addition of xylanases to cellulase preparation also decreases the cellulase dosages during the saccharification of pretreated lignocellulosic feedstock. Furthermore, the xylan chains that are trapped within or between cellulosic microfibrils also expose due to the hydrolysis of cellulosic chains by cellulases. The exposed xylan chains would be easily available for xylanase attack. This model of cellulose degradation might indicate the synergism between xylanases and cellulases during the saccharification of lignocellulosic feedstocks (Hu et al., 2015, 2013, 2011). Kumar et al. (2016b) performed the hydrolysis of pearl millet stover by crude enzyme-containing cellulases, β-glucosidase, and xylanases and reported a saccharification yield up to 64.77%. The balanced level of β-glucosidase and xylanases in the crude enzyme is helpful in achieving higher saccharification yield. Bunterngsook et al. (2018) developed a tailor-made synergistic enzyme system having cellobiohydrolase from *Talaromyces cellulolyticus*, an endoglucanase from *Thielavia terrestris*, a β-glucosidase from xylanase from *Aspergillus aculeatus* at the ratio of 0.34:0.27:0.14:0.25 respectively. This enzyme preparation released 797 mg/g of reducing sugars from alkaline-catalyzed steam explosion pretreated sugarcane bagasse that sugar yield was 17.37% higher than obtained with commercial enzyme Accellerase 1500.

5.3.4 GENERATION OF PRODUCT INHIBITORS

Biotechnological transformation of lignocellulosic feedstocks to advanced biofuels and other biochemicals requires a pretreatment to improve the digestibility of cellulose and hemicelluloses for enzymes. During the process of this pretreatment, some inhibitory compounds are generated from lignocellulosic biomass. These compounds have an inhibitory effect on enzymatic saccharification and microbial fermentation. The type of lignocellulosic material and pretreatment method determines the generation of inhibitory compounds. Different aromatic and phenolic compounds, aliphatic acids, furan aldehydes, inorganic ions, bioalcohols, or other fermentation products act as inhibitors for microorganisms and hydrolytic enzymes. Cellulases and xylanases are inhibited by their products such as sugars including cellobiose and glucose. Fermentation products like ethanol and phenolic compounds have also shown the inhibitory effect on these hydrolytic enzymes. Several phenolic compounds are generated as a result of acid-catalyzed pretreatment of lignocellulosic materials. Phenolic compounds are generated from lignins or extractives. The furan aldehydes like furfural and 5-hyroxymethyl furfural (HMF) are generated due to the dehydration reaction of pentose and hexose sugars, respectively. During fermentation, the growth of yeasts and ethanol yield is adversely affected due to the presence of furfural and HMF in the fermentation media (Jönsson et al., 2013; Jönsson and Martín, 2016; Van Dyk and Pletschke, 2012). Acetyl groups of hemicelluloses are responsible for the generation of acetic acid while formic acid is produced due to the degradation of furfural and HMF, and levulinic acid is generated by the breakdown of HMF. The problem of inhibitors can be minimized by using several strategies. Detoxification of lignocellulosic hydrolysates is one of the very useful strategies to avoid the problems caused by inhibitors. Various biological, physical, and chemical methods have been used for the process of detoxification. The approaches that have been employed for detoxification include over liming, liquid-liquid extraction, liquid-solid extraction, heating and evaporation, and treatments with microbes and microbial enzymes. Laccases and lignin peroxidases have been employed for the removal of phenolic inhibitors. Detoxification is an additional step that increases the process cost (Jönsson et al., 2013; Palmqvist and Hahn-Hägerdal, 2000; Ulbricht et al., 1984). Therefore, some researchers have isolated inhibitors tolerant ethanologenic yeasts which ferment the reducing sugars in the presence of inhibitors (Choudhary et al., 2017).

5.3.5 SIMULTANEOUS SACCHARIFICATION AND FERMENTATION (SSAF)

Presently, a cost-intensive method i.e. separate hydrolysis and fermentation, is utilized for the production of most biofuels including bioethanol. It needs separate vessels for hydrolysis of lignocellulosic biomass and fermentation. It also requires a large number of utilities that make the process costly. In separate hydrolysis and fermentation, the efficiency of cellulases and hemicellulases is lowered by end-product inhibition during the process of saccharification. Reducing sugars, disaccharides, and oligosaccharides inhibit the enzymes that are involved in the hydrolysis of lignocellulosic feedstocks. Cellobiose acts as an inhibitor for cellulases while glucose inhibits the action of β-glucosidase. Endo-xylanases are inhibited by xylose and xylobiose but only occasionally. Therefore, to overcome the problem of the problem of feedback inhibition of cellulases and xylanases Simultaneous Saccharification and Fermentation (SSAF) is utilized. In the process of SSAF reducing sugars are released by the action of cellulolytic and hemicellulolytic enzymes and immediately utilized by fermenting microorganism to produce ethanol (Table 5.3). By continuous utilization of fermentable sugars by fermentation microorganisms, the concentration of cellobiose and glucose in fermentation media is minimized to reduce the inhibitory action of reducing sugars such as cellobiose and glucose (Choudhary et al., 2017; Gupte and Madamwar, 1997; Taherzadeh and Karimi, 2007; Van Dyk and Pletschke, 2012; Wen et al., 2005). Furthermore, SSAF has been reported to give higher ethanol yields from lignocellulosic feedstocks than separate hydrolysis and fermentation and it also requires lower amounts of enzyme. The approach of SSAF has also shown other advantages such as the lower risk of contamination. The presence of ethanol during the hydrolysis of lignocellulosic feedstock minimizes the risk

TABLE 5.3

Simultaneous Saccharification and Fermentation for Biofuels Production

Source of Enzyme	Enzyme	Substrate	Fermenting Microorganism	Temperature (°C) During SSAF	Final Product	References
Aspergillus niger SH3	Holocellulase (15 FPU/ g substrate)	Paddy straw	*Candida tropicalis* Y6 *Saccharomyces cerevisiae* JRC6	40	Ethanol	Choudhary et al. (2017)
Celluclast 1.5 L (NOVO Nordisk)	Cellulolytic complex (15 FPU/ g substrate)	Poplar, Eucalyptus, wheat straw, sweet sorghum bagasse, Brassica carinata	*Kluyveromyces marxianus* CECT 10875	42	Ethanol	Ballesteros et al. (2004)
NA	Cellulase (15 FPU/ g substrate), glucosidase (30 IU/g substrate)	Oil palm empty fruit bunches	*Mucor indicus*	37	Ethanol	Christia et al. (2016)
Cellic® CTec2 Cellic® HTec2 (Novozymes)	Cellulases, hemicellulases (2.5–20 FPU/ g substrate)	Rice straw	*Mucor indicus*	37	Ethanol	Molaverdi et al. (2019)
Genencor GC220	Cellulases (10.5-15.8 FPU/ g substrate)	Sugarcane bagasse	*Bacillus coagulans* DSM2314	50	Lactic acid	Christia et al. (2016)
Advanced Enzymes	Cellulases (20 FPU/ g substrate)	Oil palm front	*Enterobacter cloacae* sp.SG1	37	2, 3-butanediol	Hazeena et al. (2019)
Acremonium cellulolyticus	Cellulase (10 FPU/ g substrate)	Oil palm empty fruit bunches	*Clostridium acetobutylicum* ATCC 824	37	Butanol	Md Razali et al. (2018)
Acidothermus cellulolyticus	Cellulases (9–36 FPU/ g substrate)	Oil palm empty fruit bunches	*Enterobacter aerogene*	37	Hydrogen	Diah et al. (2018)
Cellic CTec2 (Novozymes)	Cellulases (10 FPU/ g substrate)	Oil palm trunk	*Thermoanaerobacterium thermosaccharolyticum* KKU19	50	Hydrogen, methane	Sitthikitpanya et al. (2018)

of contamination during SSAF. Moreover, the number of vessels required for SSAF is reduced in comparison to separate hydrolysis and fermentation which results in lower capital cost of the process.

In spite of several advantages of SSAF over separate hydrolysis and fermentation, the critical problem with simultaneous saccharification and fermentation is the difference in the optimum temperature of hydrolyzing enzymes and the fermenting microorganism. The strains of *Saccharomyces* that are the most efficient producers of ethanol require an operating temperature of 35°C. On the other hand, cellulases and xylanases that are utilized for saccharification have an optimum temperature of 50°C. At lower temperatures, the rate of hydrolysis reduces substantially which is unfavorable in terms of increased processing time. A possible solution to solve this problem is the utilization of thermo-tolerant strains of yeasts instead of mesophilic yeasts, which would allow higher temperatures for simultaneous saccharification and fermentation. Therefore, the isolation of thermo-tolerant ethanogenic microorganisms is essential for SSAF processes. Several investigators have screened the thermo-tolerant strains of bacteria and yeasts for their potential use for SSAF. The thermo-tolerant microorganisms such as *Candida acidothermophilum*, *Candida tropicalis*, *Kluyveromyces marxianus*, *Kluyveromyces fragilis*, etc. have been utilized for the SSAF(Bansal et al., 2009; Choudhary et al., 2017; de Castro and de Castro, 2012; Dutta et al., 2008; Hari Krishna et al., 2001; Horn et al., 2012a, 2012b; Kádár et al., 2004; Taherzadeh and Karimi, 2007). Hazeena et al. (2019) carried out simultaneous saccharification and fermentation alkali pretreated oil palm front biomass for the production of 2,3-butanediol by *Enterobacter cloacae* sp.SG1. Recently, solid-state simultaneous saccharification and fermentation (SSSF) have been reported for the production of ethanol. Molaverdi et al. (2019) studied ethanol production by rice straw using the approach of SSSF. Rice straw was pretreated with sodium carbonate and maximum ethanol yield (89.5%) was obtained by SSSF using *Mucor indicus*.

REFERENCES

Aro, E., 2016. From first generation biofuels to advanced solar biofuels. *Ambio* 45, 24–31. https://doi.org/10.1007/s13280-015-0730-0

Bajaj, B.K., Manhas, K., 2012. Production and characterization of xylanase from Bacillus licheniformis P11(C) with potential for fruit juice and bakery industry. *Biocatal. Agric. Biotechnol.* 1, 330–337. https://doi.org/http://dx.doi.org/10.1016/j.bcab.2012.07.003

Ballesteros, M., Oliva, J.M., Negro, M.J., Manzanares, P., Ballesteros, I., 2004. Ethanol from lignocellulosic materials by a simultaneous saccharification and fermentation process (SFS) with Kluyveromyces marxianus CECT 10875. *Process Biochem.* 39, 1843–1848. https://doi.org/http://dx.doi.org/10.1016/j.procbio.2003.09.011

Bansal, P., Hall, M., Realff, M.J., Lee, J.H., Bommarius, A.S., 2009. Modeling cellulase kinetics on lignocellulosic substrates. *Biotechnol. Adv.* 27, 833–848. https://doi.org/http://dx.doi.org/10.1016/j.biotechadv.2009.06.005

Barros-Rios, J., Romani, A., Peleteiro, S., Garrote, G., Ordas, B., 2016. Second-generation bioethanol of hydrothermally pretreated stover biomass from maize genotypes. *Biomass Bioenergy* 90, 42–49. https://doi.org/https://doi.org/10.1016/j.biombioe.2016.03.029

Bastawde, K.B., 1992. Xylan structure, microbial xylanases, and their mode of action. *World J. Microbiol. Biotechnol.* 8, 353–368. https://doi.org/10.1007/bf01198746

Battan, B., Dhiman, S., Ahlawat, S., Mahajan, R., Sharma, J., 2012. Application of thermostable xylanase of *Bacillus pumilus* in textile processing. *Indian J. Microbiol.* 52, 222–229. https://doi.org/10.1007/s12088-011-0118-1

Beg, Q.K., Kapoor, M., Mahajan, L., Hoondal, G.S., 2001. Microbial xylanases and their industrial applications: A review. *Appl. Microbiol. Biotechnol.* 56, 326–338.

Bharti, A.K., Kumar, A., Kumar, A., Dutt, D., 2018. Exploitation of *Parthenium hysterophorous* biomass as low-cost substrate for cellulase and xylanase production under solid-state fermentation using *Talaromyces stipitatus* MTCC 12687. *J. Radiat. Res. Appl. Sci.* 11, 271–280. https://doi.org/10.1016/j.jrras.2018.01.003

Brigham, J., Adney, W., Himmel, M., 1996. Hemicellulases: Diversity and applications, in: Wyman, C. (Ed.), *Handbook on Bioethanol: Production and Utilization*. Taylor and Francis, Washington, DC, pp. 119–142.

Bunterngsook, B., Laothanachareon, T., Chotirotsukon, C., Inoue, H., Fujii, T., Hoshino, T., Roongsawang, N., Kuboon, S., Kraithong, W., Techanan, W., Kraikul, N., Champreda, V., 2018. Development of tailor-made synergistic cellulolytic enzyme system for saccharification of steam exploded sugarcane bagasse. *J. Biosci. Bioeng.* 125, 390–396. https://doi.org/https://doi.org/10.1016/j.jbiosc.2017.11.001

Butt, M.S., Nadeem, T.M., Zulfiqar, A., Sultan, M.T., 2008. Xylanases and their applications in baking industry. *Food Technol. Biotechnol.* 46, 22–31.

Cen, P., Xia, L., 1999. Production of cellulase by solid-state fermentation, in: Tsao, G.T., Brainard, A.P., Bungay, H.R., Cao, N.J., Cen, P., Chen, Z., Du, J., Foody, B., Gong, C.S., Hall, P., Ho, N.W.Y., Irwin, D.C., Iyer, P., Jeffries, T.W., Ladisch, C.M., Ladisch, M.R., Lee, Y.Y., Mosier, N.S., Mühlemann, H.M., Sedlak, M., Shi, N.Q., Tsao, G.T., Tolan, J.S., Torget, R.W., Wilson, D.B., Xia, L. (Eds.), *Recent Progress in Bioconversion of Lignocellulosics*. Springer, Berlin, Heidelberg, pp. 69–92. https://doi.org/10.1007/3-540-49194-5_4

Chandel, A.K., Singh, O.V., Venkateswar Rao, L., Chandrasekhar, G., Lakshmi Narasu, M., 2011. Bioconversion of novel substrate *Saccharum spontaneum*, a weedy material, into ethanol by Pichia stipitis NCIM3498. *Bioresour. Technol.* 102, 1709–1714. https://doi.org/http://dx.doi.org/10.1016/j.biortech.2010.08.016

Chandra, M., Kalra, A., Sharma, P.K., Kumar, H., Sangwan, R.S., 2010. Optimization of cellulases production by *Trichoderma citrinoviride* on marc of *Artemisia annua* and its application for bioconversion process. *Biomass Bioenergy* 34, 805–811. https://doi.org/http://dx.doi.org/10.1016/j.biombioe.2010.01.024

Chavez, R., Bull, P., Eyzaguirre, J., 2006. The xylanolytic enzyme system from the genus Penicillium. *J. Biotechnol.* 123, 413–433. https://doi.org/http://dx.doi.org/10.1016/j.jbiotec.2005.12.036

Cheng, C.-L., Lo, Y.-C., Lee, K.-S., Lee, D.-J., Lin, C.-Y., Chang, J.-S., 2011. Biohydrogen production from lignocellulosic feedstock. *Bioresour. Technol.* 102, 8514–8523. https://doi.org/https://doi.org/10.1016/j.biortech.2011.04.059

Choudhary, J., Singh, S., Nain, L., 2017. Bioprospecting thermotolerant ethanologenic yeasts for simultaneous saccharification and fermentation from diverse environments. *J. Biosci. Bioeng.* 123, 342–346. https://doi.org/https://doi.org/10.1016/j.jbiosc.2016.10.007

Christia, A., Setiowati, A.D., Millati, R., Karimi, K., Cahyanto, M.N., Niklasson, C., Taherzadeh, M.J., 2016. Ethanol production from alkali-pretreated oil palm empty fruit bunch by simultaneous saccharification and fermentation with *Mucor indicus*. *Int. J. Green Energy* 13, 566–572. https://doi.org/10.1080/15435075.2014.978004

Collins, T., Gerday, C., Feller, G., 2005. Xylanases, xylanase families and extremophilic xylanases. *FEMS Microbiol. Rev.* 29, 3–23. https://doi.org/10.1016/j.femsre.2004.06.005

de Almeida Antunes Ferraz, J.L., Souza, L.O., Soares, G.A., Coutinho, J.P., de Oliveira, J.R., Aguiar-Oliveira, E., Franco, M., 2018. Enzymatic saccharification of lignocellulosic residues using cellulolytic enzyme extract produced by *Penicillium roqueforti* ATCC 10110 cultivated on residue of yellow mombin fruit. *Bioresour. Technol.* 248, 214–220. https://doi.org/https://doi.org/10.1016/j.biortech.2017.06.048

de Castro, S., de Castro, A., 2012. Assessment of the Brazilian potential for the production of enzymes for biofuels from agroindustrial materials. *Biomass Convers. Biorefinery* 2, 87–107. https://doi.org/10.1007/s13399-012-0031-9

Delabona, P.da S., Pirota, R.D.P.B., Codima, C.A., Tremacoldi, C.R., Rodrigues, A., Farinas, C.S., 2012. Using Amazon forest fungi and agricultural residues as a strategy to produce cellulolytic enzymes. *Biomass Bioenergy* 37, 243–250. https://doi.org/http://dx.doi.org/10.1016/j.biombioe.2011.12.006

Diah, K., Joni, P., Endang, S., Sumi, H., 2018. Biohydrogen production through separate hydrolysis and fermentation and simultaneous saccharification and fermentation of empty fruit bunch of palm oil. *Res. J. Chem. Environ.* 22, 193–197.

Dutt, D., Tyagi, C.H., Singh, R.P., Kumar, A., 2012. Effect of enzyme concoctions on fiber surface roughness and deinking efficiency of sorted office paper. *Cellul. Chem. Technol.* 46, 611–623.

Dutta, T., Sahoo, R., Sengupta, R., Ray, S., Bhattacharjee, A., Ghosh, S., 2008. Novel cellulases from an extremophilic filamentous fungi *Penicillium citrinum*: production and characterization. *J. Ind. Microbiol. Biotechnol.* 35, 275–282. https://doi.org/10.1007/s10295-008-0304-2

Dwivedi, U.N., Singh, P., Pandey, V.P., Kumar, A., 2011. Structure–function relationship among bacterial, fungal and plant laccases. *J. Mol. Catal. B Enzym.* 68, 117–128. https://doi.org/https://doi.org/10.1016/j.molcatb.2010.11.002

EIA, 2013. International Energy Outlook 2013, U.S. Energy Information Administration, DOE/EIA-0484(2013). http://www.eia.gov/forecasts/ieo/pdf/0484%282013%29.pdf.

Fang, H., Zhao, C., Song, X.Y., 2010. Optimization of enzymatic hydrolysis of steam-exploded corn stover by two approaches: Response surface methodology or using cellulase from mixed cultures of *Trichoderma reesei* RUT-C30 and *Aspergillus niger* NL02. *Bioresour. Technol.* 101, 4111–4119. https://doi.org/http://dx.doi.org/10.1016/j.biortech.2010.01.078

Farinas, C.S., Loyo, M.M., Junior, A.B., Tardioli, P.W., Couri, S., 2010. Finding stable cellulase and xylanase: evaluation of the synergistic effect of pH and temperature. *N. Biotechnol.* 27, 810–816.

Garoma, T., Shackelford, T., 2014. Electroporation of Chlorella vulgaris to enhance biomethane production. *Bioresour. Technol.* 169, 778–783. https://doi.org/http://dx.doi.org/10.1016/j.biortech.2014.07.001

Gautam, A., Kumar, A., Dutt, D., 2015. Production of cellulase-free xylanase by *Aspergillus flavus* ARC-12 using pearl millet stover as the substrate under solid-state fermentation. *J. Adv. Enzym. Res.* 1, 1–9.

Gautam, A., Kumar, A., Bharti, A.K., Dutt, D., 2018. Rice straw fermentation by *Schizophyllum commune* ARC-11 to produce high level of xylanase for its application in pre-bleaching. *J. Genet. Eng. Biotechnol.* 16, 693–701. https://doi.org/10.1016/j.jgeb.2018.02.006

Gautam, R., Sharma, J., 2014. Optimization, purification of cellulase produced from *Bacillus subtilis* Subsp. inaquosorum under solid state fermentation and its potential applications in denim industry. *Int. J. Sci. Res.* 3, 1759–1763.

Ghosh, S., Chowdhury, R., Bhattacharya, P., 2017. Sustainability of cereal straws for the fermentative production of second generation biofuels: A review of the efficiency and economics of biochemical pretreatment processes. *Appl. Energy* 198, 284–298. https://doi.org/https://doi.org/10.1016/j.apenergy.2016.12.091

Gupte, A., Madamwar, D., 1997. Solid state fermentation of lignocellulosic waste for cellulase and β-glucosidase production by cocultivation of *Aspergillus ellipticus* and *Aspergillus fumigatus. Biotechnol. Prog.* 13, 166–169. https://doi.org/10.1021/bp970004g

Gusakov, A.V, 2011. Alternatives to *Trichoderma reesei* in biofuel production. *Trends Biotechnol.*29, 419–425. https://doi.org/http://dx.doi.org/10.1016/j.tibtech.2011.04.004

Hari Krishna, S., Janardhan Reddy, T., Chowdary, G.V, 2001. Simultaneous saccharification and fermentation of lignocellulosic wastes to ethanol using a thermotolerant yeast. *Bioresour. Technol.* 77, 193–196. https://doi.org/https://doi.org/10.1016/S0960-8524(00)00151-6

Hazeena, S.H., Nair Salini, C., Sindhu, R., Pandey, A., Binod, P., 2019. Simultaneous saccharification and fermentation of oil palm front for the production of 2,3-butanediol. *Bioresour. Technol.* 278, 145–149. https://doi.org/https://doi.org/10.1016/j.biortech.2019.01.042

Hölker, U., Höfer, M., Lenz, J., 2004. Biotechnological advantages of laboratory-scale solid-state fermentation with fungi. *Appl. Microbiol. Biotechnol.* 64, 175–186. https://doi.org/10.1007/s00253-003-1504-3

Horn, S., Vaaje-Kolstad, G., Westereng, B., Eijsink, V.G.H., 2012a. Novel enzymes for the degradation of cellulose. *Biotechnol. Biofuels* 5, 45.

Horn, S.J., Sorlie, M., Varum, K.M., Valjamae, P., Eijsink, V.G.H., 2012b. Measuring processivity. *Methods Enzym.* 510, 69–96.

Hu, J., Arantes, V., Saddler, J.N., 2011. The enhancement of enzymatic hydrolysis of lignocellulosic substrates by the addition of accessory enzymes such as xylanase: is it an additive or synergistic effect? *Biotechnol. Biofuels* 4, 36. https://doi.org/10.1186/1754-6834-4-36

Hu, J., Arantes, V., Pribowo, A., Saddler, J.N., 2013. The synergistic action of accessory enzymes enhances the hydrolytic potential of a "cellulase mixture" but is highly substrate specific. *Biotechnol. Biofuels* 6, 112. https://doi.org/10.1186/1754-6834-6-112

Hu, J., Chandra, R., Arantes, V., Gourlay, K., Susan van Dyk, J., Saddler, J.N., 2015. The addition of accessory enzymes enhances the hydrolytic performance of cellulase enzymes at high solid loadings. *Bioresour. Technol.* 186, 149–153. https://doi.org/https://doi.org/10.1016/j.biortech.2015.03.055

Ibrahim, C.O., 2008. Development of applications of industrial enzymes from Malaysian indigenous microbial sources. *Bioresour. Technol.* 99, 4572–4582. https://doi.org/http://dx.doi.org/10.1016/j.biortech.2007.07.040

Jönsson, L.J., Martín, C., 2016. Pretreatment of lignocellulose: Formation of inhibitory by-products and strategies for minimizing their effects. *Bioresour. Technol.* 199, 103–112. https://doi.org/https://doi.org/10.1016/j.biortech.2015.10.009

Jönsson, L.J., Alriksson, B., Nilvebrant, N.-O., 2013. Bioconversion of lignocellulose: Inhibitors and detoxification. *Biotechnol. Biofuels* 6, 16. https://doi.org/10.1186/1754-6834-6-16

Kádár, Z., Szengyel, Z., Réczey, K., 2004. Simultaneous saccharification and fermentation (SSF) of industrial wastes for the production of ethanol. *Ind. Crops Prod.* 20, 103–110. https://doi.org/https://doi.org/10.1016/j.indcrop.2003.12.015

Kogo, T., Yoshida, Y., Koganei, K., Matsumoto, H., Watanabe, T., Ogihara, J., Kasumi, T., 2017. Production of rice straw hydrolysis enzymes by the fungi *Trichoderma reesei* and *Humicola insolens* using rice straw as a carbon source. *Bioresour. Technol.* 233, 67–73. https://doi.org/10.1016/j.biortech.2017.01.075

Kovacs, K., 2009. *Production of Cellulolytic Enzymes with Trichoderma Atroviride Mutants for the Biomass-to-Bioethanol Process.* Lund University, Sweden.

Kuhad, R.C., Gupta, R., Singh, A., 2011. Microbial cellulases and their industrial applications. *Enzyme Res.* 2011, 10. https://doi.org/10.4061/2011/280696

Kumar, A., Dutt, D., Gautam, A., 2016a. Production of crude enzyme from *Aspergillus nidulans* AKB-25 using black gram residue as the substrate and its industrial applications. *J. Genet. Eng. Biotechnol.* 14, 107–118. https://doi.org/10.1016/j.jgeb.2016.06.004

Kumar, A., Dutt, D., Gautam, A., 2016b. Pretreatment and enzymatic hydrolysis of pearl millet stover by multi-enzymes from *Aspergillus nidulans* AKB-25. *Cellul. Chem. Technol.*50, 781–790.

Kumar, A., Gautam, A., Dutt, D., 2016c. Biotechnological transformation of lignocellulosic biomass in to industrial products: An overview. *Adv. Biosci. Biotechnol.* 7, 149–168. https://doi.org/10.4236/abb.2016.73014

Kumar, A., Gautam, A., Dutt, D., 2016d. Co-cultivation of Penicillium sp. AKB-24 and *Aspergillus nidulans* AKB-25 as a cost-effective method to produce cellulases for the hydrolysis of pearl millet stover. *Fermentation* 2, 1–22. https://doi.org/10.3390/fermentation2020012

Kumar, A., Minuye, N., Bezie, Y., Yadav, M., 2018. A review of factors affecting enzymatic hydrolysis of pretreated lignocellulosic biomass. *Res. J. Chem. Environ.* 22, 7.

Kumar, R., Wyman, C.E., 2009. Effects of cellulase and xylanase enzymes on the decon-struction of solids from pretreatment of poplar by leading technologies. *Biotechnol. Prog.* 25, 302–314. https://doi.org/10.1002/btpr.102

Kumar, V., Chhabra, D., Shukla, P., 2017. Xylanase production from *Thermomyces lanugino-sus* VAPS-24 using low cost agro-industrial residues via hybrid optimization tools and its potential use for saccharification. *Bioresour. Technol.* 243, 1009–1019. https://doi.org/https://doi.org/10.1016/j.biortech.2017.07.094

Ladole, M.R., Mevada, J.S., Pandit, A.B., 2017. Ultrasonic hyperactivation of cellulase immo-bilized on magnetic nanoparticles. *Bioresour. Technol.* 239, 117–126. https://doi.org/https://doi.org/10.1016/j.biortech.2017.04.096

Lynd, L.R., Weimer, P.J., van Zyl, W.H., Pretorius, I.S., 2002. Microbial cellulose utilization: Fundamentals and biotechnology. *Microbiol. Mol. Biol. R.* 66, 506–577.

Maity, C., Ghosh, K., Halder, S., Jana, A., Adak, A., Das Mohapatra, P., Pati, B., Mondal, K., 2012. Xylanase isozymes from the newly isolated Bacillus sp. CKBx1D and optimiza-tion of its deinking potentiality. *Appl. Biochem. Biotechnol.* 167, 1208–1219. https://doi.org/10.1007/s12010-012-9556-4

Manavalan, T., Manavalan, A., Heese, K., 2015. Characterization of lignocellulolytic enzymes from white-rot fungi. *Curr. Microbiol.* 70, 485–498. https://doi.org/10.1007/s00284-014-0743-0

Md Razali, A.N., Ibrahim, F.M., Kamal Bahrin, E., Abd-Aziz, S., 2018. Optimisation of simultaneous saccharification and fermentation (SSF) for biobutanol production using pretreated oil palm empty fruit bunch. *Molecules.* https://doi.org/10.3390/molecules23081944

Mohapatra, S., Padhy, S., Das Mohapatra, P.K., Thatoi, H.N., 2018. Enhanced reducing sugar production by saccharification of lignocellulosic biomass, Pennisetum species through cellulase from a newly isolated *Aspergillus fumigatus*. *Bioresour. Technol.* 253, 262–272. https://doi.org/https://doi.org/10.1016/j.biortech.2018.01.023

Molaverdi, M., Karimi, K., Mirmohamadsadeghi, S., Galbe, M., 2019. High titer ethanol production from rice straw via solid-state simultaneous saccharification and fermenta-tion by *Mucor indicus* at low enzyme loading. *Energy Convers. Manag.* 182, 520–529. https://doi.org/https://doi.org/10.1016/j.enconman.2018.12.078

Mørkeberg, A., Jørgensen, H., and Olsson, L., 2004. Screening genus Penicillium for pro-ducers of cellulolytic and xylanolytic enzymes. *Appl. Biochem. Biotechnol.* 113–116, 389–401.

Nguyen, Q.A., Keller, F.A., Tucker, M.P., Lombard, C.K., Jenkins, B.M., Yomogida, D.E., Tiangco, V.M., 1999. Bioconversion of mixed solids waste to ethanol, in: *Twentieth Symposium on Biotechnology for Fuels and Chemicals*. Humana Press, Totowa, NJ, pp. 455–472.

Niladevi, K.N., 2009. Ligninolytic enzymes, in: Nigam, P.-N., Pandey, A. (Eds.), *Biotechnology for Agro-Industrial Residues Utilisation: Utilisation of Agro-Residues*. Springer Science & Business Media, pp. 397–414.

Palmqvist, E., Hahn-Hägerdal, B., 2000. Fermentation of lignocellulosic hydrolysates. I: Inhibition and detoxification. *Bioresour. Technol.* 74, 17–24. https://doi.org/http://dx.doi.org/10.1016/S0960-8524(99)00160-1

Pandey, A., 2003. Solid-state fermentation. *Biochem. Eng. J.* 13, 81–84. https://doi.org/http://dx.doi.org/10.1016/S1369-703X(02)00121-3

Pandey, A., Soccol, C.R., Mitchell, D., 2000. New developments in solid state fermentation: I-bioprocesses and products. *Process Biochem.* 35, 1153–1169. https://doi.org/http://dx.doi.org/10.1016/S0032-9592(00)00152-7

Pareek, S.S., Ravi, I., Sharma, V., 2014. Induction of β-1,3-glucanase and chitinase in *Vigna aconitifolia* inoculated with *Macrophomina phaseolina. J. Plant Interact.* 9, 434–439.

Patel, H., Chapla, D., Shah, A., 2017. Bioconversion of pretreated sugarcane bagasse using enzymatic and acid followed by enzymatic hydrolysis approaches for bioethanol production. *Renew. Energy* 109, 323–331. https://doi.org/https://doi.org/10.1016/j.renene.2017.03.057

Pathak, P., Bhardwaj, N., Singh, A., 2014. Production of crude cellulase and xylanase from *Trichoderma harzianum* PPDDN10 NFCCI-2925 and its application in photocopier waste paper recycling. *Appl. Biochem. Biotechnol.* 172, 3776–3797. https://doi.org/10.1007/s12010-014-0758-9

Pennacchio, A., Ventorino, V., Cimini, D., Pepe, O., Schiraldi, C., Inverso, M., Faraco, V., 2018. Isolation of new cellulase and xylanase producing strains and application to lignocellulosic biomasses hydrolysis and succinic acid production. *Bioresour. Technol.* 259, 325–333. https://doi.org/https://doi.org/10.1016/j.biortech.2018.03.027

Plácido, J., Capareda, S., 2015. Ligninolytic enzymes: a biotechnological alternative for bioethanol production. *Bioresour. Bioprocess* 2, 23. https://doi.org/10.1186/s40643-015-0049-5

Prakash, D., Nawani, N., Prakash, M., Bodas, M., Mandal, A., Khetmalas, M., Kapadnis, B., 2013. Actinomycetes: A repertory of green catalysts with a potential revenue resource. *Biomed Res. Int.* 2013, 8. https://doi.org/10.1155/2013/264020

Qi, B., Luo, J., Chen, G., Chen, X., Wan, Y., 2012. Application of ultrafiltration and nanofiltration for recycling cellulase and concentrating glucose from enzymatic hydrolyzate of steam exploded wheat straw. *Bioresour. Technol.* 104, 466–472. https://doi.org/https://doi.org/10.1016/j.biortech.2011.10.049

Raghavarao, K.S.M.S., Ranganathan, T. V, Karanth, N.G., 2003. Some engineering aspects of solid-state fermentation. *Biochem. Eng. J.* 13, 127–135. https://doi.org/http://dx.doi.org/10.1016/S1369-703X(02)00125-0

Rodrigues, A.C., Leitão, A.F., Moreira, S., Felby, C., Gama, M., 2012. Recycling of cellulases in lignocellulosic hydrolysates using alkaline elution. *Bioresour. Technol.* 110, 526–533. https://doi.org/https://doi.org/10.1016/j.biortech.2012.01.140

Rosales-Calderon, O., Trajano, H.L., Posarac, D., Duff, S.J.B., 2017. Enzyme recycling by adsorption during hydrolysis of oxygen-delignified wheat straw. *ACS Sustain. Chem. Eng.* 5, 9701–9708. https://doi.org/10.1021/acssuschemeng.7b01294

Saha, B.C., Cotta, M.A., 2007. Enzymatic saccharification and fermentation of alkaline peroxide pretreated rice hulls to ethanol. *Enzyme Microb. Technol.* 41, 528–532. https://doi.org/https://doi.org/10.1016/j.enzmictec.2007.04.006

Saini, A., Aggarwal, N.K., Yadav, A., 2018. Microbial cellulases: Role in second-generation ethanol production, in: Singh, J., Sharma, D., Kumar, G., Sharma, N.R. (Eds.), *Microbial Bioprospecting for Sustainable Development.* Springer, Singapore, pp. 167–187. https://doi.org/10.1007/978-981-13-0053-0_8

Sanchez, C., 2009. Lignocellulosic residues: Biodegradation and bioconversion by fungi. *Biotechnol. Adv.* 27, 185–194. https://doi.org/http://dx.doi.org/10.1016/j.biotechadv.2008.11.001

Saritha, M., 2014. Enhanced saccharification of steam-pretreated rice straw by commercial cellulases supplemented with xylanase. *J. Bioprocess. Biotech.* 4, 1. https://doi. org/10.4172/2155-9821.1000188

Sharma, A., Thakur, V.V., Shrivastava, A., Jain, R.K., Mathur, R.M., Gupta, R., Kuhad, R.C., 2014. Xylanase and laccase based enzymatic kraft pulp bleaching reduces adsorbable organic halogen (AOX) in bleach effluents: A pilot scale study. *Bioresour. Technol.* 169, 96–102. https://doi.org/http://dx.doi.org/10.1016/j.biortech.2014.06.066

Shrestha, P., Khanal, S.K., Pometto Iii, A.L., van Leeuwen, J., 2010. Ethanol production via in situ fungal saccharification and fermentation of mild alkali and steam pretreated corn fiber. *Bioresour. Technol.* 101, 8698–8705. https://doi.org/http://dx.doi.org/10.1016/j. biortech.2010.06.089

Singh, A., Kumar, P.K., 1991. Fusarium oxysporum: status in bioethanol production. *Crit. Rev. Biotechnol.* 11, 129–147. https://doi.org/10.3109/07388559109040619

Singhania, R.R., Sukumaran, R.K., Patel, A.K., Larroche, C., Pandey, A., 2010. Advancement and comparative profiles in the production technologies using solid-state and submerged fermentation for microbial cellulases. *Enzyme Microb. Technol.* 46, 541–549. https://doi.org/http://dx.doi.org/10.1016/j.enzmictec.2010.03.010

Sitthikitpanya, S., Reungsang, A., Prasertsan, P., 2018. Two-stage thermophilic bio-hydrogen and methane production from lime-pretreated oil palm trunk by simultaneous saccharification and fermentation. *Int. J. Hydrogen Energy* 43, 4284–4293. https://doi. org/10.1016/j.ijhydene.2018.01.063

Soliman, S.A., El-Zawahry, Y.A., El-Mougith, A.A., 2013. Fungal biodegradation of agroindustrial waste, in: Ven, T. van de, Kadla, J. (Eds.), *Cellulose – Biomass Conversion.* InTech Open, pp. 75–100. https://doi.org/45306

Sunna, A., Antranikian, G., 1997. Xylanolytic enzymes from fungi and bacteria. *Crit. Rev Biotechnol.* 17, 39–67.

Taherzadeh, M.J., Karimi, K., 2007. Enzymatic-based hydrolysis processes for ethanol. *BioResources* 2, 707–738.

Techapun, C., Poosaran, N., Watanabe, M., Sasaki, K., 2003. Thermostable and alkaline-tolerant microbial cellulase-free xylanases produced from agricultural wastes and the properties required for use in pulp bleaching bioprocesses: A review. *Process Biochem.* 38, 1327–1340. https://doi.org/http://dx.doi.org/10.1016/S0032-9592(02)00331-X

Tengerdy, R.P., 1996. Cellulase production by solid state fermentation. *J. Sci. Ind. Res. (India)* 55, 313–316.

Tomei, J., Helliwell, R., 2016. Food versus fuel? Going beyond biofuels. *Land Use Policy* 56, 320–326. https://doi.org/https://doi.org/10.1016/j.landusepol.2015.11.015

Tsai, C.-T., Meyer, S.A., 2014. Enzymatic cellulose hydrolysis: Enzyme reusability and visualization of β-glucosidase iImmobilized in calcium alginate. *Molecules* 19, 19390–19406. https://doi.org/10.3390/molecules191219390

Tu, M., Chandra, R.P., Saddler, J.N., 2007. Recycling cellulases during the hydrolysis of steam exploded and ethanol pretreated lodgepole pine. *Biotechnol. Prog.* 23, 1130–1137. https://doi.org/10.1021/bp070129d

Tu, M., Zhang, X., Paice, M., MacFarlane, P., Saddler, J.N., 2009. The potential of enzyme recycling during the hydrolysis of a mixed softwood feedstock. *Bioresour. Technol.* 100, 6407–6415. https://doi.org/http://dx.doi.org/10.1016/j.biortech.2009.06.108

Ulbricht, R.J., Northup, S.J., Thomas, J.A., 1984. A review of 5-hydroxymethylfurfural (HMF) in parenteral solutions. *Fundam. Appl. Toxicol.* 4, 843–853. https://doi.org/ https://doi.org/10.1016/0272-0590(84)90106-4

Van Dyk, J.S., Pletschke, B.I., 2012. A review of lignocellulose bioconversion using enzymatic hydrolysis and synergistic cooperation between enzymes-Factors affecting enzymes, conversion and synergy. *Biotechnol. Adv.* 30, 1458–1480. https://doi.org/ http://dx.doi.org/10.1016/j.biotechadv.2012.03.002

Weiss, N., Börjesson, J., Pedersen, L.S., Meyer, A.S., 2013. Enzymatic lignocellulose hydrolysis: Improved cellulase productivity by insoluble solids recycling. *Biotechnol. Biofuels.* 6, 5. https://doi.org/10.1186/1754-6834-6-5

Wen, Z., Liao, W., Chen, S., 2005. Production of cellulase/β-glucosidase by the mixed fungi culture *Trichoderma reesei* and *Aspergillus phoenicis* on dairy manure. *Process Biochem.* 40, 3087–3094. https://doi.org/http://dx.doi.org/10.1016/j.procbio.2005.03.044

Wood, T.M., McCrae, S.I., 1978. The cellulase of *Trichoderma koningii*. Purification and properties of some endoglucanase components with special reference to their action on cellulose when acting alone and in synergism with the cellobiohydrolase. *Biochem. J.* 171, 61–72.

Woodward, J., 1991. Synergism in cellulase systems. *Bioresour. Technol.* 36, 67–75. https://doi.org/https://doi.org/10.1016/0960-8524(91)90100-X

Xin, D., Yang, M., Chen, X., Zhang, Y., Wang, R., Wen, P., Zhang, J., 2020. Improving cellulase recycling efficiency by decreasing the inhibitory effect of unhydrolyzed solid on recycled corn stover saccharification. *Renew. Energy.* 145, 215–221. https://doi.org/https://doi.org/10.1016/j.renene.2019.06.029

Xu, X., Lin, M., Zang, Q., Shi, S., 2018. Solid state bioconversion of lignocellulosic residues by *Inonotus obliquus* for production of cellulolytic enzymes and saccharification. *Bioresour. Technol.* 247, 88–95. https://doi.org/https://doi.org/10.1016/j.biortech.2017.08.192

Yoon, L.W., Ang, T.N., Ngoh, G.C., Chua, A.S.M., 2014. Fungal solid-state fermentation and various methods of enhancement in cellulase production. *Biomass Bioenergy* 67, 319–338. https://doi.org/http://dx.doi.org/10.1016/j.biombioe.2014.05.013

Zhang, X.-Z., Sathitsuksanoh, N., Zhang, Y.-H.P., 2010. Glycoside hydrolase family 9 processive endoglucanase from *Clostridium phytofermentans*: Heterologous expression, characterization, and synergy with family 48 cellobiohydrolase. *Bioresour. Technol.* 101, 5534–5538. https://doi.org/https://doi.org/10.1016/j.biortech.2010.01.152

Zhuo, R., Yu, H., Qin, X., Ni, H., Jiang, Z., Ma, F., Zhang, X., 2018. Heterologous expression and characterization of a xylanase and xylosidase from white rot fungi and their application in synergistic hydrolysis of lignocellulose. *Chemosphere* 212, 24–33. https://doi.org/https://doi.org/10.1016/j.chemosphere.2018.08.062

6 Microbiological Removal of Heavy Metals from the Environment

An Eco-Friendly Approach

Dushyant Kumar
Indian Institute of Technology Delhi

Sandeep K. Malyan
National Institute of Hydrology

Amrish Kumar and Jagdeesh Kumar
Indian Institute of Technology Roorkee

CONTENTS

6.1 INTRODUCTION

There are two most significant ways i.e. urban and industrial expansion which responsible for the environmental contaminants and continuously releases higher than 10,000 metric tons of chemicals (heavy metals (HMs), dyes, colorants, emerging pollutants etc.) to flowing water which leads to health risk to whole ecosystem (Kumar and Sharma, 2020; Lindholm-Lehto et al., 2015). The population growth at fast pace is related to the globalization and development for man-made aims have extended the aimless utilization of HMs. The process of shading operators, acid-batteries, manures, or another day to day life needs i.e. industrial production or domestic items have acquired a critical change their essence and focus in the earth (Anyanwu et al., 2018). This adjustment outcomes in storing up of some HMs at a spot outperforming the common permissible limits causing contamination noticeable in ecosystem including aquatic, and land (Malyan et al., 2019; Singh et al., 2018). HMs in even extremely low fixations are harmful and subsequently can cause cancer and mutation (Jaishankar et al., 2014; Malyan et al., 2019; Tchounwou et al., 2012). HMs is unnecessary components. When present beyond the limits into the ecosystem (soil, air, and water) they can cause overpowering impacts on the natural assorted variety of the receiving body (Ali, Khan, and Ilahi, 2019). Metals are natural occurring products. They cannot be pulverized yet can be changed from one to another; they are profoundly relentless and stay in the earth and tend to collect and amplify in an evolved way of life (Li et al., 2015; Wuana and Okieimen, 2011). A few elements are fundamental for legitimate development and advancement (sustenance); however, it could be dangerous when utilization surpasses as far as possible(Chatterjee et al., 2012; Hejna et al., 2018). All humans including other living things deal differently with ailments at the point when they are introduced to HMs by skin-contact, inward breath, and utilization of groceries mixing HMs in it (Alissa and Ferns, 2011; Anyanwu et al., 2018; Fulekar, Sharma, and Tendulkar, 2012; Jaishankar et al., 2014). Furthermore, a scientist has clarified bioaccumulation of HMs and its exchange and related well-being in dangers (Ali and Khan, 2018). Effluents containing HMs released from local sources i.e. farming, and mechanical operations, at the point when blended with flowing water, stream and land, pollutes them (Anyanwu et al., 2018). According to the reports distributed by numerous general well-being associations and examination reports, a few million human populaces of whole nature and world are experiencing HMs-related infections(Järup, 2003; Mamtani et al., 2011; WHO, 2011). Minamata ailment (mercury harming) and itai-itai (cadmium harming) are planet renowned HMs-related human well-being dangers (Kiyono et al., 2012). Different methodologies are utilized to degrade HMs that incorporate blend of physical, organic, or a blend of these strategies, yet huge numbers of these techniques are not condition agreeable and monetarily practical. Not a solitary strategy claims total degradation of HMs (Iori et al., 2013). Universally, they are water solvent and hard to isolate through physical process. Physico-chemical strategies, for example, assimilation, metal precipitation, membrane filtration, evaporation subsequently can cause secondary contamination during the treatment (Gunatilake, 2015; Kumar, Gaurav, and Sharma, 2018; Kumar et al., 2019). The usage of microorganisms and vegetative plants for the reduction or removal of HM from environment is in inspection phase, not developed yet now. These processes are not expensive and their environmental efficient properties also convert them into less harmful contaminants. (Ayangbenro and Babalola, 2017; Kumar et al., 2017, 2019; Saha et al., 2017). The secretion of microbial enzymatic activity is able to dissolve and break them into less harmful organisms (Wuana and Okieimen, 2011). Microorganisms use various components, for example, precipitation, bio-sorption, and enzymatic change to debase into a reduced destructive form that is highly steady, less portable, and latent (Ojuederie and Babalola, 2017; Saha et al., 2017). Taking this into account, this chapter researches the capacities of microbial varieties in resistance and pollution of HMs. Further, this examination embraces an appraisal of human well-being dangers related with nearness of HMs in aquatic microbial bioremediation, as an instrument to diminish the negative impact of HMs on people well-being and the earth. Additionally, ongoing practices in biotechnological instruments and strategies to investigate microbial remediation for HMs bioremediation and biodegradation are discussed.

6.2 EFFECTS OF HEAVY METALS ON THE ENVIRONMENT

It is not easy to remove the HMs from polluted organic objects owing to their non-biodegradable and resistance characters which can cause danger to global public health (Ayangbenro and Babalola, 2017). However, HMs, for example, Cobalt (Co), Copper (Cu), Iron (Fe), Manganese (Mn), and Molybdenum (Mo) are needed in little amounts for the endurance of living life forms (Kushwaha et al., 2016), although at higher levels, they could get inconvenient. Some of them are highly perilous HMs that taint the earth and unfavorably influence the class of land, and crop development (Ayangbenro and Babalola, 2017; Chaturvedi et al., 2015; Ndeddy Aka and Babalola, 2016; Pourrut et al., 2011; Skórzyn-Polit, Drazkiewicz, and Krupa, 2010; Upadhyay et al., 2017). Their concentration limit is decided by Comprehensive Environmental Response and Liability Act (Jaishankar et al., 2014; Kushwaha et al., 2016). These toxins are significant well-springs of dangerous degenerative ailments influencing people, for example, disease, Alzheimer's, infection, atherosclerosis, Parkinson's ailment, and so on (Muszyńska and Hanus-Fajerska, 2015). The poisonousness level of each metal is resolved by the length of presentation just as the consumed measurements by the living beings. Among living beings significantly influenced by substantial metal poisonousness are flora as their typical physical exercises are harshly hindered. For instance, the operation of some chemical processes are contrarily influenced by raised degrees of HMs as reported (Jadia and Fulekar, 2009; Pourrut et al., 2011). Additionally, the poison high metal restrains cytoplasmic compounds in the cells of plant and harms cell configuration because of oxidative pressure (Chibuike and Obiora, 2014; Gaur et al., 2014) which subsequently influences plant development and digestion. Introduction of the object to elevated stages of Pb might cause genuine well-being suggestions, for example, absence of coordination and loss of motion, while extreme introduction to Cd harms inner organs, for example, the kidney, liver, and heart tissues (Flora, 2012). It is extensively recognized that the intense substantial metal harming in grown-ups and youngsters (Dadzie, 2012; Tschirhart et al., 2012) and could bring about respiratory infections, for example, diminished pneumonic capacity or lung disease. A neurotoxin can hinder the speech and ability of hearing, weakness in muscles, and also affect the focal sensory system that is influenced by Hg (Lakherwal, 2014). It aggregates in the cells of small living organisms, in water frames where it transforms into methyl mercury in the small living things, and cause damage to living organisms of water systems. Utilization of water base food materials can prompt the exchange of poisonous methyl mercury to man. Because of the inconvenient impacts of these HMs, purposeful endeavors should be made to viably annihilate them from the earth and balance out the environment.

6.3 BIOREMEDIATION

Bioremediation is a method used to expel natural pollutants from the ecological/biological system. It uses the organic instruments inalienable in microorganisms and general plants to annihilate risky toxins also, reestablish the ecological system to its natural condition. The fundamental standards of bioremediation include decreasing the dissolvability of these ecological pollutants by evolving pH, the redox responses also, adsorption of pollutants from contaminated condition (Kaushik et al., 2010). Different reports were made on improving bio-sorption of pentachlorophenol (PCP) by changing the levels of pH in watery arrangements. The list of some microorganisms is given in Table 6.1. For instance, the bio-sorption capacities of *Aspergillus niger* (Mathialagan and Viraraghavan, 2009) and *Mycobacterium chlorophenolicum* (Bosso et al., 2014a, 2014b) in the expulsion of PCP from watery arrangements were accounted for to be pH-subordinate. Bossoet al.(2014b),too, assessed the impact of pH on adsorption and desorption conduct of PCP by *M. chlorophenolicum* also, detailed that pH esteems were a fundamental boundary that influenced PCP adsorption, with adsorptive limit expanding and diminished pH. The adsorption reaction by bacterium at pH 5.4, was totally irreparable, whereas total desorption was observed at pH value 7.0 and in range of pH 6–8, superior outcomes on adsorption conduct

TABLE 6.1

Summary of Various Microbes which Utilize HMs

S. No.	Organisms	Heavy Metals (HMs)
1.	*Aspergillus niger*	Cd, Zn, Ag, Th, U
2.	*Spirogyra* sp. and *Spirulina* sp.	Cr, Cu, Fe, Mn, Se, ZnAu, Cu, Ni,
3.	*Pseudomonas aeruginosa*	U, Pb, Hg, Zn
4.	*Serratia marcescens*	U
5.	*Cladophora fascicularis*	U
6.	*Saccharomyces cerevisiae*	Pb
7.	Species of *Aspergillus*, *Mucor*,	Cr, Ni, Cu, Zn
	Penicillium, and *Rhizopus*	Cd, Cu, Fe

Source: Deng et al., 2007; Fulekar, Sharma, and Tendulkar, 2012; Kumar, Bisht, and Joshi, 2011; MacHado et al., 2009; Mane and Bhosle, 2012.

of penta-chlorophenols by microbial biomass in watery arrangement were observed (Jianlong et al., 2000).

The outcomes by different creators featured the significance of utilizing the suitable pH for ideal execution of microorganisms utilized in bioremediation. Bioremediation advancements depend on redox procedure which centers around altering the science and microbiology of water by infusing chosen chemicals into polluted liquid to improve the abatement and mining of different unwanted chemicals by on-site redox chemicals responses (Kumar Tandon, and Singh, 2016; Yeung, 2010). Redox responses include artificially changing hurtful contaminants into harmless or less poisonous mixes that are steadier, less versatile or dormant (Kumar Tandon, and Singh, 2016). It assumes an indispensable job in the change of poisonousness of HMs in land and residue into lower poisonous or harmless types (Gadd, 2010; Rajapaksha et al., 2013). Redox responses in debased soil residue and groundwater are frequently influenced by the physic-chemical possession of the procedure; however, this could be controlled through expansion of natural and inorganic changes, for example, manures and biochar (Beiyuan et al., 2017; Bolan, Kunhikrishnan, and Naidu, 2013). The use of chemical substances for example, manure in metals accumulated soils, could cause changes in the availability of microbial populace in soil by evolving pH, diminishing the dissolvability of HMs, and expanding all ochthonous microbial biomass and accessible supplements (Alburquerque, De La Fuente, and Bernal, 2011; Chen et al., 2015). Biochar is form by pyrolysis of biomass, procured from sources, for example, crop buildup, compost and strong squanders that can be utilized in the direction of invigorate microorganisms for bioremediation by making the environmental conditions additionally favorable (Rizwan et al., 2016). Exhaustive audits by a few creators have portrayed the likely estimation of biochar as a viable operator in immobilization of metals and natural toxins (Ahmed et al., 2016; Mohan et al., 2014; Rizwan et al., 2016; Yuan et al., 2017). Biochar can give, acknowledge or move electrons inside their surroundings abiotically or through natural pathways (Klüpfel et al., 2014; Khoufi, Aloui, and Sayadi Ã, 2016). It was proposed by some analysts that biochar may enhance the microbial electron transport procedures because they show comparable practical qualities to soil redox-dynamic natural concern (Graber et al., 2014; Yuan et al., 2017). Biochar acts by expanding the pH of tainted soils, influencing the bioavailability of HMs for plant take-up. The portability and harmfulness of numerous components depend essentially to oxidation conditions which, thus, are constrained by the redox responses (Kumar Tandon and Singh, 2016; Tandon, Shukla, and Singh, 2013). Tandon et al. (Tandon, Shukla, and Singh, 2013) detailed another oxidative course for change of As(III) to As(V) utilizing muddy soil carry zero-valent iron nanoparticles by blending ferric nitrate in with alcohol of monetarily accessible tea. Up to 99% expulsion of As(III) from sullied water was accomplished. The viability of bioremediation relies upon a few factors, for example, the idea of the life forms used, the predominant ecological elements at the sullied site, just as the level of the poisons in that condition (Azubuike, Chikere, and Okpokwasili, 2016). Bioremediation can be accomplished with the utilization of microorganisms (microbial bioremediation) which relies upon the metabolic capability of the microorganisms to debase ecological toxins and transform them to harmless structures through oxidation-reductions forms (Rayu, Karpouzas, and Singh, 2012). It can likewise be completed by vegetation which tie, remove, and repair contaminations from nature. Bioremediation may be two types i.e. on-site or off-site. On-site bioremediation is a type of a cleanup procedure which take place on contaminated location which includes enhancing polluted soils with supplements to animate microorganisms in their capacity to dissociate contaminants, just as include new microorganisms to nature or improve the indigenous microorganisms to debase explicit contaminants utilizing hereditary building (Dinesh Mani et al., 2007;Rayu, Karpouzas, and Singh, 2012). Use of common microorganisms in the condition for on-site bioremediation is influenced by the non-accessibility of reasonable supplement stages and additionally natural situation at the contaminated area (Bani-Melhem and Smith, 2012; Lu et al., 2014). However, ex-situ bioremediation is quite a different process; it takes the polluted media from its special contaminated location to an alternate area for treatment. The treatment is dependent on the expense profundity of defilement, type of pollution and the scale of pollution load, geological region, and geography of the polluted location (Azubuike, Chikere, and Okpokwasili, 2016).

6.4 METAL TOXICITY THROUGH FOOD CHAIN CONTAMINATION

Because of non-stop anthropogenic exercises, HMs contamination turns into a significant reason for natural contamination. Various nations have diverse degrees of metal pollution. The South East Asian countries have taken a lot of concern in checking the contamination of agrarian soils and harvests by HMs. A researcher examined metals pollution in crops, soil, and water at two polluted sites – inundated with polluted water (Kapungwe, 2013). The reports showed that HMs were available in the whole ecosystem including water, soil, and crops at both examination destinations and surpassed permissible limits. The details were provided by the Agriculture Research Institute of Egypt because polluted water was utilized for the agriculture system, elements accumulated in land and vegetative plants past the most extreme acceptable values for animal utilization (Zeid, Ghazi, and Nabawy, 2013). Among various elements, it was recommended that cadmium had mainly moveable components in the land and was progressively accessible to crops. Contaminated exhaust from industrial sites in the form of air and water that is deposited on soil, or discarding of manufacturing waste on dumping sites may cause issues in the earth far from permissible limits. In India, numerous urban and highly populated urban areas with critical modern waste production were found to have polluted lands. The degrees of Pb, As, Cr, Cd, and Zn were resolved in small plants: radish leaf (*Raphanus sativus*), cress (*Lepidium sativum*), leek (*Allium ampeloprasum*), and beet leaf (*Beta vulgaris*) gathered from Iran (Nazemi, 2012). The examination demonstrated that vegetative plants developed here are at peril (Nazemi, 2012). In an area of a lead mine, the gathering of some HMs in land and vegetation were investigated (Wilberforce and Nwabue, 2013)in the consumable vegetations, for example, *Amaranthus hybridus* (Amaranth or pigweed); *Talinum triangulare* (water leaf); *Telfaria occidentalis* (fluted pumpkin); *Solmun nigrum* (gardenegg leaf) and *Vernonia amygdalina* (harsh leaf). The outcomes uncovered that HMs qualities in vegetation have been extended for As, Cd, Cr, Pb, and Zn, separately (Singh and Ward, 2014).

The degrees of As and Pb in bitter and nursery egg leaf plants have been found beyond the extreme limits of WHO. The researchers have gathered vegetation and land samples from ordinary and polluted destinations of a mining region of Iran (Karimi, Ghaderian, and Schat, 2013). Aggregate and water mixed As in the mud went from 0.007 to 2.32 mg/kg, individually. A geo-chemical investigation was completed in nearby area of Patancheru industrialized develop area to discover the degree of substance contamination in the land (Govil, Reddy, and Krishna, 2001). It was discovered that sullied soil demonstrated a few times more significant level of harmful components than ordinary and a few metals were seen as near better than average conveyance in the soil (Singh, Müller, and Singh, 2002). The polluted land specimens were gathered from Rajasthan (Krishna and Govil, 2007). The outcome was that land in the examination zone is altogether polluted with unreasonable degree of HM components. Comparative examinations were finished near Bangalore to evaluate HM pollution in small plants and land due to usage of seepage wastewater and lake water in agrarian land (Lokeshwari and Chandrappa, 2006). The outcomes indicated noteworthy measures of HMs, over the Indian Standard cutoff points in both the soil and vegetables. In 2007, samples of soil were gathered for testing from an industrial site in Surat City in 2007 (Krishna and Govil, 2007). It was discovered that land in the investigation region is altogether polluted with quite higher centralizations of HM components. Metal poisoning is concentrated close to wastewater and water destinations of Banaras City (Sharma, Agrawal, and Marshall, 2009). In this examination, land samples of main water systems located in sub-urban zones of Banaras have been taken and dissected for HMs pollution. Samples of aquatic bodies and segments of vegetables were likewise gathered. Excluded from Cd concentration, the remainder of HMs were available inside the permissible limit of Indian Standards. A report said that details of Pb and Ni are far from the acceptable point in the vegetables gathered from polluted water– inundated locally in the Dinapur site of Banaras City (Singh et al., 2010). HMs were dissected (Chauhan and Jain, 2010) in distinctive food yields, milk and meat. Humans, animals, blood samples were gathered from various age groups from highly contaminated and generally lower risk of contaminated zones: kids (1–12 years), young adults (12–18 years), grown-ups (18–45 years), and mature ages (over 45 and 55 years for men and ladies, individually). The outcomes uncovered that utilization of edible plants (foodstuff), meat, and milk have essentially expanded values of chosen HMs in the blood samples of humans. Copper, zinc, and magnesium values were altogether elevated (p/0.05) in the blood tests gathered from the dirty zone when contrasted with control region. Elderly individuals had amassed high convergences of metals when contrasted with the more youthful ones in a similar region.

6.5 MECHANISM FOR THE REMOVAL OF HEAVY METALS (HMS) FROM THE ENVIRONMENT

For millions of years we have known that microorganisms are equipped for tolerating unfavorable circumstances (Hrynkiewicz and Baum, 2014). Less harmful types of HM particles have raised legitimate concerns for some preservationists, designers, and biotechnologist for quite a while. Microorganisms have the capacity to dispose of harmful elements and radioactive elements in order to support their adjustment into lower harmful forms. Therefore, an assortment of bioremediation of HMs originating from ruin streams in nature are being anticipated. A portion of the ideas are even passed on as a guide or to modernize levels (Lin et al., 2013; Macek et al., 2008; Upadhyay et al., 2017; Volesky, May, and Holan, 1993). There are a few systems associated with bioremediation and some have not been comprehended till now. Generally, the procedure of bio-sorption is used for the bioremediation, and it is well known. It may be dependent on the reliance of the cell digestion. (Davis, Mucci, and Volesky, 2003; Neethu et al., 2015).The primary benefit of bioremediation is less expense for remediation when evaluated with other customary strategies, for example, digging (physical evacuation of the polluted silt layers), topping (covering the sullied silt surface with clean material; confining the silt) and cremation (burning of natural elements enclosed in squander matter). Bioremediation may be completed near the polluted locations, hence lessening the threat for cleanup personnel or possible transportation mishaps. Other than the favorable circumstances, bioremediation is more affordable; nearly complete removal of waste removes long-time liability; henceforth, it can be used in combination with both physical and chemical processes. The most favorable thing of bioremediation is that it is a non-intrusive process that does not disturb the ecosystem (Vidali, 2001b). Although it is very hard to predict the speed of the treatment process for a bioremediation exercise, some fundamental components are involved with choosing the fate of bioremediation. A solution is not found to date; researchers are continuously searching for the principle of guessing the speed of deprivation of a chemical contaminant in different parts of nature (Machackova et al., 2012).

6.6 FACTORS AFFECTING BIOREMEDIATION

Bioremediation mainly depends on significant parameters: (1) Supplements – supplements are not adequate for cell digestion and development of the microorganisms in the sullied locations. Since, in the polluted destinations there are more natural carbons, these might be exhausted at the time of microbial digestion. Along these lines, enhancing the supplements, for example, nitrogen (N), phosphorus (P), and potassium (K), to the sullied location can elevate the cell digestion and development of microorganisms which enhance the bioremediation. The nourishing prerequisite of carbon/nitrogen proportions (C/N) is 10/1, and carbon/phosphorous proportion is 30/1 for bioremediation. Notwithstanding, microbial development for biodegradation happened with higher numbers of C/N (25/1) proportion in polluted locations (Atagana and Haynes, 2003). (2) Nature of poisons – toxins may be (i) strong, semi-strong, fluid, unpredictable in nature, (ii) poisonous or non-harmful, natural, and inorganic poisons, (iii) HMs, (iv) polycyclic fragrant hydrocarbons, pesticides, chlorinated solvents, and so on. (3) Soil structure – the main components of soil structure are various surfaces that interacts from less to more material of sand, sediment, and mud (Figure 6.1). A granular and well-structured soil can encourage successful conveyance of atmosphere, dampen, and supplements to the microorganisms for on-site bioremediation.

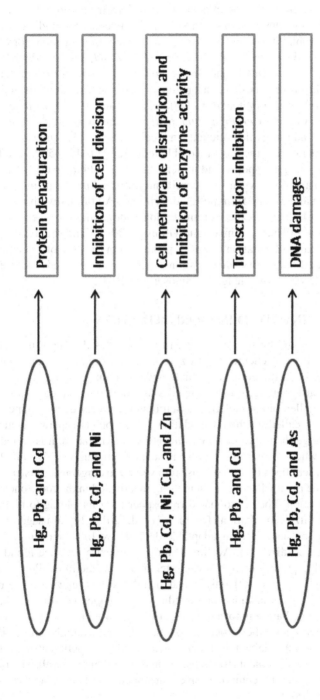

FIGURE 6.1 Heavy metal toxicity mechanism to microbes. (Modified from Khan et al., 2009.)

(4) pH – the range from 5.5–8.0 is ideal for the higher growth of organisms and to reduce the pollutants (Vidali, 2001b). (5) Moisture content – the percentage of water is the essential feature in deciding the dielectric steady of land and different channels and the dampness in soil lies between 25% and 28% (Vidali, 2001b). (6) Microbial diversity – the microbial variety of a location, for example, *Chlorobacteria, Aeromonas, Flavobacteria, Corynebacteria, Pseudomonas, Mycobacteria, Acinetobacter,* and *Streptomyces, Cyanobacteria,* and so forth. (7) Macrobenthos diverseness – consortium of water bearing animals and plants: *Anodonta woodiana, Limnodrilus hoffmeisteri* and *Eichhornia crassipes, Solenopsis molesta,* and *Ceratophyllum demersum* can possibly degrade the turbidity, organic, and inorganic pollutants in local effluents (Mangunwardoyo et al., 2013); (8) Temperature – the most favorable temp range is 15–45°C. It influences the rate of biochemical response and it increases by to fold for every 10°C change in temperature; and (9) Oxygen – predominantly utilized for the primary collapse of the hydrogen and carbon from polluted locations. Additionally, the measure of accessible air will decide if the bioremediation is done under oxygen consuming or without oxygen (Thapa, Kumar, and Ghimire, 2012). The reality, despite the fact that significant information has been picked up in the area of bioremediation through escalated laboratory investigations – the use of these advancements has not been completely acknowledged (Ramos et al., 2011).

6.7 PROCESS INVOLVED IN BIOREMEDIATION

Bioremediation is an alternative operation that provides the opportunity to wipe out, risk-free, different chemicals utilizing common organic action (Gupta and Mohapatra, 2003). At the point when bioremediation happens all alone, it is well-known as regular lessening. When it includes microorganisms along with included composts, it is called bio-stimulation; when biological destruction happens in the rhizosphere, it is called as rhizodegradation. When plants separate metals from mud and discharge them into the environment by volatilization, it is called phyto-volatilization; when plant roots retain elements from fluids, it is called rhizo-filtration; when seedlings ingest poisons from watery arrangement, it is called blast filtration; and the point when the toxin is being inactivated in the root zone is called as phytostabilization. The consolidated rhizospheric forms adding to bioremediation are called rhizo-remediation (Dhankher et al., 2011). Equal usage of various contaminations, because of their adaptable degradative limit with microbial network, are called co-digestion. Acclimatization is an extra advance found helpful in treating sludge produced from wastewater treatment facilities. Presentation of genetically modified microorganisms and metal-resistant marine microbes can likewise be utilized as advance bioremediation strategies (Poirier et al., 2013). Generally, the remediation innovations, regardless of whether set-up or off-site, do one of two things: They either expel the pollutants from the foundation (sterilization or cleanup strategies) or lessen the hazard presented by the pollutants by diminishing their publicity (adjustment strategies) (Vangronsveld et al., 2009). The bioremediation procedure can be comprehensively arranged into two groups: on-site and off-site (Vidali, 2001a).

6.7.1 In-Situ Bioremediation

On-site bioremediation is engaged in providing oxygen and nutrients by coursing fluid arrangements through sullied soils to remove normally existing microorganisms and pollutants. It is a less expensive technique which utilizes an innocuous microbial consortium to degrade the pollution load, particularly valuable for inundated soil and groundwater remediation. The procedure incorporates circumstances, for example, the invasion of water with supplements and air as electron acceptors (Chauhan and Jain 2010; Rayu, Karpouzas, and Singh, 2012; Vidali, 2001b). Furthermore, on-site bioremediation is comprehensively named 'natural bioremediation' and 'designed bioremediation'. The main methodology manages the indigenous microbial populace by taking care of them with supplements and oxygen to increment their metabolic exercises. It is un-manipulated, un-stimulated, un-enhanced organic remediation of a situation, that is, organic lessening of contaminants in nature. The second approach includes any sort of controlled, invigorated, or upgraded natural remediation of a situation (Li et al., 2010). The acquaintance of explicit microorganisms with the debased site quickens the corruption procedure by producing or upgrading favorable physicochemical conditions (Kumar et al., 2011). At the point when the situation is not appropriate, designed bioremediations are acquainted with the specific location, utilizing hereditarily designed microscopic organisms. It is accepted that free-living hereditarily designed microorganisms may have diminished degrees of endurance because of the pressured conditions. Henceforth, picking and designing the bacterial-strain quickly possibly has a higher range of bioremediation prospective without natural hazards. It will be an advancement for accomplishing a protected and feasible condition (Singh et al., 2011). The significant favorable circumstances of on-site bioremediation are: less expense, no unearthing, negligible location disturbance, negligible residue creation, and the chances of simultaneous treatment of soil and groundwater in the future. Be that as it may, the significant disadvantages are: tedious, occasional variety in the microbial action and hazardous use of treatment-added substances in the normal condition. Other than these, on-site bioremediation may not be in control and likewise not reasonable under some conditions. In such cases, either off-site bioremediation favored hereditarily designed microorganisms must be utilized; albeit invigorating original microorganisms is favored (Rayu, Karpouzas, and Singh, 2012).

6.7.2 Ex-Situ Bioremediation

Off-site bioremediation procedures include the exhuming or expulsion of sullied soil/water starting from the earliest stage. The comprehensive strategy is a frozen condition and a mixture of semi solid and liquid stages. Solid stage treatment incorporates natural waste (leaves, excrements, and horticultural squanders) and hazardous material, for instance, residential/mechanical squanders and sewage slime/metropolitan solid waste material. The treatment measures include incorporate soil biopiles, treating the soil, land cultivating, and aqua-farming (Kumar et al., 2011; Ramos et al., 2011; Rayu, Karpouzas, and Singh, 2012). A biopile is a momentary bioremediation innovation wherein uncovered soils are blended in with soil alterations, shaped into

manure heaps or cells over the ground, and encased for treatment with an air circulation system (Tchounwou et al., 2012). Biopiles – mixture of land cultivating
and treating the soil – give a positive domain to indigenous oxygen consuming and
anaerobic microorganisms. The designed units operate as air is circulated through
treated soil mass; this is common usage for the therapy of surface pollutants that
enhances loss of physical control of the pollutants by filtering and volatilization. In
the form of semi solid and liquid mixture stage, the bioremediation is a fast cycle,
polluted soil consolidates with water and different substances are added in bioreactor tank, which is blended providing ideal conditions for microorganisms to remove
the pollutants from environment. By ensuring proper sampling and maintaining
controlled conditions with collected core samples, the ex-situ bioremediation can
be accomplished (Paliwal, Puranik, and Purohit, 2012). A pilot investigation of a
half year of biopile cultivating affirmed that bio-augmentation and bio-stimulation
could upgrade the bioremediation of oil-polluted land (Cheng et al., 2015). The
ideal extent of cows excrement improved natural material decontamination and
humification measures; thus it decreased the poisonousness of elements at the time
of revolving drum composting (Singh, 2013) while assessing the bioavailability
of HMs utilizing aquatic plants (*E. crassipes*). Utilization of manures discharge
accessible supplements and vitality into the environment, particularly for microbial
movement, expanding the area of impact from the fertilizer-soil boundary (Duong,
Verma, and Penfold, 2013). An effluent treatment facility utilizing regular natural
treatment combined with micro-algae and aqua-farming was developed in a nursery
(Norström et al., 2003). A micro-algal footstep leads to lowering the phosphorus,
and the use of a sand filter to purify the liquid. The nitrification and denitrification
was reported as a reduction in nitrogen 72%, phosphorus by 47%, and COD by 90%.
A researcher assessed the morpho-physiological changes, for example, biomass creation and dividing, of nickel amassing in plants and nickel evacuation capacity, by
raising about 21-day-old *Amaranthus paniculatus L* (Iori et al., 2013). Generally,
speed and degree of biodegradation are quite prominent in a bioreactor framework
than on-site arrangements, although the restricted condition is quite sensible and
subsequently quite controllable and unsurprising. The significant downside of this
framework is that the pollutant can be taken from the dirt by means of soil washing
set in a bioreactor tank.

6.7.3 Bio-Augmentation

Bio-growth is a notable strategy for bioremediation where expansion of a regular or
laboratory synthesis microbial strain was accomplished for the remediation of contaminated natural things like water and land. This strategy is commonly utilized for
the remediation of effluent and generated sludge in bioreactors during the process.
A large portion of the microbial societies utilized in bio-expansion involve all the
essential microorganisms for biological remediation of land soil and saturated water.
Bio-expansion is used at the destinations when water and land are debased with poisonous mixes like chlorinated ethanes, to ensure that on-site organisms can totally
deteriorate these pollutants to non-harmful structures, like ethylene and chloride.
Perception of this technique is dull, because it is a delayed on-site measure and not

effective when momentum location clean-up is needed (Juwarkar et al., 2009; Kumar and Bharadvaja, 2020; Mishra, 2017).

6.7.4 BIOVENTING

Bioventing is a created and capable innovation of bioremediation. It supports the normal on-site natural debasement of HMs in non-fluid stage fluids, non-aqueous liquid phase (NAPL) which can be corrupted vigorously. Corruption of mixes is normally done inside the dirt by providing oxygen (O_2) to the effectively present soil organisms. When contrasted with soil-fume extraction soil vapor extraction (SVE), this cycle utilizes low wind stream rates to supply adequate O_2 to keep up microbial action. Generally, O_2 is given in land, utilizing contacted air infusion through wells for leftover pollution. Moreover, HMs adsorb energizes and natural mixes; for example, volatile organic compound (VOC) can be corrupted naturally as fumes move slowly through organically dynamic soil (Chauhan and Jain, 2010; Gupta et al., 2016).

6.7.5 BIO-SPARGING

This method of bioremediation is usage imbuement of compact air underneath the saturated water. It is usually used to get better combinations of saturated water oxygen and to incline the pace of natural destruction of toxins by minute creatures. Bio-sparging upgrades the contact among soil and saturated water by integrating the splashed area. This procedure is altogether versatile since it easily presents little separation across air imbuement centers and minimal effort in the structure and improvement of the system (Adams and Ready, 2003; Gupta et al., 2016; Wiltse and Dellarco, 1996).

6.7.6 BIODEGRADATION

This is a conventional expression used to depict the strategies influencing purification of natural poisons. It is found to be an improved alternative for ineffectual and costly physic-chemical remediation strategies. However, absence of data about the development and digestion of microorganisms in the contaminated condition frequently restricts its usage. The modern investigations in the comprehension of biogeochemical cycles and genomics have unlocked new viewpoints in the direction of new prospects of contamination reduction (Chauhan and Jain, 2010; Jeffries et al., 2012; Rayu, Karpouzas, and Singh, 2012; Tyagi et al., 2011). For impressive oxygen-consuming biodegradation to happen, adequate measures of broken down oxygen have to survive inside the subsurface in the form of an electron acceptor (Adams and Ready, 2003). The drawbacks of the existing bioremediation process have made it important to look for additional eco-accommodating and less expensive strategies for the removal of HMs from polluted locations. Microbially instigated calcite precipitation (MICP) gives an elective approach to remediate these types of troubles. The MICP items can emphatically adsorb HMs on their surfaces. At the time of precipitation of calcite, HMs particles with particle range near Ca^{2+}, for example,

Cd^{2+}, Pb^{2+}, Sr^{2+}, and Cu^{2+}, might be consolidated into the crystal of calcite by alternative chemical reactions (Xiangliang, 2009). A native calcified bacterial strain of *Kocuria flava* CR1 has been separated from mining zones (Achal, Pan, and Zhang, 2011). By using the MICP degradation process, the separated strain was able to destruct 95% of the copper from the polluted location. Henceforth, the capability of *K. flava* has been documented immediately as a feasible and environmentally viable tool for the remediation of copper-contaminated locations.

6.7.6.1 Application of Bioremediation

Bioremediation is a process that can possibly renovate the polluted environment (Dowarah et al., 2009) the deficiency of facts which controls the variables of the development and digestion (Kumar et al., 2017) of microorganism in dirtied situations normally restricts its implementation. Proteomics and genomic-based bioinformatics (Chauhan and Jain, 2010; Poirier et al., 2013), offers noteworthy guarantees as instruments to tackle long-lasting questions with respect to the atomic instruments engaged and to command mineralization pathways (Achal, Pan, and Zhang, 2012; Govarthanan et al., 2013; Kim and Park, 2013; Mani and Kumar, 2014). The most productive and practical remediation arrangement might be a mix of various innovations, for example, assessing the level of contamination through planned biosensors (Checa, Zurbriggen, and Soncini, 2012), unearthing the most polluted locations, planning geohydro-biological designing models (Sivakumar, 2013), followed by cleaning up the location by utilizing microorganism-supported plants (Juwarkar and Singh, 2010; Juwarkar et al., 2009; Pilon-Smits, 2005). The integrative exertion may end up being a truly outstanding ecological practice for the recovery of polluted lands.

6.7.7 BIO-STIMULATION

Microorganisms, particularly microscopic organisms and growths, are ecosystem's unique recyclers. Their ability to change organic and engineered contaminants into wellsprings of vitality (Tang et al., 2007) and resources of raw matters own development recommends that costly chemical substance or physical remediation methods must be replaced by bio-organic methods that are not expensive and quite ecologically amicable. Infusion of supplements and other advantageous elements to the local microbial populace to instigate transmission at a high rate (bio-stimulation) is one of the well-known processes for in-site bioremediation of unplanned oil spills and longstanding polluted locations around the world (Mani and Kumar, 2014; Tyagi et al., 2011). The rate of the bio-stimulation process increases microorganisms by enhancing the dose of nitrogen and phosphorus (supplements), oxygen (electron acceptors), and methane, phenol, and toluene (substrates), or adding the microorganisms with wanted synergist abilities (Baldwin et al., 2008; Mani and Kumar, 2014;).

6.7.8 BIO-MINERALIZATION

The bio-mineralization process follows the ordinary pathway for generating complex structured inorganic matter that has essential capacities in organic frameworks. The various morphologies of bio-minerals have persuaded researchers to imitate these matters through the fundamental science of bio-mineralization, which has authorized the reproduction of the extraordinary mechanical and visual assets of bio-minerals with their exceptional natural capacities, for example, route, stockpiling, and homeostasis (Govarthanan et al., 2013; Kim and Park, 2013). Nature gives stupendous instances of naturally shaped minerals for example, mesoporous silica nanoparticles (MSPs),which develop as engaging transporters for prohibited anticancer medication conveyance attributable to low harmfulness, high surface zone, and huge available pore volumes. Presently, novel techniques for bio-mineralization are being found (Achal, Pan, and Zhang, 2012) to develop natural and inorganic hybrid for exceptionally harmful components.

6.7.9 BIO-SORPTION

Bio-sorbents, the sorption of poisonous matter by organic materials, particularly from aquatic bodies such as eco-macroalgae and alginate subordinates, show higher attraction for the number of metal particles. It is developing as a possible option in contrast to the current evacuation innovations, as well as recuperation of metal particles from watery arrangements (Schiewer and Patil, 2008). The most important benefits of bio-sorption over traditional techniques incorporates ease, lower cost, high effectiveness, minimization of chemical substance, recovery of bio-sorbents, and also probability of metal recuperation (Azouaou, Sadaoui, and Mokaddem, 2008; Sud, Mahajan, and Kaur, 2008). There are more affordable than modern engineered adsorbents and grip incredible possibilities for the expulsion of harmful chemical elements from modern effluents (Fiset, Blais, and Riveros, 2008). These outcomes made a difference in the request for the sorption limit of metal particles. The group of scientists set up the capability of *Portulaca oleracea* as a potent minimal effort adsorbent for expulsion of Pb particles from water (Dubey and Shiwani, 2012). The application of biochar has been investigated as an amendment to the utisol prompting noteworthy growth in land pH and was observed with less acidic soluble properties of Cu(II) and Pb(II) after expansion of biochar (Mani and Kumar, 2014)

6.7.10 PHYTO-REMEDIATION

Phyto-remediation uses higher plants for remediation of contaminants mixed in water, soil, dregs, and other natural environments. In previous years, a few varieties of phyto-remediation were used for the remediation of poisonous HMs and other trace elements from moisture seepage, land, seepage from defuse acid mines, and also ecological samples. Phyto-remediation is sorted into different kinds, to be specific, phyto-degradation, phyto-volatilization, phyto-stabilization, phyto-extraction.

6.7.11 RHIZO-FILTRATION

Utilization of plant roots to accelerate, retain, or concentrate the mixture of poisonous HMs or radio nuclei waste from aquatic system is called rhizo-filtration. This procedure is helpful where wet soil can be shaped and all of the dirtied water is allowed to associate legitimately with plants roots. The developed vegetative roots are able to absorb high amount of HMs from water/land that has come in contact with the roots of vigorously developed vegetation (Kumar Rai, 2009; Tyagi et al., 2011).

6.8 CONCLUSION

This chapter designated the recent logical advancement in profitably applying bio-technological approaches for administrating ecosystems by using the promising area of bioremediation (done by microbes and plants) to alleviate the threat of contamination the world over. The destructive marvels have globally brought about genuine natural and social issues; these issues must be taken care of for arrangements somewhere else than the built-up physical and compound advances. The profoundly compelling microorganisms are refined for an assortment of uses; yet, at the same time the advantageous capability of microorganisms in the area of bioremediation and phyto-remediation is immense and undiscovered. The ongoing methodologies of bioremediation, for example, MICP through urease hydrolyzing microscopic organisms, arrangement of dissolvable nanoparticles as end products by microorganisms, bacterial flagging frameworks for biosensors, bio-mineralization of radio-nuclei and other harmful substantial metals, amalgamation of biomaterial or functionalized polymer–silica crossover nanoparticles, genomics and proteomics of natural medicines to the pollutants, root–microorganism communication and removal of toxins of HMs is demonstrated as talented strategies to remediate the polluted biological systems.

REFERENCES

Achal, V., X. Pan, and D. Zhang. 2012. "Bioremediation of Strontium (Sr) Contaminated Aquifer Quartz Sand Based on Carbonate Precipitation Induced by Sr Resistant Halomonas Sp." *Chemosphere* 89 (6): 764–68. https://www.sciencedirect.com/science/article/pii/S0045653512008867?casa_token=o_chsjARdHoAAAAA:UocIJaO0k7yR I4B_42rbnj9K2MM6YjQi4Um9qug7ABUZubK1m_Ou_rd91Ix7bySksKPNkSbUtQ.

Achal, Varenyam, Xiangliang Pan, and Daoyong Zhang. 2011. "Remediation of Copper-Contaminated Soil by Kocuria Flava CR1, Based on Microbially Induced Calcite Precipitation." *Ecological Engineering* 37 (10): 1601–5. https://doi.org/10.1016/j.ecoleng.2011.06.008.

Adams, Jeffrey A., and Krishna R. Ready. 2003. "Extent of Benzene Biodegradation in Saturated Soil Column During Air Sparging." *Groundwater Monitoring & Remediation* 23 (3): 85–94. https://doi.org/10.1111/j.1745-6592.2003.tb00686.x.

Ahmed, Mohammad Boshir, John L. Zhou, Huu H. Ngo, Wenshan Guo, and Mengfang Chen. 2016. "Progress in the Preparation and Application of Modified Biochar for Improved Contaminant Removal from Water and Wastewater." *Bioresource Technology*. Elsevier Ltd. https://doi.org/10.1016/j.biortech.2016.05.057.

Alburquerque, J. A., C. De La Fuente, and M. P. Bernal. 2011. "Improvement of Soil Quality after 'Alperujo' Compost Application to Two Contaminated Soils Characterised by Differing Heavy Metal Solubility." *Journal of Environmental Management* 92 (3): 722–43. https://www.sciencedirect.com/science/article/pii/S0301479710003476?casa_token=qBxqcw7LORcAAAAA:Qls_5_ca0mAkccJICVLz810kr1vKjoWfNUJzMD hhtGzbgwpcYwlhzkPd-1gjOeQzAnI5kOdM7hk.

Ali, Hazrat, and Ezzat Khan. 2018. "Bioaccumulation of Non-Essential Hazardous Heavy Metals and Metalloids in Freshwater Fish. Risk to Human Health." *Environmental Chemistry Letters.* Springer Verlag. https://doi.org/10.1007/s10311-018-0734-7.

Ali, Hazrat, Ezzat Khan, and Ikram Ilahi. 2019. "Environmental Chemistry and Ecotoxicology of Hazardous Heavy Metals: Environmental Persistence, Toxicity, and Bioaccumulation." *Journal of Chemistry* 2019. https://doi.org/10.1155/2019/6730305.

Alissa, Eman M., and Gordon A. Ferns. 2011. "Heavy Metal Poisoning and Cardiovascular Disease." *Journal of Toxicology.* https://doi.org/10.1155/2011/870125.

Anyanwu, Brilliance, Anthonet Ezejiofor, Zelinjo Igweze, and Orish Orisakwe. 2018. "Heavy Metal Mixture Exposure and Effects in Developing Nations: An Update." *Toxics* 6 (4): 65. https://doi.org/10.3390/toxics6040065.

Atagana, Harrison Ifeanyichukwu, and R. J. Haynes. 2003. "Optimization of Soil Physical and Chemical Conditions for the Bioremediation of Creosote-Contaminated Soil.". *Biodegradation.* Springer14 (4): 297–307. https://doi.org/10.1023/A:1024730722751.

Ayangbenro, Ayansina Segun, and Olubukola Oluranti Babalola. 2017. "A New Strategy for Heavy Metal Polluted Environments: A Review of Microbial Biosorbents." *International Journal of Environmental Research and Public Health Review* 94 (14): 1–16. https://doi.org/10.3390/ijerph14010094.

Azouaou, N., Z. Sadaoui, and H. Mokaddem.2008. "Removal of Cadmium from Aqueous Solution by Adsorption on Vegetable Wastes." *Journal of Applied Sciences* 8: 4638–43.

Azubuike, Christopher Chibueze, Chioma Blaise Chikere, and Gideon Chijioke Okpokwasili. 2016. "Bioremediation Techniques–Classification Based on Site of Application: Principles, Advantages, Limitations and Prospects." *World Journal of Microbiology and Biotechnology.* Springer, Netherlands. https://doi.org/10.1007/s11274-016-2137-x.

Baldwin, B.R., A.D. Peacock, M. Park, D.M. Ogles, J.D. Istok, J.P. McKinley, C.T. Resch, and D.C. White. 2008. "Multilevel Samplers as Microcosms to Assess Microbial Response to Biostimulation." *Ground Water* 46 (2): 295–304. https://doi.org/10.1111/j.1745-6584.2007.00411.x.

Bani-Melhem, K., and E. Smith. 2012. "Grey Water Treatment by a Continuous Process of an Electrocoagulation Unit and a Submerged Membrane Bioreactor System." *Chemical Engineering Journal* 198:201–10. Accessed September 20, 2018. https://www.sciencedirect.com/science/article/pii/S1385894712006377.

Beiyuan, J., Y. M. Awad, F. Beckers, D. C. Tsang, Y. S. Ok, and J. Rinkleb e.2017. "Mobility and Phytoavailability of As and Pb in a Contaminated Soil Using Pine Sawdust Biochar under Systematic Change of Redox Conditions." *Chemosphere* 178: 110–18. https://www.sciencedirect.com/science/article/pii/S0045653517303740?casa_token= oBiOzMLkhrIAAAAA:NKJcTwRK3AB6fqtwogekXIeokKY3UPj-Gmt1Inw LkAby0BRUpTcvlCgGo0Jnnz_vELnbUo5zMfA.

Bolan, Nanthi S., A. Kunhikrishnan, and R. Naidu. 2013. "Carbon Storage in a Heavy Clay Soil Landfill Site after Biosolid Application." *Science of the Total Environment* 465: 216–25. https://www.sciencedirect.com/science/article/pii/S0048969713000053?casa_ token=FVnTk2dqmEQAAAAA:2jPZOQPOLMYY2D-FqaaDxEFHNs0M8zbw_ MeUZWVQZqZpzpV1GTQrtUV7S5GSDpbbEMmBDrk7oOw.

Bosso, L., F. Lacatena, G. Cristinzio, M. Cea, M. C. Diez, and O. Rubilar.2014a. "Biosorption of Pentachlorophenol by Anthracophyllum Discolor in the Form of Live Fungal Pellets." *New Biothechnolgy* 32 (1): 21–25. https://europepmc.org/article/med/25154034.

———. 2014b. "Adsorption and Desorption of Pentachlorophenol on Cells of *Mycobacterium Chlorophenolicum* PCP-1." *New Biothechnolgy* 55 (3): 480–89. https://doi.org/10.1002/ (SICI)1097-0290(19970805)55:3<480::AID-BIT3>3.0.CO;2-8.

Chatterjee, Soumya, Lokendra Singh, Buddhadeb Chattopadhyay, Siddhartha Datta, and S. K. Mukhopadhyay. 2012. "A Study on the Waste Metal Remediation Using Floriculture at East Calcutta Wetlands, a Ramsar Site in India." *Environmental Monitoring and Assessment* 184 (8): 5139–50. https://doi.org/10.1007/s10661-011-2328-8.

Chaturvedi, Amiy Dutt, Dharm Pal, Santhosh Penta, and Awanish Kumar. 2015. "Ecotoxic Heavy Metals Transformation by Bacteria and Fungi in Aquatic Ecosystem." *World Journal of Microbiology and Biotechnology* 31 (10): 1595–1603. https://doi.org/10.1007/s11274-015-1911-5.

Chauhan, Archana, and Rakesh K. Jain. 2010. "Biodegradation: Gaining Insight through Proteomics." *Springerlink* 21: 861–879. https://doi.org/10.1007/s10532-010-9361-0.

Checa, Susana K., Matias D. Zurbriggen, Fernando C. Soncini. 2012. "Bacterial Signaling Systems as Platforms for Rational Design of New Generations of Biosensors." *Current Opinion in Biotechnology.* Elsevier Current Trends. https://doi.org/10.1016/j.copbio.2012.05.003.

Chen, Ming, Piao Xu, Guangming Zeng, Chunping Yang, Danlian Huang, and Jiachao Zhang. 2015. "Bioremediation of Soils Contaminated with Polycyclic Aromatic Hydrocarbons, Petroleum, Pesticides, Chlorophenols and Heavy Metals by Composting: Applications, Microbes and Future Research Needs." *Biotechnology Advances* 33 (6): 745–55. https://doi.org/10.1016/j.biotechadv.2015.05.003.

Cheng, Pan, Qinghe Zhang, Xiaomei Shan, Denghui Shen, Bingshuang Wang, Zhenhai Tang, Yu Jin, Chi Zhang, and Fen Huang. 2015. "Cancer Risks and Long-Term Community-Level Exposure to Pentachlorophenol in Contaminated Areas, China." *Environmental Science and Pollution Research* 22 (2): 1309–17. https://doi.org/10.1007/s11356-014-3469-4.

Chibuike, G. U., and S. C. Obiora. 2014. "Heavy Metal Polluted Soils: Effect on Plants and Bioremediation Methods." *Applied and Environmental Soil Science.* Hindwai Publishing Corporation. https://doi.org/10.1155/2014/752708.

Dadzie, Esi Saawa. 2012. "Assessment of Heavy Metal Contamination of the Densu River, Weija from Leachate," December, PhD Dissertation. http://ir.knust.edu.gh:8080/handle/123456789/5800.

Davis, T. A., A. Mucci, and B. Volesky. 2003. "A Review of the Biochemistry of Heavy Metal Biosorption by Brown Algae." *Water Research.* 37 (18): 4311–30. https://www.sciencedirect.com/science/article/pii/S0043135403002938?casa_token=8pnfflWBzhkAAAAA:TQ763XRt8Qe1V-WuBSE7ilI2JBtKdwIs0p-9-EcF5ooOR4u6jhtfy73dfUY651n6-tfoDV6btg.

Deng, Liping, Yingying Su, Hua Su, Xinting Wang, and Xiaobin Zhu. 2007. "Sorption and Desorption of Lead (II) from Wastewater by Green Algae *Cladophora fascicularis.*" *Journal of Hazardous Materials* 143 (1–2): 220–25. https://doi.org/10.1016/j.jhazmat.2006.09.009.

Dowarah, J., H. P. Deka Boruah, J. Gogoi, N. Pathak, N. Saikia, and A. K. Handique. 2009. "Eco-Restoration of a High-Sulphur Coal Mine Overburden Dumping Site in Northeast India: A Case Study." *Journal of Earth System Science* 118 (5): 597–608. https://idp.springer.com/authorize/casa?redirect_uri=https://link.springer.com/content/pdf/10.1007/s12040-009-0042-5.pdf&casa_token=VZvRiKRFTWIAAAAA:W5Qp1fZyHVIc0aeNNfNbMynnJi30_15AUYJvnlCmdC8vprMIg_7BH-yf3ELI_zlnEAeLd0a2xmSE2_Vi.

Dubey, A., and S. Shiwani. 2012. "Adsorption of Lead Using a New Green Material Obtained from Portulaca Plant." *International Journal of Environmental Science and Technology* 9 (1): 15–20. https://doi.org/10.1007/s13762-011-0012-8.

Duong, T. T. T., S. L. Verma, C. Penfold, P. Marschner. 2013. "Nutrient Release from Composts into the Surrounding Soil." *Geoderma* 195: 42–47. https://www.sciencedirect.com/science/article/pii/S0016706112004053?casa_token=6vEhcUkBgcAAAAAA:lraXJet0qNRShNUB8EZol9wy-qR1Lrnmq3VbS-vyUaQSxco-6yZb_YJbeDy3Vat-8AEoFx_YB9I.

Fiset, J. F., J. F. Blais, P. A. Riveros. 2008. "Review on the Removal of Metal Ions from Effluents Using Seaweeds, Alginate Derivatives and Other Sorbents." *Revue Des Sciences de l'Eau*. Institut National de la Research Scientifique. https://doi.org/10.7202/018776ar.

Flora, Govinder J. S. 2012. "Arsenic Toxicity and Possible Treatment Strategies: Some Recent Advancement-Indian Journals." *Current Trends in Biotechnology and Pharmacy* 6 (3): 280–89. http://www.indianjournals.com/ijor.aspx?target=ijor:ctbp&volume=6&issue=3&article=002.

Fulekar, M. H., Jaya Sharma, and Akalpita Tendulkar. 2012. "Bioremediation of Heavy Metals Using Biostimulation in Laboratory Bioreactor." *Environmental Monitoring and Assessment* 184 (12): 7299–7307. https://doi.org/10.1007/s10661-011-2499-3.

Gadd, Geoffrey Michael. 2010. "Metals, Minerals and Microbes: Geomicrobiology and Bioremediation." *Microbiology* 156 (3): 609–43. https://doi.org/10.1099/mic.0.037143-0.

Gaur, Nisha, Gagan Flora, Mahavir Yadav, and Archana Tiwari. 2014. "A Review with Recent Advancements on Bioremediation-Based Abolition of Heavy Metals." *Environmental Sciences: Processes and Impacts* 16(2): 180–93. https://doi.org/10.1039/c3em00491k.

Govarthanan, Muthusamy, Kui Jae Lee, Min Cho, Jae Su Kim, Seralathan Kamala-Kannan, and Byung Taek Oh. 2013. "Significance of Autochthonous Bacillus Sp. KK1 on Biomineralization of Lead in Mine Tailings." *Chemosphere* 90 (8): 2267–72. https://doi.org/10.1016/j.chemosphere.2012.10.038.

Govil, P. K., G. L. N. Reddy, and A. K. Krishna. 2001. "Contamination of Soil Due to Heavy Metals in the Patancheru Industrial Development Area, Andhra Pradesh, India." *Environmental Geology* 41 (3–4): 461–69. https://doi.org/10.1007/s002540100415.

Graber, E. R., L. Tsechansky, B. Lew, and E. Cohen. 2014. "Reducing Capacity of Water Extracts of Biochars and Their Solubilization of Soil Mn and Fe." *European Journal of Soil Science* 65 (1): 162–72. https://doi.org/10.1111/ejss.12071.

Gunatilake S. K.2015. "Methods of Removing Heavy Metals from Industrial Wastewater." *Journal of Multidisciplinary Engineering Science Studies (JMESS)* 1. www.jmess.org.

Gupta, Abhijit, Jyoti Joia, Aditya Sood, Ridhi Sood, Candy Sidhu, and Gaganjot Kaur. 2016. "Microbes as Potential Tool for Remediation of Heavy Metals: A Review." *Journal of Microbial and Biochemical Technology* 8 (4): 364–72. https://doi.org/10.4172/1948-5948.1000310.

Gupta, R., and H. Mohapatra. 2003. "Microbial Biomass: An Economical Alternative for Removal of Heavy Metals from Waste Water." http://nopr.niscair.res.in/handle/123456789/17155.

Hejna, M., D. Gottardo, A. Baldi, V. Dell'Orto, F. Cheli, M. Zaninelli, and L. Rossi. 2018. "Review: Nutritional Ecology of Heavy Metals." *Animal* 12 (10): 2156–70. Cambridge University Press. https://doi.org/10.1017/S175173111700355X.

Hrynkiewicz, Katarzyna, and Christel Baum. 2014. "Application of Microorganisms in Bioremediation of Environment from Heavy Metals." In *Environmental Deterioration and Human Health: Natural and Anthropogenic Determinants*, 215–27. Springer, Netherlands. https://doi.org/10.1007/978-94-007-7890-0_9.

Iori, Valentina, Fabrizio Pietrini, Alexandra Cheremisina, Nina I. Shevyakova, Nataliya Radyukina, Vladimir V. Kuznetsov, and Massimo Zacchini. 2013. "Growth Responses, Metal Accumulation and Phytoremoval Capability in Amaranthus Plants Exposed to Nickel under Hydroponics." *Water, Air, and Soil Pollution* 224 (2). https://doi.org/10.1007/s11270-013-1450-3.

Jadia, Chhotu D., and M. H. Fulekar. 2009. "Phytoremediation of Heavy Metals: Recent Techniques." *African Journal of Biotechnology* 8 (6): 921–28. http://www.academicjournals.org/AJB.

Jaishankar, Monisha, Tenzin Tseten, Naresh Anbalagan, Blessy B. Mathew, and Krishnamurthy N. Beeregowda. 2014. "Toxicity, Mechanism and Health Effects of

Some Heavy Metals." *Interdisciplinary Toxicology*. Slovak Toxicology Society. https://doi.org/10.2478/intox-2014-0009.

Järup, Lars. 2003. "Hazards of Heavy Metal Contamination." *British Medical Bulletin*. Oxford Academic. https://doi.org/10.1093/bmb/ldg032.

Jeffries, T. C., J. R. Seymour, K. Newton, R. J. Smith, L. Seuront, and J. G. Mitchell. 2012. "Increases in the Abundance of Microbial Genes Encoding Halotolerance and Photosynthesis along a Sediment Salinity Gradient." *Biogeosciences* 9 (2): 815–25. https://doi.org/10.5194/bg-9-815-2012.

Jianlong, W., Q. Yi, N. Horan, and E. Stentiford.2000. "Bioadsorption of Pentachlorophenol (PCP) from Aqueous Solution by Activated Sludge Biomass." *Bioresource Technology* 75 (2): 157–61. https://www.sciencedirect.com/science/article/pii/S0960852400000419?casa_token=0LHUhIHWdAwAAAAA:eo2-K07QSVNagMpooXUHNMDhRqfUZMvH3PT3pK-9MvmnGJdNj_WgkIMiEv-KYKNChIXCcqnXMQ.

Juwarkar, Asha A., and Sanjeev Kumar Singh. 2010. "Microbe-Assisted Phytoremediation Approach for Ecological Restoration of Zinc Mine Spoil Dump." *International Journal of Environment and Pollution* 43 (1–3): 236–50. https://doi.org/10.1504/IJEP.2010.035927.

Juwarkar, Asha A., Santosh Kumar, Yadav, P. R. Thawale, P. Kumar, S. K. Singh, T. Chakrabarti, A. A. Juwarkar, and S. K. Yadav. 2009. "Developmental Strategies for Sustainable Ecosystem on Mine Spoil Dumps: A Case of Study." *Environmental Monitoring and Assessment* 157 (1–4): 471–81. https://doi.org/10.1007/s10661-008-0549-2.

Kapungwe, Evaristo Mwaba. 2013. "Heavy Metal Contaminated Water, Soils and Crops in Peri Urban Wastewater Irrigation Farming in Mufulira and Kafue Towns in Zambia." *Journal of Geography and Geology* 5 (2). https://doi.org/10.5539/jgg.v5n2p55.

Karimi, N., S. M. Ghaderian, and H. Schat. 2013. "Arsenic in Soil and Vegetation of a Contaminated Area." *International Journal of Environmental Science and Technology* 10 (4): 743–52. https://doi.org/10.1007/s13762-013-0227-y.

Kaushik, A., H. R. Sharma, S. Jain, J. Dawra, and C. P. Kaushik. 2010. "Pesticide Pollution of River Ghaggar in Haryana, India." *Environmental Monitoring and Assessment* 160 (1–4): 61–69. https://doi.org/10.1007/s10661-008-0657-z.

Khan, Mohammad Saghir, Almas Zaidi, Parvaze Ahmad Wani, and Mohammad Oves. 2009. "Role of Plant Growth Promoting Rhizobacteria in the Remediation of Metal Contaminated Soils." *Environmental Chemistry Letters*. Springer Verlag. https://doi.org/10.1007/s10311-008-0155-0.

Khoufi, Sonia, Fathi Aloui, and Sami Sayadi Ã. 2016. "Treatment of Olive Oil Mill Wastewater by Combined Process Electro-Fenton Reaction and Anaerobic Digestion" *Water Research* 40 (2006): 2007–16. https://doi.org/10.1016/j.watres.2006.03.023.

Kim, Sungjin, and Chan Beum Park. 2013. "Bio-Inspired Synthesis of Minerals for Energy, Environment, and Medicinal Applications." *Advanced Functional Materials*. John Wiley & Sons, Ltd. https://doi.org/10.1002/adfm.201201994.

Kiyono, Masako, Yumiko Oka, Yuka Sone, Michitaka Tanaka, Ryosuke Nakamura, Masa H. Sato, Hidemitsu Pan-Hou, Kou Sakabe, and Ken ichiro Inoue. 2012. "Expression of the Bacterial Heavy Metal Transporter MerC Fused with a Plant SNARE, SYP121, in Arabidopsis Thaliana Increases Cadmium Accumulation and Tolerance." *Planta* 235 (4): 841–50. https://doi.org/10.1007/s00425-011-1543-4.

Klüpfel, Laura, Marco Keiluweit, Markus Kleber, and Michael Sander. 2014. "Redox Properties of Plant Biomass-Derived Black Carbon (Biochar)." *Environmental Science and Technology* 48 (10): 5601–11. https://doi.org/10.1021/es500906d.

Krishna, A. K., and P. K. Govil. 2007. "Soil Contamination Due to Heavy Metals from an Industrial Area of Surat, Gujarat, Western India." *Environmental Monitoring and Assessment* 124 (1–3): 263–75. https://doi.org/10.1007/s10661-006-9224-7.

Kumar A., Bisht B. S., and Joshi V. D. 2011. "Review on Bioremediation of Polluted Environment: A Management Tool." *International Journal of Environmental Sciences* 1. http://www.indianjournals.com/ijor.aspx?target=ijor:ijes&volume=1&issue=6&article=004.

Kumar, Amit, Ashish K. Chaturvedi, Kritika Yadav, et al. 2019. "Fungal Phytoremediation of Heavy Metal-Contaminated Resources: Current Scenario and Future Prospects." In *Recent Advancement in White Biotechnology Through Fungi*, 437–61. Springer, Cham. https://doi.org/10.1007/978-3-030-25506-0_18.

Kumar, Dushyant, Vivek Kumar Gaurav, and Chhaya Sharma. 2018. "Ecofriendly Remediation of Pulp and Paper Industry Wastewater by Electrocoagulation and Its Application in Agriculture." *American Journal of Plant Sciences* 9 (12): 2462–79. https://doi.org/10.4236/ajps.2018.912178.

Kumar, Dushyant, and Chhaya Sharma. 2020. "Reduction of Chlorophenols and Sludge Management from Paper Industry Wastewater Using Electrocoagulation Process." *Separation Science and Technology* 55 (15): 2844–54. https://doi.org/10.1080/01496395.2019.1646761.

Kumar, Dushyant, Chhaya Sharma, Dushyant Kumar, and Chhaya Sharma. 2019. "Remediation of Pulp and Paper Industry Effluent Using Electrocoagulation Process." *Journal of Water Resource and Protection* 11 (03): 296–310. https://doi.org/10.4236/jwarp.2019.113017.

Kumar, Lakhan, and Navneeta Bharadvaja. 2020. "Microbial Remediation of Heavy Metals." In *Microbial Bioremediation & Biodegradation*, 49–72. Springer, Singapore. https://doi.org/10.1007/978-981-15-1812-6_2.

Kumar, R, C. Acharya, and S. R. Joshi. 2011. "Isolation and Analyses of Uranium Tolerant *Serratia Marcescens* Strains and Their Utilization for Aerobic Uranium U(VI) Bioadsorption." *Journal of Microbiology* 49 (4): 568. https://doi.org/10.1007/s12275-011-0366-0.

Kumar Rai, Prabhat. 2009. "Heavy Metal Phytoremediation from Aquatic Ecosystems with Special Reference to Macrophytes." *Critical Reviews in Environmental Science and Technology* 39 (9): 697–753. https://doi.org/10.1080/10643380801910058.

Kumar, Smita S., Abudukeremu Kadier, Sandeep K. Malyan, Altaf Ahmad, and Narsi R. Bishnoi. 2017. "Phytoremediation and Rhizoremediation: Uptake, Mobilization and Sequestration of Heavy Metals by Plants." In *Plant-Microbe Interactions in Agro-Ecological Perspectives*, 367–94. Springer, Singapore. https://doi.org/10.1007/978-981-10-6593-4_15.

Kumar Tandon, Praveen, Santosh Bahadur Singh. 2016. "Redox Processes in Water Remediation." *Environmental Chemistry Letters* 14 (1): 15–25. https://doi.org/10.1007/s10311-015-0540-4.

Kushwaha, Anamika, Radha Rani, Sanjay Kumar, and Aishvarya Gautam. 2016. "Heavy Metal Detoxification and Tolerance Mechanisms in Plants: Implications for Phytoremediation." *Environmental Reviews* 24 (1): 39–51. https://doi.org/10.1139/er-2015-0010.

Lakherwal, Dimple. 2014. "Adsorption of Heavy Metals: A Review." *International Journal of Environmental Research and Development* 4. http://www.ripublication.com/ijerd.htm.

Li, Qingzhao, Huibin Liu, Mohamed Alattar, Shoufang Jiang, Jing Han, Yujiao Ma, and Chunyang Jiang. 2015. "The Preferential Accumulation of Heavy Metals in Different Tissues Following Frequent Respiratory Exposure to PM2.5 in Rats." *Scientific Reports* 5 (1): 1–8. https://doi.org/10.1038/srep16936.

Li, Xiang, Yang Zhang, Erich Gulbins, and Marcio Luis Busi Da Silva. 2010. *Handbook of Hydrocarbon and Lipid Microbiology Molecular Probes for Sphingolipids*

Metabolizing Enzymes View Project Mitochondrial Potassium Channels in Cancer View Project. https://www.researchgate.net/publication/257871939.

Lin, Hui, Junfeng Niu, Jiale Xu, Yang Li, and Yuhang Pan. 2013. "Electrochimica Acta Electrochemical Mineralization of Sulfamethoxazole by Ti/SnO_2 -Sb/Ce-PbO_2 Anode: Kinetics, Reaction Pathways, and Energy Cost Evolution." *Electrochimica Acta* 97: 167–74. https://doi.org/10.1016/j.electacta.2013.03.019.

Lindholm-lehto, Petra C., Juha S. Knuutinen, Heidi S. J. Ahkola, and Sirpa H. Herve. 2015. "Refractory Organic Pollutants and Toxicity in Pulp and Paper Mill Wastewaters." *Environmental Science and Pollution Research* 22 (9): 6473–99. https://doi.org/10.1007/s11356-015-4163-x.

Lokeshwari, H., and G. T. Chandrappa. 2006. "Impact of Heavy Metal Contamination of Bellandur Lake on Soil and Cultivated Vegetation." *Current Science* 91. https://www.jstor.org/stable/24094365.

Lu, Lu, Tyler Huggins, Song Jin, Yi Zuo, and Zhiyong Jason Ren. 2014. "Microbial Metabolism and Community Structure in Response to Bioelectrochemically Enhanced Remediation of Petroleum Hydrocarbon-Contaminated Soil." *Environmental Science & Technology* 48 (7): 4021–29. https://doi.org/10.1021/es4057906.

Macek, T., P. Kotrba, A. Svatos, M. Novakova, K. Demnerova, M. Mackova. 2008. "Novel Roles for Genetically Modified Plants in Environmental Protection." *Trends in Biotechnology* 26 (3): 146–52. https://www.sciencedirect.com/science/article/pii/S0167779908000231?casa_token=-3ZqjctIzD4AAAAA:r0MVBlhQ4c3yFnustT__3 WMr7RRU-DcqC2MlOq7riNHNBVrF8dxv0HcAuVC1mKYvl8jnePqqO0kX.

Machackova, Jirina, Zdena Wittlingerova, Kvetoslav Vlk, and Jaroslav Zima. 2012. "Major Factors Affecting In Situ Biodegradation Rates of Jet-Fuel During Large-Scale Biosparging Project in Sedimentary Bedrock." *Journal of Environmental Science and Health Part A: Toxic/Hazardous Substances and Environmental Engineering* 47 (8): 1152–65. https://doi.org/10.1080/10934529.2012.668379.

MacHado, M. D., S. Janssens, H. M.V.M. Soares, and E. V. Soares. 2009. "Removal of Heavy Metals Using a Brewer's Yeast Strain of Saccharomyces Cerevisiae: Advantages of Using Dead Biomass." *Journal of Applied Microbiology* 106 (6): 1792–1804. https://doi.org/10.1111/j.1365-2672.2009.04170.x.

Malyan, Sandeep K., Rajesh Singh, Meenakshi Rawat, Mohit Kumar, Arivalagan Pugazhendhi, Amrish Kumar, Vivek Kumar, and Smita S. Kumar. 2019. "An Overview of Carcinogenic Pollutants in Groundwater of India." *Biocatalysis and Agricultural Biotechnology* 21: 101288. https://doi.org/10.1016/j.bcab.2019.101288.

Mamtani, Ravinder, Penny Stern, Ismail Dawood, and Sohaila Cheema. 2011. "Metals and Disease: A Global Primary Health Care Perspective." *Journal of Toxicology.* https://doi.org/10.1155/2011/319136.

Mane, P. C., and A. B. Bhosle. 2012. "Bioremoval of Some Metals by Living Algae Spirogyra Sp. and Spirullina Sp. from Aqueous Solution." *International Journal of Environmental Research* 6 (2): 571–76. www.SID.ir.

Mangunwardoyo, Wibowo, Patria Mufti, Tony Sudjarwo, and M. P. Patria. 2013. "Bioremediation of Effluent Wastewater Treatment Plant Bojongsoang Bandung Indonesia Using Consorsium Aquatic Plants and Animals." *International Journal of Research and Reviews in Applied Sciences* 14 (1): 150–60. https://www.researchgate.net/publication/273756248.

Mani, D., and Chitranjan Kumar. 2014. "Biotechnological Advances in Bioremediation of Heavy Metals Contaminated Ecosystems: An Overview with Special Reference to Phytoremediation." *International Journal of Environmental Science and Technology* 11 (3): 843–72. https://doi.org/10.1007/s13762-013-0299-8.

Mani, Dinesh, A. E. Bechan, A. E. Sharma, and Chitranjan Kumar. 2007. "Phytoaccumulation, Interaction, Toxicity and Remediation of Cadmium from Helianthus Annuus L.

(Sunflower)." *Bulletin of Environmental Contamination and Toxicology* 79 (1): 71–79. https://doi.org/10.1007/s00128-007-9153-3.

Mathialagan, Thyagarajan, and Thiruvenkatachari Viraraghavan. 2009. "Biosorption of Pentachlorophenol from Aqueous Solutions by a Fungal Biomass." *Bioresource Technology* 100 (2): 549–58. https://www.sciencedirect.com/science/article/pii/S0960852408005762?casa_token=jQs0EAcJpScAAAAA:X_uNWJwcllHW3_obLqW4dsy8_2B31w-kcfVvqOKhRb-rI7lPike6lAQ4JE68MRIyVti_tlcEtg.

Mishra, Geetesh Kumar. 2017. "Microbes in Heavy Metal Remediation: A Review on Current Trends and Patents." *Ingentaconnect.Com* 11 (3): 188–96. https://doi.org/10.2174/1872208311666170120121025.

Mohan, Dinesh, Ankur Sarswat, Yong Sik Ok, and Charles U. Pittman. 2014. "Organic and Inorganic Contaminants Removal from Water with Biochar, a Renewable, Low Cost and Sustainable Adsorbent – A Critical Review." *Bioresource Technology* 160 (May): 191–202. https://doi.org/10.1016/j.biortech.2014.01.120.

Muszyńska, Ewa, and Ewa Hanus-Fajerska. 2015. "Why Are Heavy Metal Hyperaccumulating Plants so Amazing?" *Computational Biology and Bionanotechnology REVIEW PAPERS* 96 (4): 265–71. https://doi.org/10.5114/bta.2015.57730.

Nazemi, Saeid. 2012. "Cr (2.4-5.88), Zn (54.27-170.23), As (1.92-5.49),Cd (1.94-2.43) and Pb." *Journal of Applied Environmental and Biological Sciences* 2 (8): 48. www.textroad.com.

Ndeddy Aka, Robinson Junior, and Olubukola Oluranti Babalola. 2016. "Effect of Bacterial Inoculation of Strains of *Pseudomonas Aeruginosa, Alcaligenes Feacalis* and *Bacillus Subtilis* on Germination, Growth and Heavy Metal (Cd, Cr, and Ni) Uptake of *Brassica Juncea.*" *International Journal of Phytoremediation* 18 (2): 200–09. https://doi.org/10.1080/15226514.2015.1073671.

Neethu, C. S., K. M. Mujeeb Rahiman, A. V. Saramma, and A. A. Mohamed Hatha. 2015. "Heavy-Metal Resistance in Gram-Negative Bacteria Isolated from Kongsfjord, Arctic." *Canadian Journal of Microbiology* 61 (6): 429–35. https://doi.org/10.1139/cjm-2014-0803.

Norström, A., K. Larsdotter, L. Gumaelius, J. La Cour Jansen, and G. Dalhammar. 2003. "A Small Scale Hydroponics Wastewater Treatment System under Swedish Conditions." *Water Science and Technology* 48 (11–12): 161–67. Iwaponline.Com. https://iwaponline.com/wst/article-pdf/48/11-12/161/422031/161.pdf.

Ojuederie, Omena, and Olubukola Babalola. 2017. "Microbial and Plant-Assisted Bioremediation of Heavy Metal Polluted Environments: A Review." *International Journal of Environmental Research and Public Health* 14 (12): 1504. https://doi.org/10.3390/ijerph14121504.

Paliwal, Vasundhara, Sampada Puranik, and Hemant J. Purohit. 2012. "Integrated Perspective for Effective Bioremediation." *Applied Biochemistry and Biotechnology.* 166 (4): 903–24.https://doi.org/10.1007/s12010-011-9479-5.

Dhankher, Om Parkash, Sharon Lafferty Doty, Richard B. Meagher, and Elizabeth Pilon-Smits. 2011. "Biotechnological Approaches for Phytoremediation." In *Plant Biotechnology and Agriculture*, 309–28. Elsevier, Amsterdam. https://doi.org/10.1016/B978-0-12-381466-1.00020-1.

Pilon-Smits, Elizabeth. 2005. "Phytoremediation." *Annual Review of Plant Biology.* https://doi.org/10.1146/annurev.arplant.56.032604.144214.

Poirier, I., P. Hammann, L. Kuhn, M. Bertrand. 2013. "Strategies Developed by the Marine Bacterium Pseudomonas Fluorescens BA3SM1 to Resist Metals: A Proteome Analysis." *Aquatic Toxicology* 128 (215): 232. https://www.sciencedirect.com/science/article/pii/S0166445X12003256?casa_token=6TS-xgwdmXMAAAAA:ct3mkbry2gTHOqbtGPAd9sqhQ42XGvoor0IZS-wqFcZ1eZGPQlDrP0LQEg5o3PHLXPQf8a5RCyk.

Pourrut, Bertrand, Muhammad Shahid, Camille Dumat, Peter Winterton, and Eric Pinelli. 2011. "Lead Uptake, Toxicity, and Detoxification in Plants." *Reviews of Environmental Contamination and Toxicology* 213: 113–36. https://doi.org/10.1007/978-1-4419-9860-6_4.

Rajapaksha, Anushka Upamali, Meththika Vithanage, Yong Sik Ok, and Christopher Oze. 2013. "Cr(VI) Formation Related to Cr(III)-Muscovite and Birnessite Interactions in Ultramafic Environments." *Environmental Science and Technology* 47 (17): 9722–29. https://doi.org/10.1021/es4015025.

Ramos, J. L., S. Marqués, P. van Dillewijn, et al.2011. "Laboratory Research Aimed at Closing the Gaps in Microbial Bioremediation." *Trends in Biotechnology* 12 (29): 641–47. https://www.sciencedirect.com/science/article/pii/S0167779911001119?casa_token=q BHWfMBFOCwAAAAA:UnBHs3nwIeR_CHK52wHvEHZer-pLVQOGx0nmp371Ao 9vcjYnU3Qk7Z4f9bwGW9Iw7nnhdOPXRSs.

Rayu, S., D. G. Karpouzas, B. K. Singh. 2012. "Emerging Technologies in Bioremediation: Constraints and Opportunities." *Biodegradation* 23 (6): 917–26. https://doi.org/10.1007/s10532-012-9576-3.

Rizwan, M., Ali, S., Qayyum, M. F., Ibrahim, M., Zia-ur-Rehman, M., Abbas, T., and Ok, Y. S.2016. "Mechanisms of Biochar-Mediated Alleviation of Toxicity of Trace Elements in Plants: A Critical Review." *Environmental Science and Pollution Research* 23 (3): 2230–48. https://doi.org/10.1007/s11356-015-5697-7.

Saha, Jayanta K., Rajendiran Selladurai, M. Vassanda Coumar, M.L. Dotaniya, Samaresh Kundu, and Ashok K. Patra. 2017. *Soil Pollution – An Emerging Threat to Agriculture* 10. https://doi.org/10.1007/978-981-10-4274-4.

Schiewer, Silke, and Santosh B. Patil. 2008. "Pectin-Rich Fruit Wastes as Biosorbents for Heavy Metal Removal: Equilibrium and Kinetics." *Bioresource Technology* 99 (6): 1896–1903. https://doi.org/10.1016/j.biortech.2007.03.060.

Sharma, R. K., M. Agrawal, F. M. Marshall.2009. "Heavy Metals in Vegetables Collected from Production and Market Sites of a Tropical Urban Area of India." *Food and Chemical Toxicology* 47 (3): 583–91. https://www.sciencedirect.com/science/article/pii/S0278691508007126?casa_token=43vrj1_MOgwAAAAA:6faYQ_ff8JcvYAaS gRAaXPJ4EoLZOWEWbjs2jVHLbnFbyxIhi9kwExecDMoBexB5lTqtK_mq8t8.

Singh, A., and O. P. Ward2014. Biotechnology and Bioremediation—An Overview. https://books.google.co.in/books?hl=en&lr=&id=k0GuZ6VQH2YC&oi=fnd&pg=PA1&dq=Microbial+Bioremediation+%26+Biodegradation&ots=V6AuiHRrUZ&sig=1QYnBM zDHudkbRLSY2fsCRLtkDM#v=onepage&q=MicrobialBioremediation%26Biodegra dation&f=false.

Singh, A, R. K. Sharma, M. Agrawal, and F. M. Marshall.2010. "Health Risk Assessment of Heavy Metals via Dietary Intake of Foodstuffs from the Wastewater Irrigated Site of a Dry Tropical Area of India." *Food and Chemical Toxicology* 48 (2): 611–19. https://www.sciencedirect.com/science/article/pii/S0278691509005572?casa_token= 18npbSajBU0AAAAA:xP7rNC-6Dqo2y_UYalVAPwerTUBPhbDrVebhuR2NPkXgo 7p8147i96j0a2yw5xtnM6cfavPRRaQ.

Singh, J. 2013. "Assessment of Bioavailability and Leachability of Heavy Metals during Rotary Drum Composting of Green Waste (Water Hyacinth)." *Elsevier Ecological Engineering* 52: 59–69. https://www.sciencedirect.com/science/article/pii/S09258574 12004521?casa_token=YduELM0mBykAAAAA:cWDRy0acTRvzq7w9zEicgOP4 paq-9Y97JHOiOKj30dv6hkahCJqMPJa9fQjL4OtMAxU94U2fQ98.

Singh, J. S.,P. C. Abhilash, H. B. Singh, Rana P. Singh, D. P. Singh.2011. "Genetically Engineered Bacteria: An Emerging Tool for Environmental Remediation and Future Research Perspectives." *Gene* 1 (1–2): 1–9. https://www.sciencedirect.com/science/article/pii/S0378111911000916?casa_token=X7U0j46zmSAAAAAA:zqetzv1fIdvuzuq EjSiPfV-x0qVYrpiwRWbe1aeho4_NE74mH1mBnTT8iS_gi879Lek0x_8RBp0.

Singh, Kaptan, Rajesh Singh, Sandeep K. Malyan, Meenakshi Rawat, Pradeep Kumar, Sumant Kumar, M. K. Sharma, and Govind Pandey. 2018. "Health Risk Assessment of Drinking Water in Bathinda District, Punjab, India." *Journal of Indian Water Resources Society* 38 (3): 34–41.

Singh, Munendra, German Müller, and I. B. Singh. 2002. "Heavy Metals in Freshly Deposited Stream Sediments of Rivers Associated with Urbanisation of the Ganga Plain, India." *Water, Air, and Soil Pollution.* Springer Netherlands. https://doi.org/10.1023/A:1021339917643.

Sivakumar, D.2013. "Experimental and Analytical Model Studies on Leachate Volume Computation from Solid Waste." *International Journal of Environmental Science and Technology* 10 (5): 903–16. https://doi.org/10.1007/s13762-012-0083-1.

Skórzyn-Polit, Ewa, Maria Drazkiewicz, and Zbigniew Krupa. 2010. "Lipid Peroxidation and Antioxidative Response in Arabidopsis Thaliana Exposed to Cadmium and Copper." *Acta Physiologiae Plantarum* 32 (1): 169–75. https://doi.org/10.1007/s11738-009-0393-1.

Sud, D., G. Mahajan, M. P. Kaur. 2008. "Agricultural Waste Material as Potential Adsorbent for Sequestering Heavy Metal Ions from Aqueous Solutions – A Review." *Bioresource Technology.* Elsevier. https://doi.org/10.1016/j.biortech.2007.11.064.

Tandon, Praveen K., Ritesh C. Shukla, and Santosh B. Singh. 2013. "Removal of Arsenic(III) from Water with Clay-Supported Zerovalent Iron Nanoparticles Synthesized with the Help of Tea Liquor." *Industrial and Engineering Chemistry Research* 52 (30): 10052–58. https://doi.org/10.1021/ie400702k.

Tang, Chuyang Y., Q. Shiang Fu, Craig S. Criddle, and James O. Leckie. 2007. "Effect of Flux (Transmembrane Pressure) and Membrane Properties on Fouling and Rejection of Reverse Osmosis and Nanofiltration Membranes Treating Perfluorooctane Sulfonate Containing Wastewater." *Environmental Science and Technology* 41 (6): 2008–14. https://doi.org/10.1021/es062052f.

Tchounwou, Paul B., Clement G. Yedjou, Anita K. Patlolla, and Dwayne J. Sutton. 2012. "Heavy Metal Toxicity and the Environment." *EXS.* Springer, Basel. https://doi.org/10.1007/978-3-7643-8340-4_6.

Thapa, Bijay, Ajay Kumar, and Anish Ghimire. 2012. "A Review on Bioremediation of Petroleum Hydrocarbon Contaminants in Soil." *Kathmandu University Journal of Science, Engineering and Technology* 8. https://www.nepjol.info/index.php/kuset/article/view/6056.

Tschirhart, Céline, Pascal Handschumacher, Dominique Laffly, and Eric Bénéfice. 2012. "Resource Management, Networks and Spatial Contrasts in Human Mercury Contamination along the Rio Beni (Bolivian Amazon)." *Human Ecology.* Springer. https://doi.org/10.1007/s10745-012-9500-9.

Tyagi, Meenu, M. Manuela, R. Da Fonseca, and Carla C. C. R. De Carvalho. 2011. "Bioaugmentation and Biostimulation Strategies to Improve the Effectiveness of Bioremediation Processes." *Biodegradation* 22 (2): 231–41. https://doi.org/10.1007/s10532-010-9394-4.

Upadhyay, Neha, Kanchan Vishwakarma, Jaspreet Singh, Mitali Mishra, Vivek Kumar, Radha Rani, Rohit K. Mishra, Devendra K. Chauhan, Durgesh K. Tripathi, and Shivesh Sharma. 2017. "Tolerance and Reduction of Chromium (VI) by Bacillus Sp. MNU16 Isolated from Contaminated Coal Mining Soil." *Frontiers in Plant Science* 8 (May): 777. https://doi.org/10.3389/fpls.2017.00778.

Vangronsveld, Jaco, Rolf Herzig, Nele Weyens, et al. 2009. "Phytoremediation of Contaminated Soils and Groundwater: Lessons from the Field." *Environmental Science and Pollution Research* 16 (7): 765–94. https://doi.org/10.1007/s11356-009-0213-6.

Vidali, M. 2001a. "Bioremediation – An Overview." *Pure and Applied Chemistry* 73 (7): 1163–72. http://www.icontrolpollution.com/articles/bioremediation–an-overview-.php?aid=37408.

Vidali, M. 2001b. "Bioremediation. An Overview." *Pure and Applied Chemistry* 73. https://www.degruyter.com/view/journals/pac/73/7/article-p1163.xml.

Volesky, B., and H. May and Z. R. Holan.1993. "Cadmium Biosorption by Saccharomyces Cerevisiae." *Biotechnology and Bioengineering* 41 (8): 826–29. https://doi.org/10.1002/bit.260410809.

WHO. 2011. "'Adverse Health Effects of Heavy Metals in Children.' Children's Health and the Environment WHO Training Package for the Health Sector World." https://www.who.int/ceh/capacity/heavy_metals.pdf

Wilberforce, Oti, and F. I. Nwabue. 2013. "Heavy Metals Effect Due to Contamination of Vegetables from Enyigba Lead Mine in Ebonyi State, Nigeria." *Environment and Pollution* 2 (1). https://doi.org/10.5539/ep.v2n1p19.

Wiltse, Jeanette, and Vicki L. Dellarco. 1996. "U.S. Environmental Protection Agency Guidelines for Carcinogen Risk Assessment: Past and Future." *Mutation Research/Reviews in Genetic Toxicology* 365 (1–3): 3–15. https://doi.org/10.1016/S0165-1110(96)90009-3.

Wuana, Raymond A., and Felix E. Okieimen.2011. "Heavy Metals in Contaminated Soils: A Review of Sources, Chemistry, Risks and Best Available Strategies for Remediation." *ISRN Ecology* 2011: 1–20. https://doi.org/10.5402/2011/402647.

Xiangliang, P. 2009. "Micrologically Induced Carbonate Precipitation as a Promising Way to in Situ Immobilize Heavy Metals in Groundwater and Sediment." *Research Journal of Chemistry and Environment* 13: 3–4.

Yeung, Albert T. 2010. "Remediation Technologies for Contaminated Sites." In *Advances in Environmental Geotechnics*, 328–69. Springer, Berlin, Heidelberg. https://doi.org/10.1007/978-3-642-04460-1_25.

Yuan, Yong, Nanthi Bolan, Antonin Prévoteau, Meththika Vithanage, Jayanta Kumar Biswas, Yong Sik Ok, and Hailong Wang. 2017. "Applications of Biochar in Redox-Mediated Reactions." *Bioresource Technology*. Elsevier Ltd. https://doi.org/10.1016/j.biortech.2017.06.154.

Zeid, I. M., S. M. Ghazi, D. M. Nabawy. 2013. "Alleviation of Heavy Metals Toxicity in Waste Water Used for Plant Irrigation." *International Journal of Agronomy and Plant Production* 4 (5): 976–83. https://www.cabdirect.org/cabdirect/abstract/20133173147.

7 Recent Trends in Solar Photocatalytic Degradation of Organic Pollutants using TiO$_2$ Nanomaterials

Chhotu Ram
Adigrat University

Bushra Zaman
Utah State University

Rajesh Kumar Jena
Darbhanga College of Engineering
Aryabhatta Knowledge University

Amit Kumar
Debre Markos University

CONTENTS

7.1 INTRODUCTION

Globally, the increasing trend of organic pollutants contamination in water resources has been observed over the years. The reason behind these issues is mainly due to increasing world population, continuous industrial and agricultural development leading to increases in the water pollution (Koe et al., 2020). Wastewater generated from various activities contains high concentrations of biodegradable contaminants and treatment can be done by well-established biological treatments. However, most of industrial effluents contain toxic and non-biodegradable or bio-recalcitrant compounds. In this context, wastewater from various industries such as pharmaceutical (Olama et al., 2018), textile (Touati et al., 2016), and agricultural (Kushniarou et al., 2019) often contains toxic pollutants with low biodegradability. The organic pollutants such as pharmaceutical and pesticide compounds (PPCs) (Varma et al., 2020), phenol (Khan and Malik, 2014), chlorophenols (Hao et al., 2019), dyes (Kaur et al., 2017) and bisphenol A (BPA) as endocrine disruptor (Rani and Shanker, 2018) are released into the environment from the various industrial processes. The industrial effluents having these persistent organic pollutants with the complex composition mostly having low biological oxygen demand (BOD) to chemical oxygen demand (COD) ratio and hence, considered as lower biodegradability (Zhang, Wang, and Zhang, 2014). For example, textile industries produce effluents containing residual dyes, phenolic compounds and are discarded into the surrounding environment causing water pollution and affect the ecosystem drastically. Study has shown that 20% of global industrial water pollution comes from textile processing and complex aromatic structures of these dyes make them more recalcitrant to biodegradation (Khan and Malik, 2014; Lee et al., 2016). Thus, these organic compounds of wider categories are considered persistent, bioaccumulative, and toxic chemicals by the U.S. Environmental Protection Agency (EPA) (Lee et al., 2016). Moreover, more research is focusing on the endocrine disrupting compounds (EDPs), which are a potential threat to the environment and may cause serious effects in human health, including sexual precocity, embryotoxicity, and teratogenicity. One commonly released BPA is non-biodegradable in nature and is even difficult to degrade chemically in the environment (Rani and Shanker, 2018). Thus, significance of wastewater treatment, recycling management, and its disposal with the prescribed standard is increasing in modern times due to major public health concerns (Chequer et al., 2013).

Various traditional wastewater treatment methods such as biological treatment, physico-chemical, precipitation, etc. are available in markets and used to purify the industrial wastewater. However, these available techniques have limitations and difficulties in the complete degradation and mineralization of the complex structure of organic pollutants (Guo, Qi, and Liu, 2017). Also, conventional wastewater treatment consumes higher energy, which is one of the major impediments for the industry. Thus, the use of a photocatalytic degradation process as one of the advanced oxidation processes (AOPs) has progressed rapidly since the discovery of photo-electrochemical water-splitting reactions using a semiconductor by Fujishima and Honda in 1972. AOPs using semiconductor photocatalysts has proved to be effective for organic pollutant degradation and are also known as an environmental friendly, sustainable, and energy-saving technology. Further, this technique has applications for the low biodegradability, high complexity, and high organic pollutant concentrations in the wastewater (Guo et al., 2017). Several catalysts e.g. TiO_2, ZnO, CeO_2, CdS, WO_3, Fe_2O_3, and many other semiconductor nanomaterials are employed as a photocatalyst for different applications like energy, environment, and chemical synthesis (Carp, Huisman, and Reller, 2004; Rajaraman, Parikh, and Gandhi, 2020; Shah and Chang, 2018). However, titanium dioxide (TiO_2)-based photocatalysis has been recognized as an excellent material and is widely used for different applications. It has the advantages of being chemically stable, non-toxic, and also comparatively inexpensive (Kumar and Devi, 2011; Lazar, Varghese, and Nair, 2012; Rauf, Meetani, and Hisaindee, 2011). The main disadvantages of TiO_2 semiconductor materials are that they are less active under visible light sources due to a wide band gap. In fact, TiO_2 based photocatalysis mediated-AOP techniques have considerable potential for removal of various complex organic pollutants from wastewater under solar, visible, and UV light sources (Abdelhaleem and Chu, 2017; Hinojosa -Reyes et al., 2019; Sood et al., 2015; Yi et al., 2019; Zhang et al., 2015). The solar spectrum is comprised of 5% UV radiation that reaches the Earth surface in the range of 300–400 nm. Furthermore, solar energy could be a viable option in tropical countries and the process can be economically feasible (Molla et al., 2017; Xiong et al., 2013). In this context, various studies have been performed for the application of solar photocatalysis for the degradation of organic pollutants etc. (Aliste et al., 2020; Chu et al., 2020; Radhika and Thomas, 2017). Thus, this chapter is focused on the systematic review on the solar photocatalytic degradation of organic pollutants using TiO_2 nanomaterials. However, ongoing interests and developments in harnessing solar energy are expected to increase in the future for in-plant industrial applications.

7.2 ORGANIC POLLUTANT RELEASE IN THE ENVIRONMENT AND THEIR IMPACTS

The organic pollutants include pesticides such as dichlorodiphenyl-trichloroethane (DDT) and lindane, industrial chemicals such as polychlorinated biphenyls (PCBs), and substances such as dioxins, which are the unwanted by-products of manufacturing and combustion processes. The persistent organic pollutants (POPs) are resistant to most of the known environmental degradation processes. Due to their persistent nature, POPs bio-accumulate in animal tissues and biomagnify along with the food chain and food web with potential adverse impacts on the environment, human health, and wildlife.

7.2.1 Textile Dyes

The textile dyes are soluble organic compounds. The release of textile dyes in the environment, especially the water bodies, spoils the aesthetics and reduces the sunlight penetration into water that affects underwater life metabolism including the plants and organisms. The textile dyes increase the BOD and COD, impair photosynthesis, inhibit plant growth, provide recalcitrance and bioaccumulation, and may promote toxicity, mutagenicity, and carcinogenicity (Lellis et al., 2019). The textile dyes also enter the food chain and biomagnify at higher trophic levels (Sandhya, 2010).

7.2.2 Pharmaceuticals Compounds

Pharmaceutical compounds – used by humans, veterinary hospitals, agricultural business, etc. – in the form of prescriptions medicines, over the counter medicines, or as personal care products (PPCPs) are classified in nine different categories: hormones, antibiotics, lipid regulators, nonsteroidal anti-inflammatory drugs, beta blockers, antidepressants, anticonvulsants, antineoplastic and diagnostic contrast media (Hope, 2019). The pharmaceutical waste is released into the environment either directly or indirectly as shown in Figure 7.1. The direct route is through contamination of surface water by hospitals, households (flushed down the toilet, passed unmetabolized through human excretion, rinsed off during showers), industries, or wastewater treatment plants. The indirect route is through human consumption and excretion. The distribution of pharmaceutical products in the environment is

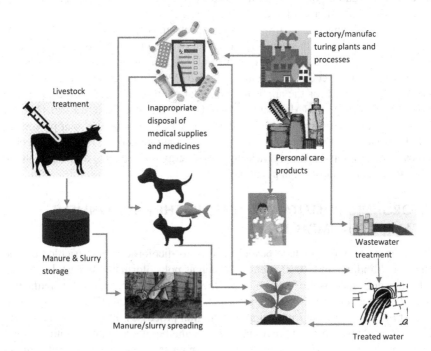

FIGURE 7.1   Release of pharmaceutical compounds into the environment.

mainly by aquatic mediums followed by food-chain dispersal due to their low vola-
tility (Nikolaou, Meric, and Fatta, 2007). Once the pharmaceutical compounds enter
the sewerage system, they have several adverse effects on humans, flora, and fauna,
sometimes fatal depending on the level of exposure.

7.2.3 PHENOLS

Phenol is synthesized and also occurs as a natural substance in nature. Phenol is the
first member of the chemical group of organic compounds with a hydroxyl group(s)
directly bonded to one or more aromatic rings. Phenol is also known as carbolic acid,
benzophenol, or hydroxybenzene with the chemical formula of C_6H_5OH (Anku,
Mamo, and Govender, 2017). Phenol is a colorless-to-white solid when pure, but the
commercial product is a liquid. Phenol evaporates more slowly than water. Phenol is
used primarily in the production of phenolic resins and in the manufacture of nylon
and other synthetic fibers. It is also used in slimicides (chemicals that kill bacteria
and fungi in slimes), as a disinfectant and antiseptic, and in medicinal preparations
(ATSDR, 2008). Phenol enters the environment through wastewater discharged by
industrial, agricultural, and domestic activities and also through the decay of natural
organic matter. Phenol usually enters into the environment in small single releases,
half of it is generally removed from air in 24 hours and remains in soil for 2 to 5 days.
In water, phenol remains for around a week. However, continuous release of phenol in the environment through various sources can result in longer residence time.
Phenol does not build or biomagnify in plants, animals, or humans. The International
Agency for Research on Cancer (IARC) and the EPA have determined that phenol is
not classifiable as to its carcinogenicity to humans.

7.2.4 PESTICIDES

In earlier days most weeds, pests, insects, and diseases were controlled using sustainable practices such as cultural, mechanical, and physical control strategies. To
increase the yield and cater to a growing population, the practice of using pesticides –
a toxic chemical substance or a biological agent – was introduced to control the
insects, weeds, rodents, fungi, or other harmful pests to avert, deter, control, and/
or kill them. Ideally, pesticides should be biodegradable and should only target the
harmful pests, but, mostly, the pesticides also kill the organisms that are harmless
and useful for the environment (Gill and Garg, 2014). The pesticides enter into the
environment through secretion into the soil and groundwater and dissemination in
the air when they are sprayed. The impact of human health depends on the type
of chemical exposure and also the duration of exposure. Pesticides also enter the
food chain through consumption of pesticide-laden food products and biomagnify at
higher trophic levels (Mostafalou and Abdollahi, 2013).

7.2.5 SURFACTANTS

Surfactants are chemicals which are widely used in cleaning products, detergents,
and personal care products. Surfactants are mainly of three types: anionic, nonionic,

and cationic (Ying, 2006). After the products are used, they are disposed of into the environment through wastewater drains, rivers, sewerage system, and contaminate soil, water, and sediments. Surfactants or surface active agents are chemicals that when added to water, reduce its surface tension and increase the wetting and spreading properties of water. Some examples of surfactants are glycerin, polysorbate 20, sodium lauryl ether sulfate, decyl glucoside, lignosulfonate, sodium stearate, alkyl polyglycolide, and sulfonate; some of them are germicides, insecticides, or fungicides, too (Augustyn, 2020). The surfactants may act as emulsifiers or demulsifiers. Literature suggests that the anionic surfactants do not harm the environment while the cationic surfactants are known to be more toxic. Surfactants as a whole do not exhibit much toxicity, however their breakdown products, nonyl and octyl phenols, are known to have carcinogenic effects (Scott and Jones, 2000).

Surfactants can be released in the environment in solid form, liquid, or as particulate matter (Figure 7.2). After the surfactant is released into the sewage drain, the solid residue is retained in the sewage sludge, sediment, and the soils. The liquid samples flow into the intakes and river, mix with natural water, sewage water, and the municipal wastewater. In the gaseous form, the surfactants appear as particulate matter, PM10 and PM2.5, and aerosols that are released by motor vehicles, lubricants, and diesel fuels, biomass burning, road dust, and construction. The discharge of surfactants into the environment causes water, air, and soil pollution. When the concentration of surfactants in water bodies increases, it effects the growth of algae and other microorganisms which subsequently affect the aquatic food chain. Literature suggests in an aquatic environment, the chronic toxicity of anionic and nonionic surfactants generally occurs at concentrations usually greater than 0.1 mg/L (Lewis, 1991). The surfactants enter into the aquatic animals through the food chain and skin penetration. Since surfactants are the by-product of detergent use, they can cause skin irritation in humans. However, if the surfactants enter the human body they can also interfere with normal physiological functions, and if the surfactants accumulate

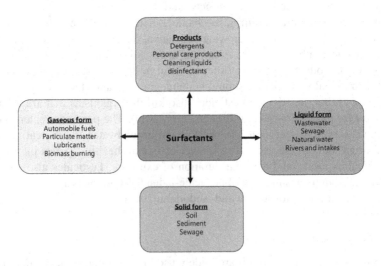

FIGURE 7.2 Surfactants release into the environment.

in the body, their toxic effects may manifest in the form of other harmful effects. Non-ionic surfactants do not cause skin irritation. However, cationic surfactants are highly toxic. One of the cationic surfactants, sodium dodecyl benzene sulfonate (SDBS), might cause liver damage and has carcinogenic and other chronic effects if absorbed through human skin (Lewis, 1991; Yuan et al., 2013).

7.2.6 HYDROCARBONS AND NITROGEN-CONTAINING COMPOUNDS

Hydrocarbons are compounds of hydrogen and carbon, like the compounds that form petroleum and natural gas. Some examples are methane, butane, propane, and hexane. The hydrocarbons and other nitrogen-containing compounds are released into the environment through oil spills, use of fossil fuels, pesticides, automobile oil, urban runoff, and toxic organic matter. The hydrocarbons can also enter into the groundwater through oil leaks associated with the exploration, production, and distribution petroleum hydrocarbons (Brown et al., 2017). The release of hydrocarbons and nitrogen-containing compounds in the environment lead to adverse effects on human health and are toxic to all forms of life. Oil spills are one phenomenon where a layer of oil is formed on the surface of water that hinders the exchange of oxygen and adversely affects the aquatic life. Arellano et al. (2015) studied the effect of hydrocarbon pollution in the Amazon forest and reported reduced chlorophyll content, high levels of foliar water content, and structural changes in leaves. The solvents used in paints, dry cleaning, and household cleaning chemicals contain dangerous hydrocarbons. These hydrocarbons can cause other effects, including comas, seizures, irregular heart rhythms, or damage to the kidneys or liver (Malaguarnera et al., 2012). Pollution due to nitrogen-containing compounds has a wide spectrum and affects global warming, leading to acid rain and eutrophication. Nitrogen is released into the environment in liquid or gaseous form and is also applied deliberately to soil as a fertilizer. Nitrates and nitrites are known to cause serious health effects. Nitrite reacts with hemoglobin in the blood and reduces the oxygen-carrying capacity of the blood (Ward et al., 2018). Nitrates can cause vitamin A deficiency and decrease the functioning of the thyroid gland. Nitrates and nitrites lead to the formation of nitro amines which are known to be one of the most common causes of cancer (Ma et al., 2018). An extensive use of fertilizers also has resulted in the release of nitrites and nitrates into the environment (Malaguarnera et al., 2012).

7.3 NANO-PHOTOCATALYSTS

Polluted waste plays a significant role in environmental pollution. There are various classical techniques that were used to decontaminate the polluted water, but the end products of these techniques need to be processed further for complete purification. Modern techniques such as biodegradation (Ram et al., 2020), fenton (Bouasla, El-Hadi Samar, and Ismail, 2010), photo-Fenton (Monteagudo et al., 2010), photocatalytic radiation (Rauf et al., 2010), sonolysis (Merouani et al., 2010), ozonation (Tehrani-Bagha, Mahmoodi, and Menger, 2010), and UV photocatalytic processes (Elmorsi et al., 2010) clearly have advantages than the earlier classical techniques mentioned because they degrade to produce harmless products. These advanced processes are better than the previously reported chemical oxidation processes, however these are much more costly. The evolution of science known as nano-science has completely replaced all the previous technologies because:

1. Nano-materials completely mineralize most of organics and are inexpensive.
2. Semiconductors remove all the organic matter from polluted water.
3. Nano-photocatalyst are non-toxic, non-corrosive, stable chemically and thermally.
4. Nano-photocatalysts are easily available, inexpensive, and stable to corrosion in the presence of water and chemicals.

7.3.1 PRINCIPLE AND MECHANISM OF PHOTOCATALYSIS

According to the conventional chemical terminology, photocatalysis is defined as a thermodynamically favored reaction in which a substance, the photocatalyst, accelerates the rate of a chemical photo-reaction, induced by the absorption of light by a catalyst or a co-existing molecule. The basic photo-physical and photochemical principles describing photocatalysis are already established and have been extensively reported (Serpone and Salinaro, 1999). When a photo-excited electron is promoted from the filled valence band of a nano-photocatalyst to the empty conduction band, the absorbed photon energy, $h\upsilon$, either equal to or exceeds the band gap of the semiconductor photocatalyst, initiates a photocatalytic reaction, which is leaving a hole in the valence band. Due to that, an electron and a hole pair ($e^- - h^+$) is generated. The mechanism of the photocatalytic degradation reactions are as follows:

$$\text{Photoexcitation: Nano-photocatalyst} + h\upsilon \rightarrow e^- + h^+ \qquad (7.1)$$

$$\text{Oxygen ionosorption: } O_2 + e^- \rightarrow O_2^{\cdot-} \qquad (7.2)$$

$$\text{Ionization of water: } H_2O \rightarrow OH^- + H^+ \qquad (7.3)$$

$$\text{Protonation of superoxides: } O_2^{\cdot-} + H^+ \rightarrow HOO\bullet \qquad (7.4)$$

The hydroperoxyl radical formed in (4) has also scavenging properties similar to O_2 thus doubly prolonging the lifetime of photohole:

$$HOO\bullet + e^- \rightarrow HO_2^- \qquad (7.5)$$

$$HOO^- + H^+ \rightarrow H_2O_2 \qquad (7.6)$$

Both the oxidation and reduction can take place at the surface of the nano-photocatalyst. Recombination between electron and hole occurs unless oxygen is available to scavenge the electrons to form superoxides ($O_2^{\cdot-}$), so after protonated form the hydroperoxyl radical ($HO_2\bullet$) and subsequently H_2O_2. These active radicals are responsible to decompose the organic waste.

7.3.2 Synthesis Techniques of Different Nano-Photocatalyst

There are several techniques that are used to synthesize the nano-photocatalysts and some are given below:

1. **High energy ball-milling:** It is a mechanical deformation process in which powder mixture placed in a ball mill is subjected to high-energy collisions from the balls (Qiu, Shen, and Gu, 2005).
2. **Hydrothermal technique:** It is a solution reaction-based approach in which nanomaterials can happen in wide temperature range from room temperature to very high temperature (J. Wang et al., 2009).
3. **Sol-gel technique:** The technique is a wet-chemical technique where either a chemical solution or colloidal particles (sol for nanoscale particle) is used to produce an integrated network (Brinker, 1994).
4. **Chemical co-precipitation:** This is the synthetic method where the supersaturation conditions necessary to induce precipitation that are usually the result of a chemical reactions (Faraji, Yamini, and Rezaee, 2010).
5. **Reverse micelle technique:** The reverse micelle technique is useful for the synthesis of nanoparticles with the desired shape and size (Santra et al., 2001).
6. **Calcination:** The method is based on the temperature (Liu et al., 2007).

7.3.3 Factors Affecting the Photocatalyst Efficiency

Photodegradation has direct influence on photocatalyst efficiency and there are a number of factors that drive the photodegradation of organic pollutants (Guillard et al., 2003; Rasoulifard, Fazli, and Eskandarian, 2015; Rauf et al., 2011):

- Light density
- Nature and concentration of substance
- Nature of photocatalyst
- Photocatalyst concentration
- pH of the solution
- Reaction temperature
- Free oxygen in solution

7.3.4 Titania (TiO_2) Versus other Photocatalysts

An ideal photocatalyst should be stable enough towards photons, and chemically and biologically inert. It should be available at low cost and moreover it should have the capability to absorb reactants under efficient photonic activation ($h\upsilon \geq Eg$) (Carp et al., 2004). Titanium dioxide has shown its importance in recent advances. It is also called titanium (IV) or *titania*, the naturally occurring oxide of titanium. The sources of titania are from ilmenite, rutile, and anatase. It has wide ranges of applications, from paint to sunscreen to food coloring. The problem associated with the existing catalysts is their high band gap. Doping one catalyst with other suitable

TABLE 7.1
Different Nano-Photocatalyst and Their Application for Degradation of Organic Pollutants

Nano-Photocatalyst	Application/Degradation	References
In VO$_4$	Organic pollutant	Zhang et al. (2006)
CdS/TiO$_2$	Phenol	Srinivasan, Wade, and Stefanakos (2006)
SiOC/ZnO	Methylene blue	Hojamberdiev et al. (2012)
LiFe(WO$_4$)$_2$	Methylene blue	Ji et al. (2011)
UV/Solar/TiO$_2$	Pharmaceuticals	Dhir, Kamboj, and Ram (2016)
UV/Solar/TiO$_2$	Chlorinated resin and fatty acids	Kumar et al. (2015)
ZnS/MMT	Eosin B	Miao et al. (2006)
UV/Solar/TiO$_2$	Procion yellow dye	Ram, Pareek, and Singh (2012)
InVO$_4$/TiO$_2$	Degradation of phenol	Yuan et al. (2010)
Au/TiO$_2$, Pd/TiO$_2$	Methyl alcohol	Ismail, Kandiel, and Bahnemann (2010)
Ag/ZnO	Organic pollutant	Shu et al. (2010)
WO$_x$-TiO$_2$	Methylene blue	Sajjad et al. (2010)
BiFeO$_3$	Rhodamine B	Madhu, Bellakki, and Manivannan (2010)

metals and metal oxides will enhance the efficiency of the photocatalyst and also makes the catalyst active in the visible region. By far, titania is the most widely employed system in photocatalysis due to its comparatively higher photocatalytic activity, low toxicity, chemical stability, and very low cost (Table 7.1). From the literature survey, it was reported that the anatase form of titania is the best combination of photo-activity and photo-stability. The minimum band gap energy required for a photon to cause a photo-generation of charge carriers over TiO$_2$ semiconductor (anatase form) is 3.2 eV corresponding to a wavelength of 388 nm (Perkowski et al., 2006). Practically, TiO$_2$ photoactivation takes place in the range of 300–388 nm. The photo-induced transfer of electrons occurring with adsorbed species on semiconductor photocatalyst depends on the band-edge position of the semiconductor and the redox potentials of the adsorbates (Fujishima, Rao, and Tryk, 2000).

7.4 OVERVIEW OF TIO$_2$ NANOMATERIAL MEDIATED PHOTOCATALYSIS

Heterogeneous photocatalysis using UV/TiO$_2$ is one of the most commonly used photocatalytic processes and is based upon the adsorption of light photons. The photocatalytic process is initiated by the absorption of light photons with energy (hυl) equal to or greater than the band gap of TiO$_2$ (3.2 eV and wavelengths lower than ~390 nm). By this process, the electron is excited to the conduction band (CB) while a positive hole is formed in the valence band (VB) at the same time (Boroski et al., 2009). Hence, an electron-hole pair is formed at the surface of TiO$_2$. This process initiates the decomposition of organic pollutants at the surface of the TiO$_2$ catalyst and used for wide application in the purification of air and water. The excited-state electrons at the conduction band and hole at the valence band can recombine and dissipate the heat adsorbed on the semiconductor surface. Both the formed hole and electron at the conduction band are considered strong oxidizing and reducing agents, respectively. The hole (h_{VB}^+) reacts with the organic molecules due to their oxidation producing CO_2 and H_2O as the final products (Equation 7.8). Further, Equation 7.9 shows the oxidation of organic compounds by reacting with water molecules to generate the hydroxyl radical ($^{\cdot}OH$) on h_{VB}^+. This step is valuable in photochemical reactions due to $^{\cdot}OH$ ion production. This radical step has the second highest oxidation potential (2.80 V) after the strongest oxidant-fluorine. The $^{\cdot}OH$ is electrophilic in nature (electron preferring) and can non-selectively oxidize almost all electron-rich organic molecules, eventually converting them into simpler water and carbon dioxide as provided in Equation 7.10. The 'R' is represented by the organic pollutant in the following equations.

$$TiO_2 + hv \ (< 387 \ nm) \rightarrow e_{CB}^- + h_{VB}^+ \tag{7.7}$$

$$h_{VB}^+ + R \rightarrow intermediate \rightarrow CO_2 + H_2O \tag{7.8}$$

$$H_2O + h_{VB}^+ \rightarrow {}^{\cdot}OH + H^+ \tag{7.9}$$

$${}^{\cdot}OH + R \rightarrow intermediates \rightarrow CO_2 + H_2O \tag{7.10}$$

$$e_{CB}^- + O_2 \rightarrow O_2^{\cdot -} \tag{7.11}$$

The semiconductor conduction band further reacts with the oxygen and forms the superoxide anion radical ($O_2^{\cdot -}$), which can be seen in Figure 7.3. Further, the reaction process can lead to the formation of hydroxyl peroxide which forms $^{\cdot}OH$ (Pirkanniemi and Sillanpää, 2002). The mechanism of the electron hole-pair

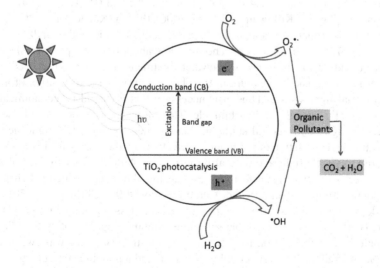

FIGURE 7.3 Photocatalytic degradation of organic pollutants using TiO_2 nanomaterials.

formation occurs when the TiO_2 is irradiated under artificial and solar light (Chong et al., 2010).The presence of dissolved oxygen is extremely important during the photocatalytic process, as it can make the recombination process on TiO_2 (e_{CB}^-/h_{VB}^+) that results in maintaining the electroneutrality of the TiO_2 particles (Boroski et al., 2009). In other words, it is important to reduce the rate of the recombination process of e_{CB}^- and h_{VB}^+, and also avoid the higher number of electrons accumulated on the conduction band. Thus, it can lead to the effective photocatalytic degradation of organic pollutants by reduction process of oxygen (Herrmann, 1999; Hoffmann et al., 1995).

7.5 ADVANCEMENTS IN SOLAR PHOTOCATALYTIC DEGRADATION FOR ORGANIC POLLUTANTS

Photochemical reactions are carried out through the process of photolysis in which chemical compounds absorb radiation energy and generate radicals. For this purpose, either solar energy or low- and medium-pressure mercury lamps can be used as a light source (Chen et al., 2010; Jiao et al., 2008). In photolysis, the hydroxyl compounds are generated by water splitting:

$$H_2O \xrightarrow{h\nu} H^{\cdot} + OH^{\cdot} \qquad (7.12)$$

However, it is well established that hydroxyl radicals generated are not sufficient enough to fully degrade refractory pollutants due to slow reactions (Kabra, Chaudhary, and Sawhney, 2004). Further, to accelerate the photolysis process, semiconductor metal salts act as catalysts in the photochemical processes (Abramovic, Sojic, and Anderluh, 2007; Herrmann et al., 1998).Titanium dioxide is an extensively used nanomaterial by many scientists and researchers for the degradation organic pollutants like benzofuran, 2,4-dichlorophenoxyaceticacid (Herrmann et al., 1998), nitrobenzene (Tayade, Kulkarni, and Jasra, 2006), thiram (Kaneco et al., 2009), benzoic acid, o-chlorobenzoic acid, o-nitrobenzoic acid, phthalic acid (Gandhi, Mishra, and Joshi, 2012) and many others. The overview about the TiO_2 photocatalysis has also been studied for industrial wastewater treatment and briefly discussed in the previous sections. The key feature in the photocatalysis is the utilization of a wide and industrial light source. Thus, it is necessary that it should be economical and environmentally friendly and sustainable. There are three options of light source: visible, UV light, and natural sunlight. Among these options, UV light mediated photocatalytic degradation of pollutants is effective, but the energy utilization is high compared to visible light and natural light sources. Another disadvantage of a UV light source is the harmful effect on skin causing skin cancer, due to their highly penetrating nature (Matsumura and Ananthaswamy, 2004; Wittlich et al., 2016).

Wide band gap TiO_2 semiconductor materials are less active under a visible light source. Hence, it was observed that solar light contains approximately 5% UV light and could be a viable option to utilize as on industrial scale for wastewater treatment. Solar irradiation is a renewable, abundant, and a pollution-free energy source at low-cost commercial applications. However, physicochemical and biological treatment systems are widely used on an industrial scale with high efficiency; the major drawback is the consumption of higher energy. Thus, the industry is moving towards the water energy nexus approach which has substantial capability to improve the performance of wastewater treatment processes, equally focusing on the two interconnected aspects i.e. water and energy resources (Crini and Lichtfouse, 2019; Maktabifard, Zaborowska, and Makinia, 2018; Scott et al., 2011). Research has started in this area by the use of solar irradiation with the market available TiO_2 for the degradation of organic pollutants. Many other researchers are working to improve the photocatalytic efficiency of TiO_2 in the presence of natural sun light. Recent study (Parangi and Mishra, 2019) reviewed the different modification strategies in TiO_2 nanoparticles for improving photocatalytic activity under visible and solar light with various synthesis methods. In this way, TiO_2 doping like metal and non-metal has involved the incorporation of atoms of dopant into the TiO_2 structure (Mogal et al., 2013). The utilization of metal dopant with TiO_2 is reported to alter the band gap energy, physical properties, and charge carrier recombination rate. The metal dopant plays a vital role in capturing of electron and helps in the separation of charge carriers. The incorporation of metal dopant has shifted the band gap edge towards the higher wavelength. It was also observed that metal-doped TiO_2 materials work much better under solar light as compared to undoped TiO_2 (Connelly, Wahab, and Idriss, 2012; Sathishkumar et al., 2011; Vega et al., 2018). Thus, the present section deals with the recently developed undoped and doped TiO_2 nanomaterial-mediated solar photocatalytic wastewater treatment for the removal of various types of complex organic pollutants.

7.5.1 UNDOPED TiO_2 SOLAR PHOTOCATALYSIS

Semiconductor photocatalysis using solar radiation as the source of photons for the activation of the catalyst has received much attention over the past several years. TiO_2 has shown the potential for use in converting solar energy to electricity and this can have a significant impact on reducing the world's dependency on fossil fuels. However, the photocatalytic properties of the TiO_2 are of greater importance, since many organic pollutants can be destroyed by oxidation and reduction processes occurring on TiO_2 surfaces (Chang et al., 2004; McCullagh et al., 2007; Ramirez et al., 2010; Wang et al., 2009). Recent work has been carried out using TiO_2 and ZnO photocatalysts for the degradation of caffeine (CAF) under solar light (290–1200 nm) irradiation. A comprehensive parametric study shows that for both photocatalysts, the optimum reaction conditions are pH 7.0, light intensity of 100 mW cm^{-2} and photocatalyst amount of 1.0 g L^{-1}, and results showed that solar light coupled with unmodified TiO_2 can be successfully degraded by caffeine in water (Ghosh et al., 2019). Author studied the use of titanium dioxide-coated glass spheres to treat 2, 4-dichlorophenoxy acetic acid (2,4-D), methyl chlorophenoxy propionic acid (MCPP), and 3, 6-dichloro-2-methoxy benzoic acid, a commercially available herbicide. Furthermore, photocatalytic degradation of sulfolane and a typical naphthenic acid (cyclopentane carboxylic acid; CPA) were also tested under ambient conditions. The results showed that 99.8% degradation of 2, 4-D, 100% degradation of both MCPP, and Dicamba in Killex® solution, and 97.4% degradation of sulfolane by capturing 3.18 MJ/m²solar energy (Heydari, Langford, and Achari, 2019). However, the mineralization of organic compounds converts to the CO_2 and mineral acids by the use of TiO_2. This process depends directly upon the catalyst dose, pH of the aqueous solution, and pollutant concentration (Barndõk et al., 2016; Chatti et al., 2007; Durán and Monteagudo, 2007). However, other operating parameters are the reaction temperature, reactor configuration, reactor material, and air sparging in the reactor solution intensity (Salgado-Tránsito et al., 2015). Literature also cites some limitations of TiO_2 as photocatalytic material such as a relative high e_{cb}^-/h_{vb}^+ recombination rate and wide band gap energy (Iliev et al., 2010; Li et al., 2001). The practical application of TiO_2 is inhibited by its low photon utilization efficiency and the need for an ultraviolet excitation source, which accounts for only a small fraction of solar light (~5%). Therefore, it is a challenging problem to develop new TiO_2 photocatalytic systems with enhanced activities under both UV and visible light irradiation(Ramos-Delgado et al., 2013).

7.5.2 Doped TiO₂ Nanomaterial

Researchers are working towards the development of efficient photocatalytic materials and it has led to significant progress in many years, thereby generating a large number of published research papers. TiO_2 has a relatively large band gap energy (~3.2 eV) that requires UV light for photoactivation and results in a very low efficiency in utilizing solar light. In fact, UV light accounts for only about the 5% of the solar spectrum and 45% of visible light (Dong et al., 2015; Szczepanik, 2017). The published literature shown many strategies employed to overcome the drawbacks of the original photocatalyst. The main strategy is modifying the TiO_2 surface with the aim of extending the wavelength of photoactivation of TiO_2 into the visible region of the spectrum, thereby increasing the utilization of solar energy. Many reports are suggesting to alter the structure of TiO_2 by doping to enhance their photocatalytic activity. It prevents the electron/hole pair recombination and thus allowing more charge carriers to successfully diffuse to the surface and increasing the absorption affinity of TiO_2 towards organic pollutants (Mogal et al., 2013). Further, there is an advantage over preventing the aggregation and agglomeration of the nano-titania particles while easing their recovery from treated water. The metal and non-metal doping utilized with TiO_2 is used to change the physical properties band gap energy and charge carrier recombination rate. Thus, doped metal TiO_2 plays a vital role in trapping of electrons, which further helps in increasing its activity by shifting their band gap energy. Previous research also cites better efficiency of metal-doped TiO_2 nanomaterials under solar and visible light as compared to undoped TiO_2 (Connelly et al., 2012; Sathishkumar et al., 2011; Vega et al., 2018). Thus, in recent years many reviews have been published focusing on the organic pollutants removal and eliminating the original TiO_2 limitations (Dong et al., 2015; Low, Cheng, and Yu, 2017).

7.5.2.1 Metal-Doped TiO₂ Nanomaterial

In order to extend the spectral response of TiO_2 in the visible region and to enhance its photocatalytic efficiency, several strategies have been developed. Recently several studies report doping with various metals to lower the band gap of TiO_2. However, the photocatalytic activity of metal-doped TiO_2 depends, to a large extent, upon the nature of the dopant ion and its nature, method used, type of TiO_2 used, and reaction for which the catalyst is used under various reaction conditions (Kuvarega and Mamba, 2017). It is usually observed that doped TiO_2 results shifts in the absorption spectrum to longer wavelengths due to the overlap of Ti3d orbitals with the d levels of metals that favors the use of visible light to photoactivate the TiO_2 (Moma and Jeffrey, 2018). In the metal-doped TiO_2 mechanism, there is a shifting of fermi levels towards the positive value of the conduction band as compared to the TiO_2. It is observed to increase the absorption of higher wavelengths from emitted light sources (Varma et al., 2020). For this purpose, doping of TiO_2 with transition metals such as Cr (Inturi et al., 2014), Co (Inturi et al., 2014), Fe (Birben et al., 2017; Inturi et al., 2014; Sood et al., 2015), Ni (Inturi et al., 2014; Yadav et al., 2014), Mn (Inturi et al., 2014; Tripathi et al., 2015), V (Khan and Berk, 2014), Cu (Choudhury, Choudhury, and Borah 2015), Ni (Yadav et al., 2014) and Zn (Kaviyarasu et al., 2017), has been studied by different research groups. Some of these reports indicated that TiO_2 metal doping improves the photocatalytic activity for the degradation of environmental pollutants. This is mainly attributed by the change in the electronic structure resulting in the absorption region being shifted from UV to visible light. Previous work investigated the comparison of the doping of TiO_2 nanoparticles with the Fe, Cr, V, Mn, Ce, Mo, Cu, Ni, Y, and Zr and it was observed that Fe, Cr, and V showed the improved conversions in the visible region while other metal incorporated (Mo, Mn, Ce, Cu, Ni, Y, and Zr) exhibited an inhibiting effect on the photocatalytic activity (Inturi et al., 2014).

Further, noble metals such as Au, Ag, Pt, and Pd played a wider role in improving photocatalytic activity under visible light illuminated by the electron trapping and by delaying the charge carrier recombination rate (Kim, Hwang, and Choi, 2005; Li and Li, 2001; Mogal et al., 2013; Pelaez et al., 2012). Table 7.2 represents the various organic contaminants degradation by using doped materials under solar irradiation. Hence, one study investigated the photocatalytic activity of a series of noble metal Me/TiO_2 photocatalysts (Me: Pt, Ru, Pd, Rh) for the degradation of BPA under solar irradiation. The Pt/TiO_2 catalyst exhibited the higher efficiency under solar irradiation. The Rh/TiO_2 catalyst exhibited increased photoactivity for the title reaction in the presence of humic acid (20 mg/L), while Pt/TiO_2 catalyst exhibits the higher efficiency under solar light without humic acid (Repousi et al., 2017).

Other researchers (Hu et al., 2015; Wang et al., 2017) found similar findings with the Pt nanoparticle metal and improved photocatalytic performance by further increasing the electron transfer rate to the oxidant. The photocatalytic hydrogen generation was influenced by the various parameters such as platinum loading (wt.%) on TiO_2, pH solution, and light (UV, visible, and solar) intensities. Ibuprofen (25 mg/L) was degraded by the synthesized $Bi-TiO_2$ and $Ni-TiO_2$ by the metal doped nanomaterial (Bhatia and Dhir, 2016). In the case of $Bi-TiO_2$, 89% degradation of ibuprofen was used when metal doped around 0.25 wt.% at pH 6 and photocatalyst (2 g/L) after 5 h of irradiation under solar light. $Ni-TiO_2$ degradation efficiency was achieved round 78% under optimized conditions. Few reports focused on the degradation of textile dyes with $Mo-TiO_2$ and $Mg-TiO_2$ synthesized nanophotocatalysts

TABLE 7.2
Photocatalytic Degradation of Organic Pollutants using Metal Doped TiO$_2$Nanomaterial under Natural Solar Light

Doped TiO$_2$ Nanomaterial	Targeted Pollutant	Irradiation Time	Experimental Conditions	Removal Efficiency	References
WO$_3$/TiO$_2$	Malathion (12 mg/L)	160 min.	2 wt.% WO$_3$; pH: 7; catalyst dose: 1 g/L	100%	Ramos-Delgado et al. (2013)
Rh/TiO$_2$	Bisphenol A (260 µg/L)	30 min.	0.5 wt.% Rh/TiO2 ; 100 mg/L catalyst dose; 20 mg/L humic acid	100%	Repousi et al. (2017)
Bi/TiO$_2$	Ibuprofen (25 mg/L)	360 min.	0.25 wt.% Bi; pH: 6; photocatalyst: 2 g/L	89%	Bhatia and Dhir (2016)
Ni/TiO$_2$	Ibuprofen (25 mg/L)	360 min.	0.5 wt.% Ni; pH: 6; photocatalyst: 2 g/L	78%	Bhatia and Dhir (2016)
Fe/TiO$_2$	Carbendazim (8 mg/L)	300 min.	2 wt.% Fe	98%	Kaur et al. (2016)
Pd/TiO$_2$	Sulfamethoxazole (1 mg/L)	10 min.	1 wt.% Pd; 50 mg/L catalyst dose; natural pH	100%	Borowska et al. (2019)
Mo/TiO$_2$	Methylene Blue (10 ppm)	120 min.	0.9 wt.%; 1 g/L catalyst dose; pH: 7	98%	Chaudhari et al. (2018)
Pt/TiO$_2$	Sulfamethoxazole (1 mg/L)	30 min.	1 wt. % Pt; 50 mg/L catalyst dose	90%	Borowska et al. (2019)
NiO/TiO$_2$	Methylene Blue (20 ppm)	30 min.	300 W Xe-lamp; 200 mg/L catalyst dose	80%	Liu et al. (2018)
Mg/TiO$_2$	Brilliant blue G 250 (0.01M)	300 min.		64%	Shivaraju et al. (2017)
	Methyl violet (0.01 M)	300 min.	0.8 wt.%; 1 g/L catalyst dose;	100%	
	Methyl red (0.01 M)	300 min.	natural pH	80%	
	Methyl orange (0.01 M)	300 min.		79%	
	Brilliant green (0.01 M)	300 min.		73%	

(Chaudhari et al., 2018; Shivaraju et al., 2017). Thus, the methylene blue was degraded with the Mo-TiO$_2$ under the solar irradiation with experimental conditions i.e. pH 7, catalyst dose of 2 g/L, and time duration of 2 h. However, the Mg-TiO$_2$ doped catalyst investigated to remove the five different dyes i.e. brilliant blue, methyl violet, methyl red, methyl orange, and brilliant green. The removal efficiency was observed from 64% to 100% under different experimental conditions (Table 7.2). Moreover, various organic pollutants such as malathion, carbendazim, and sulfamethoxazole were also studied by using WO_3-TiO_2, Fe-TiO_2, and Pd-TiO_2 doped nanomaterials under natural sunlight (Borowska et al., 2019; Kaur et al., 2016; Ramos-Delgado et al., 2013). The degradation efficiency and various optimized parameters of the investigation are provided in the Table 7.2.

7.5.2.2 Non-Metal-Doped TiO_2 Nanomaterial

Non-metal-doped TiO_2 nanomaterials are synthesized by the various dopant materials N, C, S, F, I, B, and even graphene as nanocomposite. It is revealed from previous studies that non-metal-doped TiO_2 nanomaterials are more appropriate in shifting of photocatalytic activity into the visible region as compared to metal dopant (Jaiswal et al., 2015; Lee et al., 2018). The mechanism of non-metal doping indicates the alteration of valence band by creating at new valence band at upward direction as shown in Figure 7.3 and its application in organic pollutant degradation. Observation shows the shifting of valence band to conduction band in higher electron transfers in solar and visible light as compared to UV light. The higher surface area and lower crystalline size are achieved during non-metal doping on TiO_2 as compared to bare TiO_2 (Marschall and Wang, 2014; Tan, Wong, and Mohamed, 2011). The effective separation of electron and hole occurs that further enhances the degradation performance of organic pollutants. The photochemical process generates the superoxide ions ($O_2^{-\bullet}$) and hydroxyl radical ($HO^{\bullet}$) that are engaged in the degradation process ND converts complex compound into CO_2, H_2O and other intermediate products (Bakar, Byzynski, and Ribeiro, 2016; Zhang et al., 2012). Table 7.3 systematically presents a few examples of non-metal doping nanomaterials for organic pollutants degradation under solar irradiation. Authors investigated the nitrogen doped TiO_2 nanomaterial for the degradation of acephate and cefazolin organic pollutants. N-doped TiO_2 photocatalysts were developed by an incipient wet impregnation method and exhibited substantial photocatalytic activity under direct solar irradiation with 80% of cefazolin degradation achieved in 30min (Gurkan et al., 2012). Another study investigated the degradation of the insecticide acephate under the visible light photocatalytic activity of N-doped TiO_2nanotube arrays and observed a high degradation rate of 84% within 2 h (Zhang et al., 2015). Various researches have been performed to investigate the important role of other non-metals, i.e. boron, carbon, activated carbon, sulfur, graphene, and polypyrrole and their degradation efficiency for relevant model compounds (Table 7.3).

7.5.3 Co-Doping

Co-doping is a process when two or more dopants are added on the TiO_2 surface. However, single metal and non-metal doped TiO_2 nanomaterials exhibited excellent performance in decreasing the electrons and holes recombination, but there are limitations due to thermal stability and losing a number of dopants during laboratory synthesis process (Abdullah et al., 2017).Thus, co-doping is gaining much interest with TiO_2 electronic structure alteration done by co-doping, with further formation of new doping levels inside its band gap and suggested expansion in the TiO_2 absorption range(Abdullah et al., 2017; Jin et al., 2017). However, limited work

TABLE 7.3

Photocatalytic Degradation of Organic Pollutants using Non-Metal Doped TiO$_2$Nanomaterial under Natural Solar Light

Doped TiO$_2$Nanomaterial	Targeted Pollutant	Irradiation Time, Min.	Experimental Conditions	Removal Efficiency	References
N-TiO$_2$	Acephate (8×10^{-4} mol/L)	120	Calcined at 500 °C	84%	Zhang et al. (2015)
N-TiO$_2$	Cefazolin (1.0×10^{-2} mol/L)	30	pH:6.4; Catalyst dose: 2 g/L	80%	Gurkan et al. (2012)
C-TiO$_2$	Methylene blue (20 mg/L)	60	Neutral pH; Catalyst dose: 1 g/L	Greater than 0.1 C$_t$/C$_o$	Xiao et al. (2008)
AC-TiO$_2$	4-Chlorophenol (20 mg/L)	At 2000 min Wm^{-2}	No pH adjustment; Catalyst dose: 0.2 g/L	100%	Herrmann et al. (1999)
Polypyrrole-TiO$_2$	Methylene orange (10 mg/L)	160	No pH adjustment; 0.45 g/L	~90%	Wang et al. (2008)
Graphene-TiO$_2$	Phenol (40 ppm)	60	Neutral pH, TiO$_2$-G dose: 1.25 g/L; Solar radiation: 100 mW/cm^2	100%	Malekshoar et al. (2014)
B-TiO$_2$	Pentachlorophenol (20 ppm)	360	B amount in TiO$_2$ 188 mg L^{-1}; Visible range (400-620 nm); Intensity: 7.5 mW cm^{-2}, natural pH	57.20%	Lu et al. (2007)
S-TiO$_2$	Rhodamine B (22.5 mg L^{-1}	120	30 mW cm^{-2} (Indoor sunlight); Catalyst dose: 20mg/L; natural pH	71.60%	Zhu et al. (2015)
S-TiO$_2$	Methylene blue (20 mg L^{-1})	120	30 mW cm^{-2} (Indoor sunlight); Catalyst dose: 20 mg/L; natural pH	87.30%	Zhu et al. (2015)

is reported in the literature for the combination of either metals or non-metals. Co-doping applications for the degradation of organic pollutants has been checked by various researchers worldwide. The photo-degradation efficiency of a synthesized Mn-N-co-doped TiO$_2$ photocatalyst was evaluated for the degradation of organophosphate pesticides such as quinalphos and 2-chlorophenol under different solar spectrums. The adsorption energies of quinalphos and 2-chlorophenol were compared with the N-doped TiO$_2$ and Mn-N-co-doped TiO$_2$ photocatalysts, and they found the probable positions on the surface of photocatalysts where pesticide molecules could adsorb and degrade efficiently (Sharotri, Sharma, and Sud, 2019).

Another study also found the similarly enhanced photocatalytic findings for the degradation of rhodamine B textile dye using the Mn-N-co-doped catalyst (Quan et al., 2014). Researchers compared the laboratory prepared visible light active rhodium and antimony co-doped TiO_2 (Rh-Sb: TiO_2) and copper doped (Cu_xO/ A-Rh-Sb: TiO_2) nanorod for the degradation of bisphenol A organic pollutant. Cu_xO/A-Rh-Sb: TiO_2 nanorod photocatalytic activity was observed to be higher due to the synergistic effect of acid treatment in Rh-Sb:TiO_2 and dispersed Cu_xO nanoparticles improved the charge transfer at the conduction band of the photocatalyst (Dhandole et al., 2018). Jin et al. (2017) compared the degradation efficiency of prepared cobalt and sulfur co-doped TiO_2 (2%Co-5%S-TiO_2) for rhodamine B; they exhibited improved degradation as compared to undoped TiO_2. The observation showed the decrease in grain size with the increase in Co and S doping up to a certain range that leads to the photo-response threshold increasing significantly to about 760 nm. Recent study synthesized the hybrid nano-photocatalyst (TiO_2-S/rGO), by doping TiO_2 with sulfur and incorporating reduced graphene oxide (rGO). The TiO_2-S/rGO nanohybrid showed the higher photocatalytic activity for the degradation of organic dye methyl blue under simulated sunlight. This attribute is mainly due to the sulfur doping which is effective in narrowing the band gap of TiO_2, and the rGO addition helped to expedite the electron transfer thus making a more efficient separation of electron-hole pair (Wang et al., 2017). Another work (Park et al., 2016)cites the novel C/N-doped TiO_2 photoelectrodes preparation, properties testing, and their role under the visible light irradiation. Results showed the high surface area, active charge transfer, and decreased charge recombination as optical responses. Visible-light-driven photodegradation was carried out for rhodamine B and phenol by nitrogen and sulfur co-doped TiO_2 photocatalysts (NST). The dopant elements were observed to shift from the adsorption edge shoulder from UV to visible region. Moreover, a red shift of the absorption edge in the visible-light region is expected to be the formation of new energy levels near to the conduction bands because of the incorporation of dopant elements (N and S) into the bandgap of the TiO_2 crystal lattice (Bakar, Ribeiro, and Logo, 2016). The photodegradation of diclofenac from an aqueous medium under UV-A light conditions has been investigated by using co-doped with manganese and silver TiO_2 photocatalysts. A high surface specific area of 165 m^2/g, low band gap energy (2.7 eV) and effective charge separations were obtained for the 0.5%Ag-0.6% Mn/TiO_2 photocatalyst. It was found that TiO_2 co-doped with Mn and Ag exhibited the maximum diclofenac removal (86%) after 4 h irradiation (Tbessi et al., 2019). It can be concluded that metal dopants tend to suffer from thermal instability causing an increase in photo-induced carrier recombination centers, which decreases the lifetime of the electron-hole pairs. However, non-metal doping is more appropriate for the enhancement of TiO_2 photocatalytic activity in the visible region. This is mainly due to the presence of impure states near the valence band, localized oxygen vacancies below the conduction band that contributes to a low electron mobility in TiO_2 (Cronemeyer, 1959; Feng et al., 2013; Rumaiz et al., 2009). Thus, metal and non-metal dopants are capable of enhancing the absorption of visible light to a certain range and, more important, can be developed as an efficient photocatalyst for the degradation of organic pollutants.

7.6 CONCLUSION

Rapid industrialization has a major challenge in mitigating the water pollution and associated problems with the entire world. However, major concerns for the commonly used wastewater treatment methods are the efficiency in terms of complete removal of organics and energy consumption. Thus, advanced wastewater treatment methods conserve the water resources and are widely used for degradation of organic compounds in both aspects of pollutant removal and using solar irritations with energy efficient systems. TiO_2 photocatalysis generates the hydroxyl radicals which have high oxidizing potential during photochemical reactions and can completely mineralize into the simpler compounds with zero secondary waste generation. TiO_2 got significant attention due to its outstanding properties including strong oxidizing capability, corrosion resistance, chemical and mechanical stability, and non-toxicity. The present chapter reviewed the industrial organic pollutants generation and their complete degradation with the undoped TiO_2 and modified TiO_2 photocatalytic treatment under natural solar light. Various reports represent the modification on the surface of TiO_2 nanomaterials by pursuing metal, non-metal, and co-doped and its applications. However, it was observed that doped TiO_2 nanomaterial-based photocatalytic treatments have high potential due to their physical and chemical stability, reusability, and broad spectrum activities under solar and visible light sources. Hence, considerable research is going on to identify the more efficient nanocomposite industrial materials that have wide applications under different light sources and comparable high degradation potential against bio-recalcitrant or toxic compounds.

REFERENCES

Abdelhaleem, A. and W. Chu. 2017. "Photodegradation of 4-Chlorophenoxyacetic Acid under Visible LED Activated N-Doped TiO_2 and the Mechanism of Stepwise Rate Increment of the Reused Catalyst." *Journal of Hazardous Materials* 338:491–501.

Abdullah, H., M. M. R. Khan, H. R. Ong, and Z. Yaakob. 2017. "Modified TiO_2 Photocatalyst for CO_2 Photocatalytic Reduction: An Overview." *Journal of CO2 Utilization* 22:15–32.

Abramovic, B., D. Sojic, and V. Anderluh. 2007. "Visible-Light-Induced Photocatalytic Degradation of Herbicide Mecoprop in Aqueous Suspension of TiO_2."*Acta Chimica Slovenica* 54:558–564.

Aliste, M., G. Pérez-Lucas, N. Vela, I. Garrido, J. Fenoll, and S. Navarro. 2020. "Solar-Driven Photocatalytic Treatment as Sustainable Strategy to Remove Pesticide Residues from Leaching Water." *Environmental Science and Pollution Research* 27(7):7222–33.

Anku, W. W., M. A. Mamo, and P. P. Govender. 2017. "Phenolic Compounds in Water: Sources, Reactivity, Toxicity and Treatment Methods." In: *Phenolic Compounds: Natural Sources, Importance and Applications.* edited by M. Soto-Hernandez, M. Palma-Tenango, and M. R. Garcia-Mateos. IntechOpen, USA.

Arellano, P., K. Tansey, H. Balzter, and D. S. Boyd. 2015. "Detecting the Effects of Hydrocarbon Pollution in the Amazon Forest Using Hyperspectral Satellite Images." *Environmental Pollution* 205:225–39.

ATSDR. 2008. *Agency for Toxic Substances and Disease Registry. Toxicological Profile for Phenol.* U.S. Department of Health and Human Services, Public Health Service, Atlanta, GA.

Augustyn, A. 2020. "Surfactant." In: *Encyclopædia Britannica. T. E. o. E. Britannica, Encyclopædia Britannica, Inc.* https://www.britannica.com/science/surfactant

Bakar, S. A., G. Byzynski, and C. Ribeiro. 2016. "Synergistic Effect on the Photocatalytic Activity of N-Doped TiO2 Nanorods Synthesised by Novel Route with Exposed (110) Facet." *Journal of Alloys and Compounds* 666: 38–49.

Bakar, S. A., C. Ribeiro, and O. Logo. 2016. "Prospective Aspects of Preferential {001} Facets of N,S-Co-Doped TiO2 Photocatalysts for Visible-Light-Responsive Photocatalytic Activity." *RSC Advances* 6(92):89274–87.

Barndõk, H., D. Hermosilla, C. Han, D. D. Dionysiou, C. Negro, and Á. Blanco. 2016. "Degradation of 1,4-Dioxane from Industrial Wastewater by Solar Photocatalysis Using Immobilized NF-TiO$_2$ Composite with Monodisperse TiO$_2$ Nanoparticles." *Applied Catalysis B: Environmental* 180:44–52.

Bhatia, V. and A. Dhir. 2016. "Transition Metal Doped TiO$_2$ Mediated Photocatalytic Degradation of Anti-Inflammatory Drug under Solar Irradiations." *Journal of Environmental Chemical Engineering* 4(1):1267–73.

Birben, N. C., C. S. Uyguner-Demirel, S. S. Kavurmaci, Y. Y. Gürkan, N. Turkten, Z. Cinar, and M. Bekbolet. 2017. "Application of Fe-Doped TiO$_2$ Specimens for the Solar Photocatalytic Degradation of Humic Acid." *Catalysis Today* 281:78–84.

Boroski, M., A. C. Rodrigues, J. C. Garcia, L. C. Sampaio, J. Nozaki, and N. Hioka. 2009. "Combined Electrocoagulation and TiO$_2$ Photoassisted Treatment Applied to Wastewater Effluents from Pharmaceutical and Cosmetic Industries." *Journal of Hazardous Materials* 162(1):448–54.

Borowska, E., J. F. Gomes, R. C. Martins, R. M. Quinta-Ferreira, and H. Horn. 2019. "Solar Photocatalytic Degradation of Sulfamethoxazole by TiO$_2$ Modified with Noble Metals." *Catalysts* 9:1–19.

Bouasla, C., M. El-Hadi Samar, and F. Ismail. 2010. "Degradation of Methyl Violet 6B Dye by the Fenton Process." *Desalination* 254(1):35–41.

Brinker, C. Jeffrey. 1994. "Sol–Gel Processing of Silica." In: *The Colloid Chemistry of Silica.* Vol. 234, American Chemical Society, Washington, D.C., pp. 18–361.

Brown, D. M., B. Matthijs, R. Gill, J. Dawick, and P. J. Boogaard. 2017. "Heavy Hydrocarbon Fate and Transport in the Environment." *Quarterly Journal of Engineering Geology and Hydrogeology* 50:333–46.

Carp, O., C. L. Huisman, and A. Reller. 2004. "Photoinduced Reactivity of Titanium Dioxide." *Progress in Solid State Chemistry* 32(1):33–177.

Chang, C. N., Y. S. Ma, G. C. Fang, A. C. Chao, M. C. Tsai, and H. F. Sung. 2004. "Decolorizing of Lignin Wastewater Using the Photochemical UV/TiO$_2$ Process." *Chemosphere* 56(10):1011–17.

Chatti, R., S. S. Rayalu, N. Dubey, N. Labhsetwar, and S. Devotta. 2007. "Solar-Based Photoreduction of Methyl Orange Using Zeolite Supported Photocatalytic Materials." *Solar Energy Materials and Solar Cells* 91(2):180–90.

Chaudhari, S. M., P. M. Gawal, P. K. Sane, S. M. Sontakke, and P. R. Nemade. 2018. "Solar Light-Assisted Photocatalytic Degradation of Methylene Blue with Mo/TiO$_2$: A Comparison with Cr- and Ni-Doped TiO$_2$." *Research on Chemical Intermediates* 44(5):3115–34.

Chen, B., W. Lee, P. K. Westerhoff, S. W. Krasner, and P. Herckes. 2010. "Solar Photolysis Kinetics of Disinfection Byproducts." *Water Research* 44(11):3401–9.

Chequer, F. M. D., G. A. R. Oliveira, E. R. A. Ferraz, J. C. Cardoso, M. V. B. Zanoni, and D. P. Oliveira. 2013. "Textile Dyes: Dyeing Process and Environmental Impact." In: *Eco-Friendly Textile Dying and Finishing,* edited by Melih Günay. IntechOpen, USA.

Chong, M. N., B. Jin, C. W. K. Chow, and C. Saint. 2010. "Recent Developments in Photocatalytic Water Treatment Technology: A Review." *Water Research* 44(10):2997–3027.

Choudhury, B., A. Choudhury, and D. Borah. 2015. "Interplay of Dopants and Defects in Making Cu Doped TiO$_2$ Nanoparticle a Ferromagnetic Semiconductor." *Journal of Alloys and Compounds* 646:692–98.

Chu, L., J. Zhang, Z. Wu, C. Wang, Y. Sun, S. Dong, and J. Sun. 2020. "Solar-Driven Photocatalytic Removal of Organic Pollutants over Direct Z-Scheme Coral-Branch Shape Bi2O$_3$/SnO$_2$ Composites." *Materials Characterization* 159:110036.

Connelly, K., A. K. Wahab, and H. Idriss. 2012. "Photoreaction of Au/TiO$_2$ for Hydrogen Production from Renewables: A Review on the Synergistic Effect between Anatase and Rutile Phases of TiO$_2$." *Materials for Renewable and Sustainable Energy* 1(1):3.

Crini, G. and E. Lichtfouse. 2019. "Advantages and Disadvantages of Techniques Used for Wastewater Treatment." *Environmental Chemistry Letters* 17(1):145–55.

Cronemeyer, D. 1959. "Infrared Absorption of Reduced Rutile TiO$_2$ Single Crystals." *Physical Review* 113(5):1222–26.

Dhandole, L. K., S. G. Kim, Y. S. Seo, M. A. Mahadik, H. S. Chung, S. Y. Lee, S. H. Choi, M. Cho, J. Ryu, and J. S. Jang. 2018. "Enhanced Photocatalytic Degradation of Organic Pollutants and Inactivation of Listeria Monocytogenes by Visible Light Active Rh–Sb Codoped TiO2 Nanorods." *ACS Sustainable Chemistry & Engineering* 6(3):4302–15.

Dhir, A., M. Kamboj, and C. Ram. 2016. "Studies on the Use of Calcium Hypochlorite in the TiO2 Mediated Degradation of Pharmaceutical Wastewater." *Environmental Engineering and Management Journal* 15(8):1713–20.

Dong, Haoran, Guangming Zeng, Lin Tang, Changzheng Fan, Chang Zhang, Xiaoxiao He, and Yan He. 2015. "An Overview on Limitations of TiO$_2$-Based Particles for Photocatalytic Degradation of Organic Pollutants and the Corresponding Countermeasures." *Water Research* 79:128–46.

Durán, A. and J. M. Monteagudo. 2007. "Solar Photocatalytic Degradation of Reactive Blue 4 Using a Fresnel Lens." *Water Research* 41(3):690–98.

Elmorsi, Taha M., Yasser M. Riyad, Zeinhom H. Mohamed, and Hassan M. H. Abd El Bary. 2010. "Decolorization of Mordant Red 73 Azo Dye in Water Using H$_2$O$_2$/UV and Photo-Fenton Treatment." *Journal of Hazardous Materials* 174(1):352–58.

Faraji, M., Y. Yamini, and M. Rezaee. 2010. "Magnetic Nanoparticles: Synthesis, Stabilization, Functionalization, Characterization, and Applications." *Journal of the Iranian Chemical Society* 7(1):1–37.

Feng, N., Q. Wang, A. Zheng, Z. Zhang, J. Fan, S. B. Liu, J. P. Amoureux, and F. Deng. 2013. "Understanding the High Photocatalytic Activity of (B, Ag)-Codoped TiO$_2$ under Solar-Light Irradiation with XPS." *Journal of American Chemical Society* 135(4):1607–16.

Fujishima, Akira, Tata N. Rao, and Donald A. Tryk. 2000. "Titanium Dioxide Photocatalysis." *Journal of Photochemistry and Photobiology C: Photochemistry Reviews* 1(1):1–21.

Gandhi, V. G., M. K. Mishra, and P. A. Joshi. 2012. "A Study on Deactivation and Regeneration of Titanium Dioxide during Photocatalytic Degradation of Phthalic Acid." *Journal of Industrial and Engineering Chemistry* 18(6):1902–7.

Ghosh, M., K. Manoli, X. Shen, J. Wang, and A. K. Ray. 2019. "Solar Photocatalytic Degradation of Caffeine with Titanium Dioxide and Zinc Oxide Nanoparticles." *Journal of Photochemistry and Photobiology A: Chemistry* 377:1–7.

Gill, H. K. and H. Garg. 2014. "Pesticides: Environmental Impacts and Management Strategies." In: *Pesticides – Toxic Aspects*, edited by S. Soloneski. IntechOpen, USA.

Guillard, Chantal, Hinda Lachheb, Ammar Houas, Mohamed Ksibi, Elimame Elaloui, and Jean-Marie Herrmann. 2003. "Influence of Chemical Structure of Dyes, of PH and of Inorganic Salts on Their Photocatalytic Degradation by TiO$_2$ Comparison of the Efficiency of Powder and Supported TiO$_2$." *Journal of Photochemistry and Photobiology A: Chemistry* 158(1):27–36.

Guo, Y., P. S. Qi, and Y. Z. Liu. 2017. "A Review on Advanced Treatment of Pharmaceutical Wastewater." In: IOP Conference Series: Earth and Environmental Science, Volume 63, 2017 International Conference on Environmental and Energy Engineering (IC3E 2017). 22–24 March 2017, Suzhou, China, pp. 1–6.

Gurkan, Y. Y., N. Turkten, A. Hatipoglu, and Z. Cinar. 2012. "Photocatalytic Degradation of Cefazolin over N-Doped TiO_2 under UV and Sunlight Irradiation: Prediction of the Reaction Paths via Conceptual DFT." *Chemical Engineering Journal* 184:113–24.

Hao, Q., Y. Liu, T. Chen, G. Qingfeng, W. Wei, and B. J. Ni. 2019. "Bi_2O_3@Carbon Nanocomposites for Solar-Driven Photocatalytic Degradation of Chlorophenols." *ACS Applicaiton Nano Material* 2:2308–16.

Herrmann, J. M. 1999. "Heterogeneous Photocatalysis: Fundamentals and Applications to the Removal of Various Types of Aqueous Pollutants." *Catalysis Today* 53:115–29.

Herrmann, J. M., J. Disdier, P. Pichat, S. Malato, and J. Blanco. 1998. "TiO_2-Based Solar Photocatalytic Detoxification of Water Containing Organic Pollutants. Case Studies of 2,4dichlorophenoxyaceticacid (2,4-D) and of Benzofuran." *Applied Catalysis: B* 17:15–23.

Herrmann, J. M., J. Matos, J. Disdier, C. Guillard, J. Laine, S. Malato, and J. Blanco. 1999. "Solar Photocatalytic Degradation of 4-Chlorophenol Using the Synergistic Effect between Titania and Activated Carbon in Aqueous Suspension." *Catalysis Today* 54(2):255–65.

Heydari, G., C. H. Langford, and G. Achari. 2019. "Passive Solar Photocatalytic Treatment of Emerging Contaminants in Water: A Field Study." *Catalysts* 9:1045–59.

Hinojosa -Reyes, M., R. Camposeco -Solis, Facundo Ruiz, V. Rodríguez- González, and E. Moctezuma. 2019. "Promotional Effect of Metal Doping on Nanostructured TiO_2 during the Photocatalytic Degradation of 4-Chlorophenol and Naproxen Sodium as Pollutants." *Materials Science in Semiconductor Processing* 100:130–39.

Hoffmann, M. R., S. T. Martin, W. Choi, and D. W. Bahnemann. 1995. "Environmental Applications of Semiconductor Photocatalysis." *Chemical Reviews* 95(1):69–96.

Hojamberdiev, Mirabbos, Ravi Mohan Prasad, Koji Morita, Marco Antônio Schiavon, and Ralf Riedel. 2012. "Polymer-Derived Mesoporous SiOC/ZnO Nanocomposite for the Purification of Water Contaminated with Organic Dyes." *Microporous and Mesoporous Materials* 151:330–38.

Hope, S. 2019. *The Effects of Ultraviolet Filters and Sunscreen on Corals and Aquatic Ecosystems: Bibliography*. NOAA Central Library, Silver Spring, MD.

Hu, Y., X. Song, S. Jiang, and C. Wei. 2015. "Enhanced Photocatalytic Activity of Pt-Doped TiO2 for NOx Oxidation Both under UV and Visible Light Irradiation: A Synergistic Effect of Lattice Pt4+ and Surface PtO." *Chemical Engineering Journal* 274:102–12.

Iliev, V., D. Tomova, S. Rakovsky, A. Eliyas, and G. L. Puma. 2010. "Enhancement of Photocatalytic Oxidation of Oxalic Acid by Gold Modified WO_3/TiO_2 Photocatalysts under UV and Visible Light Irradiation." *Journal of Molecular Catalysis A: Chemical* 327(1):51–57.

Inturi, S. N. R., T. Boningari, M. Suidan, and P. G. Smirniotis. 2014. "Visible-Light-Induced Photodegradation of Gas Phase Acetonitrile Using Aerosol-Made Transition Metal (V, Cr, Fe, Co, Mn, Mo, Ni, Cu, Y, Ce, and Zr) Doped TiO_2." *Applied Catalysis B: Environmental* 144:333–42.

Ismail, Adel A., Tarek A. Kandiel, and Detlef W. Bahnemann. 2010. "Novel (and Better?) Titania-Based Photocatalysts: Brookite Nanorods and Mesoporous Structures." *Journal of Photochemistry and Photobiology A: Chemistry* 216(2):183–93.

Jaiswal, R., J. Bharambe, N. Patel, A. Dashora, D. C. Kothari, and A. Miotello. 2015. "Copper and Nitrogen Co-Doped TiO_2 Photocatalyst with Enhanced Optical Absorption and Catalytic Activity." *Applied Catalysis B: Environmental* 168–169:333–41.

Ji, Fei, Chaolin Li, Jiahuan Zhang, and Lei Deng. 2011. "Heterogeneous Photo-Fenton Decolorization of Methylene Blue over $LiFe(WO_4)_2$ Catalyst." *Journal of Hazardous Materials* 186(2):1979–84.

Jiao, S., S. Zheng, D. Yin, L. Wang, and L. Chen. 2008. "Aqueous Photolysis of Tetracycline and Toxicity of Photolytic Products to Luminescent Bacteria." *Chemosphere* 73(3):377–82.

Jin, Q., C. Nie, Q. Shen, Y. Xu, and Y. Nie. 2017. "Cobalt and Sulfur Co-Doped TiO_2 Nanostructures with Enhanced Photo-Response Properties for Photocatalyst." *Functional Materials Letters* 10(05):1750061.

Kabra, K., R. Chaudhary, and R. L. Sawhney. 2004. "Treatment of Hazardous Organic and Inorganic Compounds through Aqueous-Phase Photocatalysis: A Review." *Industrial & Engineering Chemistry Research* 43(24):7683–96.

Kaneco, S., N. Li, K. -k. Itoh, H. Katsumata, T. Suzuki, and K. Ohta. 2009. "Titanium Dioxide Mediated Solar Photocatalytic Degradation of Thiram in Aqueous Solution: Kinetics and Mineralization." *Chemical Engineering Journal* 148(1):50–56.

Kaur, S., S. Sharma, A. Umar, S. Singh, S. K. Mehta, and S. K. Kansal. 2017. "Solar Light Driven Enhanced Photocatalytic Degradation of Brilliant Green Dye Based on ZnS Quantum Dots." *Superlattices and Microstructures* 103:365–75.

Kaur, T., A. Sraw, A. P. Toor, and R. K. Wanchoo. 2016. "Utilization of Solar Energy for the Degradation of Carbendazim and Propiconazole by Fe Doped TiO_2." *Solar Energy* 125:65–76.

Kaviyarasu, K., N. Geetha, K. Kanimozhi, C. M. Magdalane, S. Sivaranjani, A. Ayeshamariam, J. Kennedy, and M. Maaza. 2017. "In Vitro Cytotoxicity Effect and Antibacterial Performance of Human Lung Epithelial Cells A549 Activity of Zinc Oxide Doped TiO_2 Nanocrystals: Investigation of Bio-Medical Application by Chemical Method." *Materials Science and Engineering: C* 74:325–33.

Khan, H. and D. Berk. 2014. "Synthesis, Physicochemical Properties and Visible Light Photocatalytic Studies of Molybdenum, Iron and Vanadium Doped Titanium Dioxide." *Reaction Kinetics, Mechanisms and Catalysis* 111(1):393–414.

Khan, S. and A. Malik. 2014. "Environmental and Health Effects of Textile Industry Wastewater." In: *Environmental Deterioration and Human Health: Natural and Anthropogenic Determinants*, edited by A. Malik, E. Grohmann. Springer, Dordrecht, pp. 55–71.

Kim, S., S. J. Hwang, and W. Choi. 2005. "Visible Light Active Platinum-Ion-Doped TiO2 Photocatalyst." *The Journal of Physical Chemistry B* 109:24260–67.

Koe, W. S., J. W. Lee, W. C. Chong, Y. L. Pang, and L. C. Sim. 2020. "An Overview of Photocatalytic Degradation: Photocatalysts, Mechanisms, and Development of Photocatalytic Membrane." *Environmental Science and Pollution Research* 27(3):2522–65.

Kumar, P., S. Kumar, N. K. Bhardwaj, and C. Ram. 2015. "Removal of Chlorinated Resin and Fatty Acids from Paper Mill Wastewater through Photocatalysis." In: *Proceedings of International Multi-Track Conference on Sciences, Engineering and Technical Innovations (IMCT 2015)*. May 22–23, Jalandhar, India, pp. 260–64.

Kumar, S. G. and L. G. Devi. 2011. "Review on Modified TiO_2 Photocatalysis under UV/Visible Light: Selected Results and Related Mechanisms on Interfacial Charge Carrier Transfer Dynamics." *The Journal of Physical Chemistry A* 115(46):13211–41.

Kushniarou, A., I. Garrido, J. Fenoll, N. Vela, P. Flores, G. Navarro, P. Hellín, and S. Navarro. 2019. "Solar Photocatalytic Reclamation of Agro-Waste Water Polluted with Twelve Pesticides for Agricultural Reuse." *Chemosphere* 214:839–45.

Kuvarega, A. T. and B. B. Mamba. 2017. "TiO2 -Based Photocatalysis: Toward Visible Light-Responsive Photocatalysts through Doping and Fabrication of Carbon-Based Nanocomposites." *Critical Reviews in Solid State and Materials Sciences* 42(4):295–346.

Lazar, M. A., S. Varghese, and S. S. Nair. 2012. "Photocatalytic Water Treatment by Titanium Dioxide: Recent Updates." *Catalysts* 2:572–601.

Lee, H. C., H. S. Park, S. K. Cho, K. M. Nam, and A. J. Bard. 2018. "Direct Photoelectrochemical Characterization of Photocatalytic H, N Doped TiO_2 Powder Suspensions." *Journal of Electroanalytical Chemistry* 819:38–45.

Lee, K. M., C. W. Lai, K. S. Ngai, and J. C. Juan. 2016. "Recent Developments of Zinc Oxide Based Photocatalyst in Water Treatment Technology: A Review." *Water Research* 88:428–48.

Lellis, B., C. Z. Fávaro-Polonio, J. A. Pamphile, and J. C. Polonio. 2019. "Effects of Textile Dyes on Health and the Environment and Bioremediation Potential of Living Organisms." *Biotechnology Research and Innovation* 3(2):275–90.

Lewis, M. A. 1991. "Chronic and Sublethal Toxicities of Surfactants to Aquatic Animals: A Review and Risk Assessment." *Water Research* 25(1):101–13.

Li, X. Z. and F. B. Li. 2001. "Study of Au/Au_3+ -TiO_2 Photocatalysts toward Visible Photooxidation for Water and Wastewater Treatment." *Environmental Science Technology* 35:2381–87.

Li, X. Z., F. B. Li, C. L. Yang, and W. K. Ge. 2001. "Photocatalytic Activity of WOx-TiO_2 under Visible Light Irradiation." *Journal of Photochemistry and Photobiology A: Chemistry* 141(2):209–17.

Liu, J., Y. Li, J. Ke, S. Wang, L. Wang, and H. Xiao. 2018. "Black NiO-TiO_2 Nanorods for Solar Photocatalysis: Recognition of Electronic Structure and Reaction Mechanism." *Applied Catalysis B: Environmental* 224:705–14.

Liu, Jianjun, Xinping Li, Shengli Zuo, and Yingchun Yu. 2007. "Preparation and Photocatalytic Activity of Silver and TiO_2 Nanoparticles/Montmorillonite Composites." *Applied Clay Science* 37(3):275–80.

Low, J., B. Cheng, and J. Yu. 2017. "Surface Modification and Enhanced Photocatalytic CO2 Reduction Performance of TiO_2: A Review." *Applied Surface Science* 392:658–86.

Lu, N., X. Quan, J. Y. Li, S. Chen, H. T. Yu, and G. H. Chen. 2007. "Fabrication of Boron-Doped TiO_2 Nanotube Array Electrode and Investigation of Its Photoelectrochemical Capability." *The Journal of Physical Chemistry C* 111(32):11836–42.

Ma, L., L. Hu, X. Feng, and S. Wang. 2018. "Nitrate and Nitrite in Health and Disease." *Aging and Disease* 9(5): 938–45.

Madhu, C., Manjunath B. Bellakki, and V. Manivannan. 2010. "Synthesis and Characterization of Cation-Doped $BiFeO_3$ Materials for Photocatalytic Applications." *Indian Journal of Engineering & Material Sciences* 17:131–39.

Maktabifard, M., E. Zaborowska, and J. Makinia. 2018. "Achieving Energy Neutrality in Wastewater Treatment Plants through Energy Savings and Enhancing Renewable Energy Production." *Reviews in Environmental Science and Bio/Technology* 17(4):655–89.

Malaguarnera, G., E. Cataudella, M. Giordano, G. Nunnari, G. Chisari, and M. Malaguarnera. 2012. "Toxic Hepatitis in Occupational Exposure to Solvents." *World Journal of Gastroenterology* 18 (22). Baishideng Publishing Group Co., Limited: 2756–66.

Malekshoar, G., K. Pal, Q. He, A. Yu, and A. K. Ray. 2014. "Enhanced Solar Photocatalytic Degradation of Phenol with Coupled Graphene-Based Titanium Dioxide and Zinc Oxide." *Industrial & Engineering Chemistry Research* 53(49):18824–32.

Marschall, R. and L. Wang. 2014. "Non-Metal Doping of Transition Metal Oxides for Visible-Light Photocatalysis." *Catalysis Today* 225:111–35.

Matsumura, Y. and H. N. Ananthaswamy. 2004. "Toxic Effects of Ultraviolet Radiation on the Skin." *Toxicology and Applied Pharmacology* 195(3):298–308.

McCullagh, C., J. M. C. Robertson, D. W. Bahnemann, and P. K. J. Robertson. 2007. "The Application of TiO_2 Photocatalysis for Disinfection of Water Contaminated with Pathogenic Micro-Organisms: A Review." *Research on Chemical Intermediates* 33(3):359–75.

Merouani, Slimane, Oualid Hamdaoui, Fethi Saoudi, and Mahdi Chiha. 2010. "Sonochemical Degradation of Rhodamine B in Aqueous Phase: Effects of Additives." *Chemical Engineering Journal* 158(3):550–57.

Miao, Shiding, Zhimin Liu, Buxing Han, Haowen Yang, Zhenjiang Miao, and Zhenyu Sun. 2006. "Synthesis and Characterization of ZnS-Montmorillonite Nanocomposites and Their Application for Degrading Eosin B."*Journal of Colloid and Interface Science* 301(1):116–22.

Mogal, S. I., M. Mishra, V. G. Gandhi, and R. J. Tayade. 2013. "Metal Doped Titanium Dioxide: Synthesis and Effect of Metal Ions on Physico-Chemical and Photocatalytic Properties." *Materials Science Forum* 734:364–78.

Molla, M. A. I., I. Tateishi, M. Furukawa, H. Katsumata, T. Suzuki, and S. Kaneco. 2017. "Photocatalytic Decolorization of Dye with Self-Dye-Sensitization under Fluorescent Light Irradiation." *ChemEngineering* 1(2):1–8.

Moma, J. and B. Jeffrey. 2018. "Modified Titanium Dioxide for Photocatalytic Applications." In: *Photocatalysts Applications and Attributes*, edited by S. B. Khan and K. Akhtar. IntechOpen, USA.

Monteagudo, J. M., A. Durán, I. San Martín, and M. Aguirre. 2010. "Catalytic Degradation of Orange II in a Ferrioxalate-Assisted Photo-Fenton Process Using a Combined UV-A/C–Solar Pilot-Plant System." *Applied Catalysis B: Environmental* 95(1):120–29.

Mostafalou, S. and M. Abdollahi. 2013. "Pesticides and Human Chronic Diseases: Evidences, Mechanisms, and Perspectives." *Toxicology and Applied Pharmacology* 268(2):157–77.

Nikolaou, A., S. Meric, and D. Fatta. 2007. "Occurrence Patterns of Pharmaceuticals in Water and Wastewater Environments." *Analytical and Bioanalytical Chemistry* 387(4):1225–34.

Olama, N., M. Dehghani, and M. Malakootian. 2018. "The Removal of Amoxicillin from Aquatic Solutions Using the TiO_2/UV-C Nanophotocatalytic Method Doped with Trivalent Iron." *Applied Water Science* 8(4):97.

Parangi, T. and M. K. Mishra. 2019. "Titania Nanoparticles as Modified Photocatalysts: A Review on Design and Development." *Comments on Inorganic Chemistry* 39(2):90–126.

Park, J. Y., C. S. Kim, K. Okuyama, H. M. Lee, H. D. Jang, S. E. Lee, and T. O. Kim. 2016. "Copper and Nitrogen Doping on TiO_2 Photoelectrodes and Their Functions in Dye-Sensitized Solar Cells." *Journal of Power Sources* 306:764–71.

Pelaez, M., N. T. Nolan, S. C. Pillai, M. K. Seery, P. Falaras, A. G. Kontos, P. S. M. Dunlop, J. W. J. Hamilton, J. A. Byrne, K. O'Shea, M. H. Entezari, and D. D. Dionysiou. 2012. "A Review on the Visible Light Active Titanium Dioxide Photocatalysts for Environmental Applications." *Applied Catalysis B* 125:331–49.

Perkowski, J., S. Bzdon, A. Bulska, and W. K. Jóźwiak. 2006. "Decomposition of Detergents Present in Car-Wash Sewage by Titania Photo-Assisted Oxidation." *Polish Journal of Environmental Studies* 15(3):457–65.

Pirkanniemi, K. and M. Sillanpää. 2002. "Heterogeneous Water Phase Catalysis as an Environmental Application: A Review." *Chemosphere* 48(10):1047–60.

Qiu, Jianxun, Haigen Shen, and Mingyuan Gu. 2005. "Microwave Absorption of Nanosized Barium Ferrite Particles Prepared Using High-Energy Ball Milling." *Powder Technology* 154(2):116–19.

Quan, F., Y. Hu, X. Zhang, and C. Wei. 2014. "Simple Preparation of Mn-N-Codoped TiO_2 Photocatalyst and the Enhanced Photocatalytic Activity under Visible Light Irradiation." *Applied Surface Science* 320:120–27.

Radhika, S. and J. Thomas. 2017. "Solar Light Driven Photocatalytic Degradation of Organic Pollutants Using ZnO Nanorods Coupled with Photosensitive Molecules." *Journal of Environmental Chemical Engineering* 5(5):4239–50.

Rajaraman, T. S., S. P. Parikh, and V. G. Gandhi. 2020. "Black TiO_2: A Review of Its Properties and Conflicting Trends." *Chemical Engineering Journal* 389:123918.

Ram, C., R. K. Pareek, and V. Singh. 2012. "Photocatalytic Degradation of Textile Dye by Using Titanium Dioxide Nanocatalyst." *International Journal Of Theoretical & Applied Sciences* 4 (2): 82–88.

Ram, C., P. Rani, K. A. Gebru, and M. G. Abrha. 2020. "Pulpand Paper Industry Wastewater Treatment: Use of Microbes and Their Enzymes." *Physical Sciences Reviews* 1–13.

Ramirez, A. M., K. Demeestere, N. B. De, T. Mäntylä, and E. Levänen. 2010. "Titanium Dioxide Coated Cementitious Materials for Air Purifying Purposes: Preparation, Characterization and Toluene Removal Potential." *Building and Environment* 45(4):832–38.

Ramos-Delgado, N. A., L. Hinojosa-Reyes, I. L. Guzman-Mar, M. A. Gracia-Pinilla, and A. Hernández-Ramírez. 2013. "Synthesis by Sol–Gel of WO_3/TiO_2 for Solar Photocatalytic Degradation of Malathion Pesticide." *Catalysis Today* 209:35–40.

Rani, M. and U. Shanker. 2018. "Insight in to the Degradation of Bisphenol A by Doped ZnO@ZnHCF Nanocubes: High Photocatalytic Performance." *Journal of Colloid and Interface Science* 530:16–28.

Rasoulifard, Mohammad Hossein, Mostafa Fazli, and Mohammad Reza Eskandarian. 2015. "Performance of the Light-Emitting-Diodes in a Continuous Photoreactor for Degradation of Direct Red 23 Using $UV-LED/S_2O_8^{2-}$ Process." *Journal of Industrial and Engineering Chemistry* 24:121–26.

Rauf, M. A., M. A. Meetani, and S. Hisaindee. 2011. "An Overview on the Photocatalytic Degradation of Azo Dyes in the Presence of TiO2 Doped with Selective Transition Metals." *Desalination* 276(1):13–27.

Rauf, Muhammad A., Mohammed A. Meetani, A. Khaleel, and Amal Ahmed. 2010. "Photocatalytic Degradation of Methylene Blue Using a Mixed Catalyst and Product Analysis by LC/MS." *Chemical Engineering Journal* 157(2):373–78.

Repousi, V., A. Petala, Z. Frontistis, M. Antonopoulou, I. Konstantinou, D. I. Kondarides, and D. Mantzavinos. 2017. "Photocatalytic Degradation of Bisphenol A over Rh/TiO_2 Suspensions in Different Water Matrices." *Catalysis Today* 284:59–66.

Rumaiz, A. K., J. C. Woicik, E. Cockayne, H. Y. Lin, G. H. Jaffari, and S. I. Shah. 2009. "Oxygen Vacancies in N Doped Anatase TiO_2: Experiment and First-Principles Calculations." *Applied Physics Letters* 95:262111.

Sajjad, Ahmed Khan Leghari, Sajjad Shamaila, Baozhu Tian, Feng Chen, and Jinlong Zhang. 2010. "Comparative Studies of Operational Parameters of Degradation of Azo Dyes in Visible Light by Highly Efficient WOx/TiO_2 Photocatalyst." *Journal of Hazardous Materials* 177(1):781–91.

Salgado-Tránsito, I., A. E. Jiménez-González, M. L. Ramón-García, C. A. Pineda-Arellano, and C. A. Estrada-Gasca. 2015. "Design of a Novel CPC Collector for the Photodegradation of Carbaryl Pesticides as a Function of the Solar Concentration Ratio." *Solar Energy* 115:537–51.

Sandhya, S. 2010. "Biodegradation of Azo Dyes under Anaerobic Condition: Role of Azoreductase." In: *The Handbook of Environmental Chemistry*, Vol. 9, edited by H.A. Erkurt.Springer, Berlin, Heidelberg, pp. 39–57.

Santra, Swadeshmukul, Rovelyn Tapec, Nikoleta Theodoropoulou, Jon Dobson, Arthur Hebard, and Weihong Tan. 2001. "Synthesis and Characterization of Silica-Coated Iron Oxide Nanoparticles in Microemulsion: The Effect of Nonionic Surfactants." *Langmuir* 17(10):2900–2906.

Sathishkumar, P., S. Anandan, P. Maruthamuthu, T. Swaminathan, M. Zhou, and M. Ashokkumar. 2011. "Synthesis of Fe3+ Doped TiO2 Photocatalysts for the Visible Assisted Degradation of an Azo Dye." *Colloids and Surfaces A: Physicochemical and Engineering Aspects* 375(1):231–36.

Scott, C. A., S. A. Pierce, M. J. Pasqualetti, A. L. Jones, B. E. Montz, and J. H. Hoover. 2011. "Policy and Institutional Dimensions of the Water–Energy Nexus." *Energy Policy* 39(10):6622–30.

Scott, M. J. and M. N. Jones. 2000. "The Biodegradation of Surfactants in the Environment." *Biochimica et Biophysica Acta (BBA) – Biomembranes* 1508(1):235–51.

Serpone, Nick and Angela Salinaro. 1999. "Terminology, Relative Photonic Efficiencies and Quantum Yields in Heterogeneous Photocatalysis. Part I: Suggested Protocol." *Pure and Applied Chemistry* 71(2):303–20.

Shah, K. J. and P. C. Chang. 2018. "Shape-Control Synthesis and Photocatalytic Applications of CeO_2 to Remediate Organic Pollutant Containing Wastewater: A Review." In: *Photocatalytic Nanomaterials for Environmental Applications*, edited by V. Tayade, R. J. Gandhi. Materials Research Forum LLC, USA.

Sharotri, N., D. Sharma, and D. Sud. 2019. "Experimental and Theoretical Investigations of Mn-N-Co-Doped TiO_2 Photocatalyst for Visible Light Induced Degradation of Organic Pollutants." *Journal of Materials Research and Technology* 8(5):3995–4009.

Shivaraju, H. P., G. Midhun, K. M. Anil Kumar, S. Pallavi, N. Pallavi, and S. Behzad. 2017. "Degradation of Selected Industrial Dyes Using Mg-Doped TiO_2 Polyscales under Natural Sun Light as an Alternative Driving Energy." *Applied Water Science* 7(7):3937–48.

Shu, Huoming, Jimin Xie, Hui Xu, Huaming Li, Zheng Gu, Guangsong Sun, and Yuanguo Xu. 2010. "Structural Characterization and Photocatalytic Activity of $NiO/AgNbO_3$." *Journal of Alloys and Compounds* 496(1):633–37.

Sood, S., A. Umar, S. K. Mehta, and S. K. Kansal. 2015. "Highly Effective Fe-Doped TiO_2 Nanoparticles Photocatalysts for Visible-Light Driven Photocatalytic Degradation of Toxic Organic Compounds." *Journal of Colloid and Interface Science* 450:213–23.

Srinivasan, S. S., J. Wade, and E. K. Stefanakos. 2006. "Visible Light Photocatalysis via CdS/ TiO2 Nanocomposite Materials." *Journal of Nanomaterials* 1–7.https://doi.org/10.1155/ JNM/2006/87326.

Szczepanik, B. 2017. "Photocatalytic Degradation of Organic Contaminants over Clay-TiO2 Nanocomposites: A Review." *Applied Clay Science* 141:227–39.

Tan, Y. N., C. L. Wong, and A. R. Mohamed. 2011. "An Overview on the Photocatalytic Activity of Nano-Doped-TiO2 in the Degradation of Organic Pollutants." Edited by P. Sánchez and D. Chicot. *ISRN Materials Science* 2011:261219.

Tayade, R. J., R. G. Kulkarni, and R. V. Jasra. 2006. "Photocatalytic Degradation of Aqueous Nitrobenzene by Nanocrystalline TiO_2." *Industrial & Engineering Chemistry Research* 45(3):922–27.

Tbessi, I., M. Benito, J. Llorca, E. Molins, S. Sayadi, and W. Najjar. 2019. "Silver and Manganese Co-Doped Titanium Oxide Aerogel for Effective Diclofenac Degradation under UV-A Light Irradiation." *Journal of Alloys and Compounds* 779:314–25.

Tehrani-Bagha, A. R., N. M. Mahmoodi, and F. M. Menger. 2010. "Degradation of a Persistent Organic Dye from Colored Textile Wastewater by Ozonation." *Desalination* 260(1):34–38.

Touati, A., T. Hammedi, W. Najjar, Z. Ksibi, and S. Sayadi. 2016. "Photocatalytic Degradation of Textile Wastewater in Presence of Hydrogen Peroxide: Effect of Cerium Doping Titania." *Journal of Industrial and Engineering Chemistry* 35:36–44.

Tripathi, A. K., M. C. Mathpal, P. Kumar, M. K. Singh, M. A. G. Soler, and A. Agarwal. 2015. "Structural, Optical and Photoconductivity of Sn and Mn Doped TiO2 Nanoparticles." *Journal of Alloys and Compounds* 622:37–47.

Varma, K. S., R. J. Tayade, K. J. Shah, P. A. Joshi, A. D. Shukla, and V. G. Gandhi. 2020. "Photocatalytic Degradation of Pharmaceutical and Pesticide Compounds (PPCs) Using Doped TiO_2 Nanomaterials: A Review." *Water-Energy Nexus* 3:46–61.

Vega, M. P. B., M. Hinojosa-Reyes, A. Hernández-Ramírez, J. L. G. Mar, V. Rodríguez-González, and L. Hinojosa-Reyes. 2018. "Visible Light Photocatalytic Activity of Sol–Gel Ni-Doped TiO_2 on p-Arsanilic Acid Degradation." *Journal of Sol-Gel Science and Technology* 85(3):723–31.

Wang, D., Y. Wang, X. Li, Q. Luo, J. An, and J. Yue. 2008. "Sunlight Photocatalytic Activity of Polypyrrole–TiO_2 Nanocomposites Prepared by 'in Situ' Method." *Catalysis Communications* 9(6):1162–66.

Wang, F., R. J. Wong, J. H. Ho, Y. Jiang, and R. Amal. 2017. "Sensitization of Pt/TiO_2 Using Plasmonic Au Nanoparticles for Hydrogen Evolution under Visible-Light Irradiation." *ACS Applied Materials & Interfaces* 9(36):30575–82.

Wang, Jingying, Fenglian Ren, Ran Yi, Aiguo Yan, Guanzhou Qiu, and Xiaohe Liu. 2009. "Solvothermal Synthesis and Magnetic Properties of Size-Controlled Nickel Ferrite Nanoparticles." *Journal of Alloys and Compounds* 479(1):791–96.

Wang, W., Z. Wang, J. Liu, Z. Luo, S. L. Suib, P. He, G. Ding, Z. Zhang, and L. Sun. 2017. "Single-Step One-Pot Synthesis of TiO_2 Nanosheets Doped with Sulfur on Reduced Graphene Oxide with Enhanced Photocatalytic Activity." *Scientific Reports (Nature. Com)* 7(1):1–9.

Wang, Y., Y. Huang, W. Ho, L. Zhang, Z. Zou, and S. Lee. 2009. "Biomolecule-Controlled Hydrothermal Synthesis of C–N–S-Tridoped TiO_2 Nanocrystalline Photocatalysts for NO Removal under Simulated Solar Light Irradiation." *Journal of Hazardous Materials* 169(1):77–87.

Ward, M.H., R.R. Jones, J.D. Brender, T.M. de Kok, P.J. Weyer, B.T. Nolan, C.M. Villanueva, and S.G. van Breda. 2018. "Drinking Water Nitrate and Human Health: An Updated Review." *International Journal of Environmental Research and Public Health* 15 (7). MDPI:1557.

Wittlich, M., S. Westerhausen, P. Kleinespel, G. Rifer, and W. Stöppelmann. 2016. "An Approximation of Occupational Lifetime UVR Exposure: Algorithm for Retrospective Assessment and Current Measurements." *Journal of the European Academy of Dermatology and Venereology* 30(S3):27–33.

Xiao, Q., J. Zhang, C. Xiao, Z. Si, and X. Tan. 2008. "Solar Photocatalytic Degradation of Methylene Blue in Carbon-Doped TiO2 Nanoparticles Suspension." *Solar Energy* 82(8):706–13.

Xiong, J., G. Cheng, F. Qin, R. Wang, H. Sun, and R. Chen. 2013. "Tunable BiOCl Hierarchical Nanostructures for High-Efficient Photocatalysis under Visible Light Irradiation." *Chemical Engineering Journal* 220:228–36.

Yadav, H. M., S. V. Otari, R. A. Bohara, S. S. Mali, S. H. Pawar, and S. D. Delekar. 2014. "Synthesis and Visible Light Photocatalytic Antibacterial Activity of Nickel-Doped TiO2 Nanoparticles against Gram-Positive and Gram-Negative Bacteria." *Journal of Photochemistry and Photobiology A: Chemistry* 294:130–36.

Yi, C., Q. Liao, W. Deng, Y. Huang, J. Mao, B. Zhang, and G. Wu. 2019. "The Preparation of Amorphous TiO_2 Doped with Cationic S and Its Application to the Degradation of DCFs under Visible Light Irradiation." *Science of the Total Environment* 684:527–36.

Ying, G. G. 2006. "Fate, Behavior and Effects of Surfactants and Their Degradation Products in the Environment." *Environment International* 32(3):417–31.

Yuan, C., Z. Z. Xu, M. X. Fan, H. Y. Liu, Y. Xie, and T. Zhu. 2013. "Study on Characteristics and Harm of Surfactants." *Journal of Chemical and Pharmaceutical Research* 6(7):2233–37.

Yuan, Jin, Yong-Kang Lü, Yu Li, and Jun-Ping Li. 2010. "Synthesis and Characterization of Magnetic TiO_2/SiO_2/$NiFe_2O_4$ Composite Photocatalysts." *Chemical Research in Chinese Universities* 26(2):278–82.

Zhang, Liwu, Hongbo Fu, Chuan Zhang, and Yongfa Zhu. 2006. "Synthesis, Characterization, and Photocatalytic Properties of InVO4 Nanoparticles." *Journal of Solid State Chemistry* 179(3):804–11.

Zhang, T., X. Wang, and X. Zhang. 2014. "Recent Progress in TiO_2-Mediated Solar Photocatalysis for Industrial Wastewater Treatment." Edited by Q. Wang. *International Journal of Photoenergy* 2014: 1–12.

Zhang, X., J. Zhou, Y. Gu, and D. Fan. 2015. "Visible-Light Photocatalytic Activity of N-Doped TiO2 Nanotube Arrays on Acephate Degradation." Edited by S. Zhou. *Journal of Nanomaterials* 2015:1–6.

Zhang, Y., P. Zhang, Y. Huo, D. Zhang, G. Li, and H. Li. 2012. "Ethanol Supercritical Route for Fabricating Bimodal Carbon Modified Mesoporous TiO$_2$ with Enhanced Photocatalytic Capability in Degrading Phenol." *Applied Catalysis B: Environmental* 115–116:236–44.

Zhu, M., C. Zhai, L. Qiu, C. Lu, A. S. Paton, Y. Du, and M. C. Goh. 2015. "New Method to Synthesize S-Doped TiO$_2$ with Stable and Highly Efficient Photocatalytic Performance under Indoor Sunlight Irradiation." *ACS Sustainable Chemistry & Engineering* 3(12):3123–29.

8 Recent Advancement in Phytoremediation for Removal of Toxic Compounds

Yilkal Bezie
Bahir Dar University
Debre Markos University

Mengistie Taye
Bahir Dar University

Amit Kumar
Debre Markos University

CONTENTS

8.1 INTRODUCTION

The environmental pollution is the usual consequence of people's search for a better life and economic development through rapid industrialization (Adu and Denkyirah, 2017; Paul *et al.*, 2014), expansion of urbanization, rapidly growing human population, mining, and agricultural development. It is the contamination of the environment with pollutants above the permissible limit. Industries have mainly contributed to the generation of these pollutants that cause objectionable effects, impairing the welfare of the environment, reducing the quality of life and leading to death (Bekele, 2018). A pollutant is anything that exists in the environment more than its normal concentration. Waste produced by anthropogenic activity is widely diverse, which is difficult to classify them adequately. Pollutants that create nuisance in the environment are usually industrial wastes, municipal solid wastes, agricultural runoffs, leachates (organic contaminants), and radioactive wastes (Bauddh et al., 2017). The organic pollutants, heavy metals, and radioactive wastes are the potential risks inducing contaminants (Qiao et al., 2020). They can cause multiple issues directly affecting plants as well as animals including human beings, sometimes by changing the ecological balance (Figure 8.1) (Chirakkara et al., 2016).

Pollutants are different groups that include biodegradables, xenobiotics, toxic organic pollutants, and heavy metals which can be great health and environmental burdens (Bauddh et al., 2015; Rissato et al., 2015; Weyens et al., 2009; Worku et al., 2018). The biodegradable pollutants are not a problem since they can be cleared by microbial degradation. However, the removal of heavy metals from the environment is found to be difficult. Different activities such as mining, agriculture, metallurgy, combustion of fossil fuels, faulty waste disposal, and military operations have released an enormous amount of toxic heavy metals and metalloids into the environment (Kotrba et al., 2009). Heavy metals are contaminants that pose a great environmental burden as they are hazardous to humans, animals, plant health, and the environment at large (Francis, 2017). Due to their elemental character, heavy metals cannot be chemically degraded, and their detoxification in the environment mostly exists either in stabilization *in situ* or their removal from the matrix. For this purpose, the application of plants for the restoration of the polluted environment has been developed as a promising green alternative to traditional physical and chemical methods (Saxena et al., 2020).

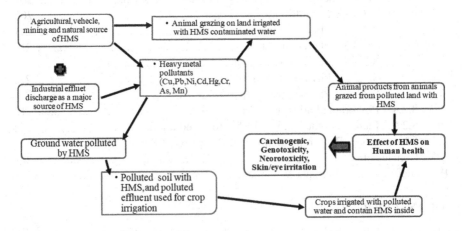

FIGURE 8.1   The transfer of toxic heavy metals from soil directly to humans and organism to humans. (Modified from Saxena et al., 2020.)

Phytoremediation is a method that uses a plant's ability to get rid of pollutants from the environment or make toxic compounds harmless (Liu *et al.*, 2018). Phytoremediation is a novel technology that takes advantage of the natural ability of plants to extract chemicals from water, soil, and air using energy from sunlight (Doty, 2008). Both edible and inedible plants could be used for phytoremediation and create a clean environment. The plants with huge biomass are ideal for phytoremediation; however, edible crops with less biomass are less efficient for the absorption of toxic elements to create a safe environment (Rahman *et al.*, 2016; Rai *et al.*, 2020). In the process of phytoremediation, plants are grown in a contaminated matrix to remove environmental contaminants by facilitating sequestration and/or degradation (detoxification) of the pollutants (Awa and Hadibarata, 2020). Plants are unique organisms equipped with remarkable metabolic and absorption capabilities as well as a transport system that can take up nutrients or contaminants selectively from the growth matrix, soil, or water (Zhang *et al.*, 2018). Phytoremediation technology has public acceptance and is important to restore the balance of a stressed environment. The long term implications of phytoremediation in removing environmental contaminants must be addressed deeply and it is important to follow with caution (Rai *et al.*, 2020).

The phytoremediation techniques such as phytostabilization, phytovolatilazation, phytodegradation, phytoextraction, phytofiltration, and phytodetoxification are the used for the removal of heavy metals from the matrix through their uptake by plants (Awa and Hadibarata, 2020; Rahman *et al.*, 2016). The process has a considerable advantage over the traditional techniques because of its cost effectiveness, potential treatment of multiple heavy metals at the same time, no need for the excavation of polluted soil, good public acceptance, and an easy follow-up of the processing of the biomass produced. For an efficient phytoremediation process, plants that are naturally hyperaccumulating, fast-growing, and producing higher biomass could be targeted (Rastogi and Nandal, 2020). Increased heavy metal concentrations in the soil have become a significant problem in the modern world due to several anthropogenic activities (Zhong *et al.*, 2017). Heavy metals are non-biodegradable and have long biological half-lives, thus, once entered into the food chain, their concentration ultimately poses a threat to human life (Zare *et al.*, 2020). The one captivating solution for this problem is to use green plants for heavy metal removal from soil and render the soil harmless and reusable (Thakur *et al.*, 2016). Many research findings reported and proved that heavy metal pollutants exist in both developed and developing countries and thus signifying it as a global problem. The high concentration of toxic heavy metals (Cd, As, Fe, Cr, Zn, Cu, Mn, Pb, Ni, Hg, and others) found in the soil, surface, and groundwater have been reported in different parts of the world (Oluoch, 2018; Rastogi and Nandal, 2020).

The use of plants in conjunction with plant-associated bacteria (rhizosphere or endophytic) offers greater potential for efficient bioremediation of organic compounds, and in some cases inorganic pollutants, than using plants alone in bioremediation (Truu *et al.*, 2015). This approach is an emerging technology for cleanup, being an aesthetically pleasing and affordable solution for water pollution by organic pollutants. The utilization of plants and their associated microorganisms to remove pollutants and producing the safe environment contributes to human well-being (Worku *et al.*, 2018). The remediation of polluted water involves the planting arrangements: constructing wetlands, floating plant systems, and numerous other configurations. Combined strategies are also used for the removal of organic pollutants from water and also removing inorganic substances like nitrogen, phosphorus, and others (Alemu, 2019; Amare *et al.*, 2018b). The treatment components in the form of vegetation, filter beds, and microorganisms contribute both directly and indirectly to the removal of pollutants from wastewater (Haynes, 2015). Some pollutants are heavy metals complexes with organic compounds (mercury trichloroethylene) which are very difficult to remediate using wild plants or endophytes; in this case, transgenic plants may be required with targeted action (Basharat *et al.*, 2018). Most plant growths promoting endophytes can help their host plant to overcome contaminant induced stress responses, thereby providing improved plant growth (Ahemad, 2019). During phytoremediation of organic pollutants, plants can further benefit from endophytes involving appropriate degradation pathways and metabolic capabilities, leading to more efficient pollutant degradation and the reduction of both phytotoxicity and phytovolatilization of pollutants (Weyens *et al.*, 2009). For phytoremediation of toxic metals, endophytes possessing a metal resistance/sequestration system can lower metal phytotoxicity and affect metal translocation to the above ground plant parts. Furthermore, endophytes that can degrade organic pollutants, improve the extraction of the metals, and offer a better way to improve phytoremediation of mixed pollution (Ma *et al.*, 2016).

Land, surface, and groundwater worldwide are highly affected by continuous waste contamination from industries, research experiments, militaries, and agricultural activities either due to ignorance, lack of vision, carelessness, or high cost of waste disposal and treatment (Misra and Misra, 2019). The rapid buildup of toxic pollutants (metals, radionuclide, and organic contaminants) in the soil, surface water, and groundwater not only affects natural resource but also is a major strain on the ecosystems (Zhang *et al.*, 2018). Interest in phytoremediation as a means to solve environmental contamination has been growing rapidly. The green technology that involves tolerant plants has been utilized to clean up heavy metals and other toxic organic compounds from soil and groundwater (Paz-Alberto and Sigua, 2013). Hence, the purpose of this chapter is to address all the technologies of phytoremediation employed for removal of hazardous pollutants in the soil and wetland environments.

8.2 PHYTOREMEDIATION OF POLLUTANTS

The environment could be polluted by a mix of pollutants that are organically biodegradable and non-degradable. The biodegradables are easy to eliminate from the environment through microbial degradation; however, the big challenge is related to toxic heavy metals, organic pollutants, and xenobiotics. Since they never be simplified by microbes, they could exist for a longer time in the environment and could create health burdens to humans, animals, plants, and ecosystems (Awa and Hadibarata, 2020; Rastogi and Nandal, 2020) (Figure 8.2). Heavy metal pollutants cause potential ecological risks. Metals present in a high concentration in soil affects the growth and metabolism of plants, and the bioaccumulation of such toxic metals in the plants poses a risk to human and animal health (Saxena *et al.*, 2020). In plant-based technology the success of phytoremediation is inherently dependent upon proper plant selection (Odoh *et al.*, 2019; Patra *et al.*, 2020). Many metal-tolerant plant species, particularly

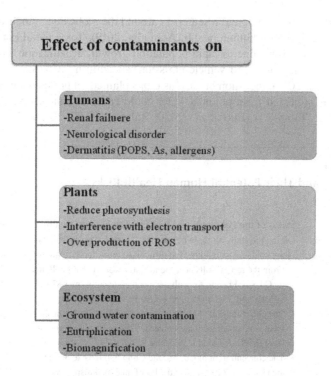

FIGURE 8.2 The consequence of heavy metal to human health, plant physiology, and environment health. (Modified from Bauddh *et al.*, 2017.)

grass species, escape toxicity through an exclusion mechanism, and are therefore better suited for phytoremediation (Subhashini *et al.*, 2013). Moreover, commonly tested plants for phytoremediation are *Ricinus communis* L., *Brassica napus*, *Pteris vittata*, sunflower *(Helianthus annuus)*, Indian mustard (*Brassica juncea*), *Vetiveria zizanioide*, tobacco (*Nicotiana tabacum*), willow tree (*Salix caprea*) and polar tree for both toxic metals and organic pollutants (Clay and Pichtel, 2019; Dimitrova *et al.*, 2018; Pan *et al.*, 2019; Paz-Alberto and Sigua, 2013; Rissato *et al.*, 2015; Siyar *et al.*, 2020).

8.2.1 TOXIC POLLUTANTS SOURCES AND THEIR RISK TO HUMANS, PLANTS, AND THE ECOSYSTEM

Heavy metals are elements that have an atomic number of more than 20 and may have variable valiancy also and present naturally in soil (Chirakkara *et al.*, 2016). Many of the metals are important for the growth and nourishment of soil flora. Heavy metals are non-biodegradable and might be poisonous depending on the metal type, its oxidation state, pH, concentration, duration and so forth, leading to the death of organisms.. Elements like Zn, Cu, Mn, Ni, Co are important for the normal function of their physiology, thereby, the plants' survival (Morkunas *et al.*, 2018). However, the role of some metals such as lead and mercury is not known (Bauddh *et al.*, 2017). Toxic metal contamination in the urban soil gets great attention from the public and also environmental researchers around the globe due to their toxicity and close relationship with human health (Adimalla, 2019). Toxic metal contamination increased in the world due to rapid population growth, urbanization, expansion of industrial sectors, increased vehicle emission, and mining (Christian and Beniah, 2019; Girmay, 2019). Ingestion of various crops planted on polluted soil and aquatic food that is collected from polluted water could cause human, animal, or wildlife health risks (Table 8.1) (Luo *et al.*, 2020; Rahman *et al.*, 2013). Moreover, toxic

TABLE 8.1
Pollutants and Their Potential Human Health Risks

Pollutants Studied	Status of the Pollutants and Risk Level to Human Health	Reference
Co, Cd, Zn, Cr, Ni, Mn, and Cu	During the study all these heavy metals reported more than the threshold and were found too difficult to reuse for agriculture	Amare *et al.* (2018a, 2018b)
Cd, Cu, Pb, and As	• From the report, soils near the industry were heavily polluted by Cd, Cu, Pb, and As with the mean concentration 4.87, 195.26, and 35.84mg/kg, respectively • The concentration of Cd and As in vegetable samples taken from smelter-affected area exceeded the maximum permissible level (MPL) for food in china by 82% and 39%, respectively • The concentration of Cd, As, and Cu in fish muscle in affected area MPL by 72%, 41%, and 24% of analyzed respectively and the health risk for local children was raised 30.25 times than the acceptable level	Cai *et al.* (2019)
Cd, Pb, and Hg	• Cd could cause cancer, a neurotoxin, and a mutagenic effect • Pb causes learning disability, seizures, and death • Hg also causes neurotoxin, affects liver CNS and kidneys, cancer developmental disorder and damage of vision, hearing, speech leading paralysis • Arsenic in water also causes cancer, keratosis, kidney, lung, and bladder disease when the element is taken by human being greater than the concentration of 50µg/litter	Adimalla (2019)
Cr	Cr(VI) is extremely toxic compared to Cr(III) and cause a carcinogenic, teratogenic, and mutagenic effect	Alemu (2019)
Zn, Cu, Cr, and Ni	• The carcinogenic risk value for Ni and Cr were found higher than the safe value (1×10^{-6}) suggesting that all receptors (especially wheat) in Kermanshah province might have significant and acceptable potential health risk because of exposure to Ni and Cr. • The carcinogenic effect for adults and children has a descending order of Ni > Cr in wheat	Doabi *et al.* (2018)
Persistent organic pollutants	• The contamination level of organic pollutants in the human body increases as their age increases • Exposure to such pollutants could bring various serious health risks such as hormone disruption, cancer, cardiovascular diseases, reproductive failure, and neurological problems	Alharbi *et al.* (2018).

metals could affect the food relationship in the ecosystem (Figure 8.1) (Christian and Beniah, 2019).

Heavy metals which are toxic whenever the hazard index is above the threshold (HI>1); they become carcinogenic. Toxic heavy metals such as As, Pb, Cu, Zn, Hg, Cr, Ni, Cd, Mn, Co, Fe, Se could cause health problems when they are above the threshold (Amare *et al.*, 2017, 2018b; Basharat *et al.*, 2018; Basu *et al.*, 2018; Bauddh *et al.*, 2017).

8.2.2 POTENTIALS AND MERITS OF PHYTOREMEDIATION

Removal of pollutants using plants is another option besides bioremediation of wastes using microbes. Phytoremediation has the potential to remove toxic pollutants from the environment and to create safe environment (Paul *et al.*, 2014). Naturally occurring plants do have a capacity to absorb toxic pollutants from soil or from water and sequestrate into their tissues. Plants are adaptive for phytoremediation in having fibrous roots, high biomass yielding in short period of time, tolerant to toxic pollutants, and having the ability to accumulate pollutants (Hauptvogl *et al.*, 2019). Moreover, the plant microbe synergy or transformed plants are efficient for phytoremediation process and resolves the burden of toxic pollutants in the environment.

Pollutants for phytoremediation are three types, the first are toxic heavy metals that are worst to health and environment, the second type of pollutants are toxic organic pollutants and the third pollutants are complex type toxics metals found with organic pollutants in the same area (Kong and Glick, 2017; Luo *et al.*, 2016; Rastogi and Nandal, 2020). By these pollutants soil or water could be polluted. To remove the pollutants from the soil or water phytoremediation could be designed as per the requirement. Plants could be used for remediation of either polluted soil polluted water (Christian and Beniah, 2019). Plants involved for phytoremediation could be natural (wild) plants; plants associated with microbes or transformed plants (Kong and Glick, 2017; Reddy *et al.*, 2019; Saxena *et al.*, 2020). Natural plants do have a special physiology and higher biomass for sequestration of the pollutants. Many natural plants are repeatedly used for the phytoremediation program in different corners of the world (Girmay, 2019). Still, few plants are assisted by endophytic bacteria and used for the clearance of environmental pollutants (Ijaz *et al.*, 2016). Another approach is the utilization of genetically transformed plants for phytoremediation by introducing genes from bacteria or other sources (Koźmińska *et al.*, 2018). Currently, the plants altered through gene editing using nucleases are also utilized to remove the pollutants from the environment (Basharat *et al.*, 2018; Basu *et al.*, 2018; Zarrin and Azra, 2018).

Phytoremediation has shown potential for the treatment of wastewater and decontamination of the terrestrial environment. Finding suitable green technology for remediation of the environment has emerged among researchers worldwide. Phytoremediation is going to be utilized as one of the suitable and recognized treatment methods for a wide range of anthropogenic pollution (Farraji *et al.*, 2016). Several remediation techniques for pollutants have been employed like physical method (remediation of contaminated area by soil washing, vitrification encapsulation of contaminated areas by impermeable vertical and layers, permeable barrier system, electro-kinetic, and thermal desorption),, and chemical remediation for soils contaminated by heavy toxic metals. However, these technologies are found to be costly, less effective, and less environment-friendly.

Phytoremediation is a cost-effective, efficient, and environmentally friendly technology to tackle both toxic heavy metals and toxic organic pollutants. Hence, these days phytoremediation received greater attention by researchers around the globe (Awa and Hadibarata, 2020; Basharat et al., 2018). In the past few years, bioremediation, or phytoremediation, is referring to microbial- or plant-based cleanup of hazardous compounds. Plants absorb heavy metals from the polluted matrix through the roots and accumulate them either in their roots or transfer to their shoots through xylem vessels where they accumulate in vacuoles. Vacuoles help to reduce the excess metallic ions from the cytoplasm and resolves their interactions with other metabolic processes due to low metabolic activity (Saxena et al., 2020). In addition to the removal of toxic pollutants from the soil, phytoremediation gives other advantages, like resolving soil erosion and controlling solubilized pollutants by hydraulic activity. Growing plant species in contaminated soil gives ample advantages, including sequestering carbon, producing biomass or biofuel, and maintaining biodiversity. However, plant-based phytoremediation is limited by several environmental extremes, such as the toxicity of pollutants and the influence of various environmental factors, such as soil texture, soil pH, vegetation reduction, and rhizosphere diversity (Basu et al., 2018).

8.2.3 PHYTOREMEDIATION MADE BY WILD PLANTS

Phytoremediation involves different processes which have been considered as one of the most appropriate technique to encounter the problem of pollutants from the environment (Rastogi and Nandal, 2020). It is believed to be a more efficient method of pollutant removal using plants compared to the traditional methods. Today, in addition to the potential of naturally existing plants, transformed plants are also involved for phytoremediation of different pollutants. The mechanisms used with both plant groups are extraction, stabilization, accumulation, degradation, and detoxification (Bauddh et al., 2017; Vazquez-Nunez et al., 2018). Different techniques that have been used for phytoremediation are discussed next.

8.2.3.1 Rhizofiltration

Rhizofiltration involves the method of using plant roots, which absorb and sequester toxic pollutants from contaminated land surfaces or groundwater (Tiwari et al., 2019). The technique of phytofiltration can be of two types: one is through the roots, known as rhizofiltration, and the second type is balstofiltration where the seedling of the plant accumulates the contaminants from the soil or the polluted water (Bauddh et al., 2017). Principally for rhizofiltration, plants must have roots that are fast-growing and more efficient in the accumulation of contaminants over a longer period. The toxic contaminants form a precipitate over the root surface. The performance of this technique depends on different factors such as the genotype of the plant and the types of contaminants. Plants for rhizofiltration should qualify with a high growth rate, high biomass production, ease of cultivation, best uptake, translocation and

FIGURE 8.3   Vetiver grass in the field with very dense biomass. (A) Rapid growing above ground biomass for rapid accumulation of contaminats; (B) very fibrous roots with high surface contact to the contaminated soil to adsorb or translocate pollutants to the above part of the plant. (Adapted with permission from Darajeh *et al.*, 2019.)

accumulation ability, high tolerance to adverse conditions, highly available in the ecosystem, and adaptable to the agro-ecology. One best example for this technique could be Vetiver grass (Figure 8.3) (Darajeh *et al.*, 2019; Siyar *et al.*, 2020).

Several studies have been done to address phytoremediation through phytofiltration using different plant genotypes in different countries. Hence, those research works have been reviewed to better understand phytofiltration and its potential for the removal of heavy metals and other organic pollutants to create a safer environment. Research was carried out in Egypt to remove heavy metal pollutants by the use of the plant, *Pistia stratiotes*. Using this plant the removal of Cr, Zn, Cu, Pb, Ni, Co, and Fe from polluted soil was evaluated and the removal through rhizofiltration was ordered as, Fe > Mn > Cr > Pb > Zn > Ni > Co > Cu > Cd (Table 8.2). The bioconcentration factor (BCF) of most studied heavy metals, except Cr and Pb, was greater than 1000, while the translocation factor (TF) of most studied metals, except Pb and Cu, did not exceed 1. The rhizofiltration potential (RP) of heavy metals was higher than 1000 for Fe, and 100 for Cr, Pb, and Cu (Galal *et al.*, 2017a). *Arundo donax* L. root cells blended with other polymers like chitosan (Cs), Gelatin (GP) and polyvinyl pyrrolidone (polymer ratio 3:1:1, respectively) were used to evaluate the efficiency of the fabricated mix for *in vitro* testing of rhizofiltration of dyes and it has been found effectively adsorbent (El-Aassar *et al.*, 2018).

The University of Huila in Coahuila, Mexico, reported removal of heavy metals like Cr, Hg, and Pb using *Zea mays* and the results indicated the mortality potential of the heavy metals Cr at 10mg/L and Hg at 4.0mg/L was reported as 40% death rate, while the mortality potential of lead (Pb) at 10mg/L reported the death rate as 20%.Hence, mercury was more toxic at lower concentrations followed by Cr at 10mg/L being more toxic than lead with the same concentration. In the phytofiltration case, *Z. mays* acted as bioaccumulater to these heavy metals. *Zea mays* adsorbed

TABLE 8.2

Phytoremediation of Pollutants Using Wild Plants

Techniques Employed by Plants	Plants Name used for Phytoremediation	Pollutants Targeted to be Removed	Concentration of Pollutants mg/kg/%Order of Remediation	Place of Study (Country)	Source
Rhizofiltration	*P. stratiotes*	Mn, Cr, Zn, Cu, Pb, Ni,Co, Fe	Root removal order of heavy metals, Fe > Mn > Cr > Pb > Zn > Ni > Co > Cu > Cd	Egypt	Galal *et al.* (2017a)
	A. donax	Heavy metals, toxins, and dyes from the environment	Dye removal	Egypt	El-Aassar *et al.* (2018)
	Z. mays	Cr, Hg, and Pb	Absorption of Hg by 88% (1.758mg/L from 2.0mg/L), Cr 68% (3.42mg/L from 5.0mg/L)but lead absorption was less (30%)	Corhuila	Benavides *et al.* (2018)
	Cattail(*T. latifolia*); sedge (*C. blanda*,); sunflower (*H. annuus*); Indian mustard (*B. juncea*)	Cr, Cu, and Cd	Cattail root had the highest BCF for Cr 1156, and Cu (2911) and Cd (6047) Mustard roots had a high BCF for Cd (3485)	India	Clay and Pichtel (2019)
	Jancus sp., *Tamarix* sp., and *Suaeda* sp.	Pb, Cd, Cr, Zn, Fe, Mn, and Cu	Most depletion for *Juncus* sp. followed by *Tamarix* sp and *Suaeda* sp. *Suada* sp.(153.42) <*Tamarix* sp. (206.30)<*Jancus* sp.(258.90), which are much lower than the bare land water (660.27)	Iran	Zare *et al.* (2020)
Phytostabilization	*B. pilosa* and *P. lanceolata*	Cu	Shoot (142 mgkg^{-1}) roots (964mgkg^{-1})	Brazil	Andreazza *et al.* (2015)

TABLE 8.2 *(Continued)*

Phytoremediation of Pollutants Using Wild Plants

Techniques Employed by Plants	Plants Name used for Phytoremediation	Pollutants Targeted to be Removed	Concentration of Pollutants mg/kg/%Order of Remediation	Place of Study (Country)	Source
	V. cuspidata	Cr, Pb, Cu, Zn, and Cd	Highest accumulation Cr, Cu, and Pb in the root	Egypt	Galal *et al.* (2017b)
	R. sceleratus	Cu, Pb, Ni, Cd, and Mn	The high concentration of Cu and Pb (27.7 and 9.9mgkg⁻¹) while the toxic concentration of Mn (2508mgkg⁻¹) is in their roots compared to their shoots	Egypt	Farahat and Galal (2017)
	Lemongrass (*Cymbopogon flexuousus*)	Cr^{+6}	Increased accumulation of chromium (Cr) from both roots and shoots within 60 days	India	Patra *et al.* (2018)
	Soybean, *M. circinelloides* and three amendments organic fertilizer, rice husk, biochar, camp site at a combination of 1:1:2:1 ratio	Cu, Zn, Pb, and Mn	Removal of heavy metals from the soil in the order of Pb > Cd > Cu > Zn > Mn	China	Li *et al.* (2019)
	L. stolonifera	Cu, Pb, Zn, Cr, Cd, and Ni	Bioaccumulation factor that exceeded 1, the translocation factor of the investigated metals was <1	Egypt	Galal et al. (2019)

(Continued)

TABLE 8.2 *(Continued)*
Phytoremediation of Pollutants Using Wild Plants

Techniques Employed by Plants	Plants Name used for Phytoremediation	Pollutants Targeted to be Removed	Concentration of Pollutants mg/kg/%Order of Remediation	Place of Study (Country)	Source
	Four aquatic macrophytes (*E. crassipes* (Mart.) Solms, *L. stolonifera* (Guill. and Perr.) P.H. Raven (*E. stagnina* [Retz.] P.), Beauv. and (*P. australis* [Cav.] Trin. ex Steud.)	Cd, Ni, and Cu	The four species had bioaccumulation factors(BAFs) greater than one, while their translocation factors (TFs) were less than 1	Egypt	Eid *et al.* (2020)
	A. donax	As	20mg/L when BC or plant growth promoting bacteria are present	Italy	Guarino *et al.* (2020)
	Indigenous weed (*S. pumila*), energy plant (*P. sinese*), Cd tolerant *Sedum plumbizincicola*) and Cu tolerant plant (*E. splendens*)	Cd, Cu	*P. sinese* treatments decreased DGT extractable Cu and Cd by 52.1% and 40.5% than *S. pumila* treatment	China	Cui *et al.* (2020)
Phytoextraction	Amaranth (*Amaranthus paniculatus*), Indian mustard(*B. juncea*),and sunflower (*H. annuus*)	Pb and Cu	In the shoot, sunflower removed significantly higher Pb(50–54%) and Cu (34–38%) compared to amaranth and Indian mustard	Malaysia	Rahman *et al.* (2013)

TABLE 8.2 *(Continued)*

Phytoremediation of Pollutants Using Wild Plants

Techniques Employed by Plants	Plants Name used for Phytoremediation	Pollutants Targeted to be Removed	Concentration of Pollutants mg/kg/%Order of Remediation	Place of Study (Country)	Source
	S. vera Forssk. Ex J.F. Gmel	Cu, Cr, Ni, Zn, and Pb	According to biological accumulator coefficient, BAC=C_{plant}/C_{soil}=0.1 to 1.0: *S. vera* is moderate accumulator plant for the metals according to the following order Cu>Zn>Ni>Cr	Libya	Bader *et al.* (2018)
	P. ensiformis, B. nivea, A. prorerus, and *H. sibthorpioides*	As, Cd, Pb, and Zn	*P. ensiformis* accumulated 1091 mg kg^{-1} As in the shoot, and its translocation factor (TF) was greater than 1, suggesting the potential capacity for As phytoextraction, *B. nivea, A. prorerus,* and *H. sibthorpioides* showed potential for phytoextraction Cd in shoots (490.3, 175.4, and 128.5 mg kg^{-1}, respectively)	China	Pan *et al.,* (2019)
	B. juncea, H. annus, and *Z. mays*	As	Additives K_2HPO_4 or $(NH4)S_2O_3$ for the mobilization of Asphytoextraction up to 80%	Italy	Franchi *et al.* (2019)

(Continued)

TABLE 8.2 *(Continued)*
Phytoremediation of Pollutants Using Wild Plants

Techniques Employed by Plants	Plants Name used for Phytoremediation	Pollutants Targeted to be Removed	Concentration of Pollutants mg/kg/%Order of Remediation	Place of Study (Country)	Source
	H. scoparia and *H. strobilaceum*	Cu, Zn, Cr, and Fe	Both plants were found to be the moderate extractor	Libya	Bader *et al.* (2020)
	Napier grasses (*P. purpureum* 'purple') and variegated giant reed (*A. donax* var versicolor)	Cd and Zn	109.3% and 55.4%, respectively	China	Hou *et al.* (2020)
	Poplar plant (*P. deltoids* X *nigra*)	Toluene	Peak-season toluene mass removal rate ranging from 313 to 743µg/day	Canada	BenIsrael *et al.* (2020)
	Three Cardoon caltivars (Sardo, Siciliano, and Spagnolo)	As, Cd, Cu, Pb, and Sb	Pb content in the rhizosphere especially in the Sardo from about 67,000 mg kg^{-1} pre remediation soil to about 35,000 mgkg^{-1} after two round growth	Italy	Capozzi *et al.* (2020)
Phytovolatilization	As hyperaccumulating plant, *P. vittata*	As	Percentage of arsenic component per sample was 37% for arsenite and 63% for arsenate *P. vittata* is effective volatilizing of As, it removed about 90% of the total uptake of As from As-contaminated soil	Japan	Sakakibara *et al.* (2010)

TABLE 8.2 *(Continued)*
Phytoremediation of Pollutants Using Wild Plants

Techniques Employed by Plants	Plants Name used for Phytoremediation	Pollutants Targeted to be Removed	Concentration of Pollutants mg/kg/%Order of Remediation	Place of Study (Country)	Source
	Willow (*Salix* sp.) and hybrid poplar (*Populus* sp.)	TCE, PCE	The transpired gases after 7 days of exposure, the fraction of transpired TCE to the TCE taken up by plant was 70–90%	U.S.	Limmer and Burken (2016)
Phytodegradation	Three endophytic bacteria augmented to two kinds of grass, *L. fusca,* and *B. mutica*	Petroleum hydrocarbons	Oil-contaminated soil (46.8 oil kg^{-1} soil) and maximum oil degradation (80%) was achieved with *B. mustica* plant augmented with the endophytes	Pakistan	Fatima *et al.* (2018)
	H. vertuculata (L.F)	Phenol	*H. verticillata* efficient degraded phenol in solutions with initial concentration lowers than 200mgL^{-1}	China	Chang *et al.* (2020)
Phytodetoxification	The grass, *V. zizanioides* (L.) Nash	Genotoxicity with heavy metals	In the research report, a significant reduction of the concentration of heavy metals and decreases genotoxic potential was observed	India	Ghosh *et al.* (2014)

(Continued)

TABLE 8.2 *(Continued)*
Phytoremediation of Pollutants Using Wild Plants

Techniques Employed by Plants	Plants Name used for Phytoremediation	Pollutants Targeted to be Removed	Concentration of Pollutants mg/ kg/%Order of Remediation	Place of Study (Country)	Source
	S. nigrum mediated by endophytic fungi	Cd	RSF-6L inoculation decreased uptake of Cd in roots and above ground parts, as evidenced by a low BCF and improved tolerance index (TI)	Republic of Korea	Khan *et al.* (2017)
	Willow tree (*S. caprea*)	Iron cyanide	Young leaves 15.197% and old leaves accumulates more due to longer exposure time	Germany	Dimitrova *et al.* (2018)

mercury and decreased it by 88% (1.758mg/L) compared to the initial concentration (2.0mg/L) (Table 8.2); likewise, the lead metal absorption potential was also found high with a reduction of 68% compared to the initial concentration of 5.0 mg/L. The absorption potential of chromium was not as significant as for the other metals, the reduction percentage was found to be 30% (1.51mg/L) (Benavides *et al.*, 2018).

There are four plant species evaluated for rhizofiltration of heavy metals available in synthetic produced water including cattail (*Typha latifolia*), sage (*Carex blanda*), sunflower (*H. annuus*), and Indian mustard (*B. juncea*). All plants evaluated for absorption of heavy metals from polluted water and soil from wetlands proved that they accumulated more metals in their roots than their shoots. Cattail root had the highest BCF with Cr (1156), Cu (2911), and Cd (6047), so this plant is proved to be a potential resource for removal of heavy metals. Moreover, Mustard root had a high BCF with Cd (3485) (Table 8.2). Mustard, cattail, and sage had TF values < 1, indicating their potential as metal excluders of produced water (Clay and Pichtel, 2019). Another research performed in Iran for the removal of heavy metals (Pb, Cd, Cr, Zn, Fe, Mn, and Cu) using plants *Jancus.* sp., *Tamarix* sp., and *Suaeda* sp.; maximum depletion was reported by *Juncus* sp., followed by *Tamarix* sp. It was found *Suaeda* sp. (153.42) <*Tamarix* sp. (206.30)<*Jancus* sp.(258.90), which were much lower than the bare land water (660.27) (Zare *et al.*, 2020). It is concluded from this discussion that mesophytes plants are ideal for rhizofiltration because these plants have extensive and fibrous root systems than hydrophytes. It can be applied for the absorption of radioactive elements from contaminated areas (Patra *et al.*, 2020). The benefit of this technique is to relocate metals from the rhizospheric site and their subsequent translocation to aerial parts of the plants.

8.2.3.2 Phytostabilization

Plants immobilize or inactivate metal pollutants at their place involving absorption by plant roots, adsorb on the roots, precipitate (Usman *et al.*, 2018). For this purpose, certain plant species specialize in immobilizing contaminants in the soil or groundwater itself. Such plants adsorb onto the root surface or zone thereby preventing migration of contaminants in the soil or away by soil erosion (Bauddh *et al.*, 2017). Several research works have been performed recently in different countries by using this technique to resolve environmental toxicity (Cui *et al.*, 2020; Eid *et al.*, 2020; Guarino *et al.*, 2020). Two indigenous plants (*Bidens pilosa* and *Plantago lanceolata*) for the phytostabilization of copper in two copper mining areas of Brazil were used. *Plantago lanceolata* plants grown in the inceptisol showed the copper accumulation in the shoots (142mg kg $^{-1}$), roots (964 mg kg $^{-1}$), and the entire plant (1106mg kg^{-1}). The high levels of Cu were phyto-accumulated from the inceptisol by *B. pilosa* and *P. lanceolata* with 3500 and 2200g ha^{-1}, respectively (Andreazza *et al.*, 2015). The plant *Vossia cuspidata* was tested for the phytostabilization of heavy metals and accumulated the highest concentration of Cr, Cu, and Pb in their roots during spring season compared to other seasons. This plant resulted in a bioaccumulation factor (BF) more than one and TF less than one, therefore, this plant has shown potential for phytostabilization of these metals (Galal *et al.*,2017b). Another plant, *Ranunculus sceleratus*, was evaluated for phytostabilization of several heavy metals. The BF for the studied heavy metals was greater than one and in the order of Ni (27.1) > Zn (20.0) > Cd (16.4) > Cu (7.7) > Mn (3.9) > Pb (3.6). The TF of all studied metals was less than one. This indicated *R. sceleratus* as a potential source for phytostabilization (Farahat and Galal, 2017). The phytostabilization potential of Lemongrass for Cr^{+6} was evaluated, and accumulation of chromium (Cr) in both roots and shoots was increased within 60 days with increasing proline and antioxidant secondary metabolites, while protein and chlorophyll content was decreased (Patra *et al.*, 2018). Since proline is a stress-tolerant amino acid, it may give tolerance to the plant.

Moreover, phytostabilization was tested using the synergy of soybean, microorganisms (*Mucor circinelloides*), and A3 amendment (having organic fertilizer, rice husk, biochar, and the ceramsite in the ratio of 1:1:2:1). It was proven as an efficient approach for immobilization of heavy metals in the plant roots. The removal rate of the metals indicated in the order Pb > Cd > Cu > Zn >Mn and the optimized protocol was proved effective to remediate toxic environmental pollutant metals (Li *et al.*, 2019). The researchers from China utilized floating macrophyte *Ludwigia stolonifera* for accumulation of heavy metals through its roots with seasonal variations. The highest concentration of Al and Cu was found in spring while the maximum concentration of Fe, Mn, and Ni was observed during summer. Furthermore, the maximum phytostabilization of Cd and Zn was achieved during autumn and Cr and Pb during winter. The BF of all metals was greater than one while the TF was less than one (Galal et al., 2019).

Phytostablization was evaluated using three plants: indigenous weed (*Setaria pumila*), energy plant (*Pennisetum sinese*), and *Elsholtzia splendens* amended with (0.1%) limestone, for removal of Cd and Cu from the soil. *Pennisetum sinese* decreased the soil availability of Cu and Cd by 40.5% and 2.1%, respectively, compared to *E. splendens* treatment. Moreover, *P. sinese* had the lowest phytotoxicity among the tested plants (Cui *et al.*, 2020).Recently, phytostabilization was performed by *A. donax* assisted with plant growth promoting bacteria (PGPB) consortium consisted of two strains of *Stenotrophomonas maltophilia* and one of *Agrobacterium* sp. to enhance plant growth and bioaccumulation of heavy metals by *A. donax* (Guarino *et al.*, 2020). The phytostabilization study in Egypt was carried out using four aquatic macrophytes: *Eichhornia crassipes* (Mart.) Solms, *L. stolonifera* (Guill. and Perr.), P.H. Raven (*Echinochloa stagnina* (Retz.) P. Beauv), and (*Phragmites australis* [Cav.] Trin.ex Steud.). As compared to the other species, *P. australis* accumulated the highest concentration of Cd and Ni while *E. stagnin* accumulated the highest concentration of Pb in its tissues (Eid *et al.*, 2020). In the phytostabilization technique, plants tend to arrest the pollutants and restrict the migration of the pollutants either into the roots or to the rhizosphere (Khan *et al.*, 2020). In the phytostabilization technique, the BF should exceed one while the TF should be less than one (Ng *et al.*, 2020).

8.2.3.3 Phytoextraction

The plants that are used for phytoextraction should have some unique properties like tolerance towards a particular pollutant, efficiently translocate pollutants to aerial as well as harvestable parts of the plant, and the ability of plant survival in stressed condition like soil salinity, water content, and pest resistant (Bauddh *et al.*, 2017). Soil polluted by industrial wastes from Malaysia was amended with N, fertilizer, and phytoextraction using sunflower, amaranth and Indian mustard. The highest accumulation of Pb and Cu (Pb: 10.1–15.5 mgkg^{-1}, Cu: 11.6–16.8 mgkg^{-1}) in the shoots was two- to fourfold compared to the roots of the three plants. Sunflower was found to be effective to remove Pb (50–54%) and Cu (34–38%) followed by amaranth and Indian mustard (Rahman *et al.*, 2013). The contaminated river water and soil sample were analyzed for heavy metals and showed Cd (0.077 mg/L), Cr (0.90 mg/L), Pb (<0.001 mg/L) and Ni (0.034mg/L) in the river and in the soil sample Cr (153.22mg/L), Ni (3071mg/L) and Pb (9.55mg/L) were detected. Three plants *A. donax*, *R. communis*, and *Vernonia amygdalina*, were evaluated for phytoremediation. These plants showed the potential for removing Cd, Cr, and Pb in different concentrations while Ni was not detected from the leaves. *Arundo donax* accumulated Cr (80.90mg/L), Pb (37.30mg/L), and Cd (25.98mg/L), *V. amygdalina* accumulated Cr (83.59mg/L), Cd (44.46mg/L) and Pb (14.49mg/L) and *R. communis* accumulated Cr (62.06mg/L), Cd(16.64mg/L and Pb (16.64mg/L). These selected plants found to be good for phytoremediation of Cr, Pb, and Cd (Amberber and Kifle, 2016).

Phytoextraction of heavy metals was carried out by *Suaeda vera* according to biological accumulator coefficient BAC=C_{plant}/C_{soil}=0.1 to 1.0; *S. vera* was found to be a moderate accumulator plant for the metals in the order of Cu > Zn > Ni > Cr, whereas Pb was not detected in the plant as well as in the soil (Bader *et al.*, 2018). Pan *et al.* (2019) performed phytoextraction of heavy metals using four plants: *Pteris ensiformis*, *Boehmeria nivea*, *Aster prorerus*, and *Hydrocotyle sibthorpioides*.

Among them, *P. ensiformis* accumulated 1091 mg kg⁻¹ of As in the shoot, showing the TF greater than 1. *Boehmeria nivea, A. prorerus*, and *H. sibthorpioides*
showed phytoextraction of Cd by 490.3, 175.4, and 128.5 mg kg⁻¹, respectively, in
shoots (Pan *et al.*, 2019). Phytoextraction of As using three different plants with
the supplementation of additives K_2HPO_4 or $(NH_4)S_2O_3$ and PGPB enhanced As
uptake in the shoots. *Brassica juncea* was found to be the most effective with the
largest uptake (up to 140%) of As (Franchi *et al.*, 2019). Bader *et al.* (2020) reported
the phytoextraction of heavy metals using *Hammada scoparia* and *Halocnemum
strobilaceum*. Biological absorption coefficient (BAC), BCF, and TF of both plants
indicated a moderate extraction of Zn, Cr, and Fe (Bader *et al.*, 2020). Moreover,
Cd and Zn phytoextractions were reported using Napier grasses (*Pennisetum purpureum 'purple'*) and variegated giant reed (*A. donax* varversicolor), and due to the
largest biomass, Napier grasses (*P. purpureum 'purple'*) accumulated Cd and Zn in
shoots, up to 109.3% and 55.4% high, respectively. *Arundo donax* (varversicolor)
was best phytoextractor for Zn while Napier grasses (*P. purpureum 'purple'*) was
found to be the best for both phytoextraction and phytostabilization (Hou *et al.*,
2020). Poplar plant (*Populus deltoids X nigra*) supported with soil bacteria used for
phytoextraction of toluene showed toluene mass removal rates ranging from 313 to
743μg/day (BenIsrael *et al.*, 2020). Phytoextraction using three cardoon cultivars
(Sardo, Siciliano, and Spagnolo) to remediate heavy metals (As, Cd, Cu, Pb, and Sb)
was performed. The Pb content in the rhizosphere especially in the Sardo was about
67,000 mgkg⁻¹ in pre-remediation soil, while after two rounds of growth resulted in
about 35,000 mgkg⁻¹ (Capozzi *et al.*, 2020).

During phytoextraction, plants absorb pollutants either from soil or water and
translocate to the aboveground part of the plant tissue and accumulate the pollutants in the harvestable part of the plant body (Ashraf *et al.*, 2019). To do so, both the
plant's BF and the TF should exceed one. In this chapter, most research works done
in different countries of the world used different plant species for the phytoextraction
of heavy metal pollutants (Lajayer *et al.*, 2019). Merits of the technology are (1) it
does not alter the site, (2) helps in restoration of mining sites, and (3) it is the princi-
pal technique for remediation of pollutants. However, there are still some limitations
of this technology such as:

    i. bioavailability of metals in the rhizospheric region is less; and
   ii. toxic metals are held in the roots of the plants (Patra *et al.*, 2020).

8.2.3.4 Phytovolatalization

Some toxic contaminants exist in the atmosphere in gaseous form, for example, toxic
metals like Hg, As, and Se. Metals adsorbed into the plant tissue in their elemental
form and then converted into gaseous species by bacteria-assisted plants is known as
biomethylation. This creates volatile molecules that are released into the atmosphere
(Bauddh *et al.*, 2017). The phytovolatilization of As and organic pollutants such
as trichloroethylene (TCE) and tetra-chloroethylene (PCE) were studied (Limmer
and Burken, 2016; Sakakibara *et al.*, 2010). Arsnic (As) in the trap was measured
by induced coupled plasma mass spectrometer (ICP-MS), and speciation of As
was analyzed using high-performance liquid chromatography (HPLC/ICP-MS).

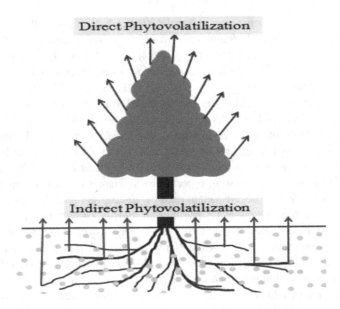

FIGURE 8.4   Phytovolatilization of pollutants. (Modified from Limmer and Burken, 2016.)

The percentage of As component per sample was 37% for arsenite and 63% for arsenate. *Pteris vittata* was found effective for volatilization of As; it removed about 90% of the total uptake of As from As-contaminated soil (Sakakibara *et al.*, 2010). *Salix* sp. and hybrid poplar (*Populus* sp.) were evaluated for the phytovolatilization of TCE and PCE. After seven days of exposure, the fraction of transpired TCE to the TCE taken up by both plants was 70–90% (Figure 8.4) (Limmer and Burken, 2016).

8.2.3.5 Phytodegradation

In the phytodegradation process, plants/microbes release enzymes that are used to metabolize and degrade contaminants in the soil, groundwater, sludges, or surface water. Polar tree metabolized TNT to 4-amino-2,6-dinitrotoluene (4-ADNT), 2-amino-4,6-dinitrotoluene (2-ADNT), and other undefined compounds (Misra and Misra, 2019). Phytodegradation using three endophytic bacteria augmented to two kinds of grass, *Leptochloa fusca* and *Brachiaria mutica*, which was reported for degradation of petroleum hydrocarbons. Phytodegradation of oil-contaminated soil (46.8 g oil kg^{-1}soil) was performed with *B. mustica* plant augmented with the endophytes, and maximum oil degradation was reported by 80% (Fatima *et al.*, 2018). *Hydrilla vertuculata* (L.F) plant was utilized for phytodegradation of phenol that showed efficiently degraded phenol in solutions with initial concentration lowers than 200mgL^{-1} (Chang *et al.*, 2020).

8.2.3.6 Phytodetoxification

In phytodetoxification, the toxic metal is not remediated directly rather it reduces/detoxifies the toxicity of toxic metals to the organisms. There are some microbes such as bacteria, phytoplankton, and fungi that are important for phytodetoxification.

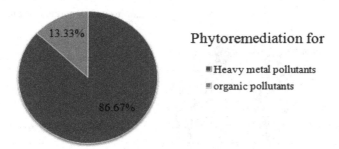

FIGURE 8.5 Research work frequencies on phytoremediation of heavy metals and organic pollutants using wild plants from Table 8.2.

The grass *Veteveria zizanioides* (L.) Nash was evaluated to remove genotoxicity of heavy metals and indicated a significant reduction in the concentration of heavy metals and lower genotoxic effects (Ghosh *et al.*, 2014). Another research reported phytodetoxification of Cd using *Solanum nigrum* assisted by endophytic fungi. It decreased the uptake of Cd in roots and aboveground parts, as evidenced by a lower BCF and improved tolerance index (TI). Application of appropriate fungal inoculation can improve tolerance mechanisms of hyper-accumulation and reduce Cd uptake. It can be recommended for phytostabilization/immobilization of heavy metals in the crop field (Khan *et al.*, 2017).

In most cases, phytoremediation of pollutants removal becomes better whenever the plant-microbe association is employed rather than the plant alone. The plant-microbe synergy has been proven more efficient in removing pollutants from the contaminated area. In the phytodegradation process, mostly organic molecules are degraded, removed, and detoxified by changing the oxidation level of the elements using enzymes. A large percentage of recent research(86.67%)has focused on heavy metal phytoremediation and 13.33% of the research was performed on organic pollutants (Figure 8.5). The coverage of heavy metal pollutants is wider than other pollutants because of industrialization, mining, and urbanization. Therefore, more attention should be given to heavy metal pollutants to resolve the problem around the globe.

Moreover, among the different techniques of phytoremediation, phytostabilization (30%) and phytoextraction (30%) techniques were dominantly employed followed by rhizofiltration (Figure 8.6). The least-used technique for phytoremediation was found to be phytodegradation and phytodetoxification (6.67% each). Most metal-tolerant plants are effective in phytostabilization or phytoextraction by either immobilizing or translocating the pollutants and sequestrating in the plant tissue, whereas phytodegradation and phytodetoxification need plant-microbe synergism to break the pollutants through their physiology.

8.2.4 PHYTOREMEDIATION USING TRANSGENIC PLANTS

Transgenic plants produced through genetic transformation are also use for the phytoremediation pollutants. The transferred gene must qualify that property when expressed in the recipient plant. The possibility of using transgene in

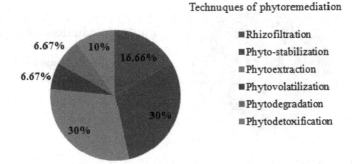

FIGURE 8.6 The application of different phytoremediation techniques using wild plants and their distribution from Table 8.2.

the phytoremediation depends on the availability of gene sequences, from different organisms, which can improve the wild plant for remediation of pollutants by conferring the plant phenotype which was not present while it was a natural plant. Modified plants for such purpose obtained genes mostly from bacteria and employed for the degradation of mostly organic pollutants (Maestri and Marmiroli, 2011). A lot of research findings declare that many potential transgenic plants are developed for remediation of the environment from pollutants (Vazquez-Nunez *et al.*, 2018). Transgenic plant strategy is based on enzyme production like peroxidase or overexpression of such enzymes for mainly organic pollutant degradation and removal from the contaminated environment (Talano *et al.*, 2012). Transgenic tobacco plants were produced by introducing the PCS gene from *Ceratophyllum demersum* cv. L. (*CdPCS1*), a submerged rootless aquatic macrophyte, which is considered a potential accumulator of heavy metals, thereby for detoxification. The *CdPCS1* cDNA of 1757bp encodes a polypeptide of 501 amino acid residues that are different from other known PCS concerning the presence of many cysteine residues known for their interaction with heavy metals. Modified tobacco plants expressing *CdPCS1* showed a large increase in PC content and non-protein thiols with enhanced bioaccumulation of Cd and As without a noticeable decrease in plant growth (Shukla *et al.*, 2012). Research on another comparative study between transgenic tobacco (*N. tabacum*) and wild tobacco plants reported phytodegradation of organic pollutants. Double transgenic (DT) tobacco plants expressing *TPX1* and *TPX2* genes for peroxidases production that is used for degradation of 2, 4, dichlorophemol (2, 4, DCP) pollutant remediation. The double transgenic line showed higher tolerance to 2, 4-DCP at an early stage of development, since their germination index was higher than that of wild-type seedlings exposed to 25mg/L pollutants (Table 8.3). Moreover, the double line transgenic tobacco was reported as efficient with up to 98% removal of 2, 4-dichlorophenol pollutants without showing toxicity after remediation (Talano *et al.*, 2012).

Phytoextraction/phytostabilization using the transgenic fern (*P. vittata*) was proven as hyperaccumulator of As. During the transformation *PvACR3* gene was taken from *Arabidopsis thaliana* and regulated by CaMV 35S promoter. In response

TABLE 8.3
Phytoremediation of Pollutants using Genetically Transformed Plants

Transformed Plants for Phytoremediation	Gene Introduced and Source of the Gene	Techniques Employed by Plants	Pollutants Targeted to Clean	Country	Source
Transgenic tobacco (*N. tabacum*)	Phytochelatin synthase (PCS) gene from *C. demersum* cv. L.(CdPCS1), a submerged rootless aquatic macrophyte which is the potential accumulator	detoxification	Cd and As	India	Shukla *et al.* (2012)
Comparative study b/n Transgenic tobacco (*N. tabacum*) and wild tobacco plants	Double transgenic (DT) tobacco plants expressing TPX1 and TPX2 genes for peroxidases production and used for degradation of 2,4,dichlorophemol(2,4,DCP) pollutant removal	Phytodegradation	2,4,Dichlorophenol	Argentina	Talano *et al.* (2012)
As hyperaccumulator fern (*P. vittata*)	*PvACR3* gene from *A. thaliana* by the CaMV 35S promoter	Phytoextraction and volatilization	As	China	Chen *et al.* (2013)
Polar tree(*Populus alba* X *P. tremula var. glandulosa*)	transgenic poplar trees which expressYCF1, a yeast ABC transporter which transports toxic metals into the vacuole	Phytoextraction and stabilization	Cd and Pb	Korea	Shim *et al.* (2016)
Transgenic alfalfa plants	*rhlA* gene, responsible for the biosynthesis of biosurfactant–rhamnolipids were obtained	Phytodegradation	Oil	Russia	Stepanova *et al.* (2016)
Transgenic Nettle (*Urtica dioica*)	The most common constitutive promoter, CaMV 35S in *U. dioica*, the CUP and bphC transgenes	Phytodegradation	Zn, Pb, Cd, Aspolychlorinated biphenyls (PCBs)	Czech	Viktorova *et al.* (2016)

(Continued)

TABLE 8.3 *(Continued)*

Phytoremediation of Pollutants using Genetically Transformed Plants

Transformed Plants for Phytoremediation	Gene Introduced and Source of the Gene	Techniques Employed by Plants	Pollutants Targeted to Clean	Country	Source
Transgenic *A. thaliana* plants	The P450 isozymes CYP1A2 expressed in *A. thaliana* were examined regarding the herbicide simazine (SIM)	Phytodegradation	Herbicide simazine (SIM) pollutant	Egypt	Azab *et al.* (2016)
Double genetically modified symbiotic system (*M. truncatula* plant and undulating bacteria)	Metallothionein gene mt4a from *A. thaliana* in roots which is Cu tolerant was used for plant inoculation Genetically modified *E. medicae* strain expressing Cu resistance genes *copAB* from pseudomonas fluorescent drive by a nodulation promoter, *nifHp* was used for inoculation	Phytostabilization	Cu	Spain	Pérez-Palacios *et al.* (2017)
Tobacco (*N. tabacum*)	*nfsI* gene, Nitro reductase gene from bacteria, plastic engineering		TNT	USA	Zhang *et al.* (2017)
Tobacco (*N. tabacum,* var *sumsun*)	*AtACR2* gene (arsenic reductase 2) of *A. thaliana*	Phytoextraction	As	India	Nahar *et al.* (2017)
Transgenic poplars	*CYP2E1* gene	Phytodegradation	Trichloroethylene (TCE)	Washington, U.S.	Legault *et al.* (2017)
Western wheatgrass (*P. smithii*)	The bacterial genes, *xplA* and *xplB*, confer the ability to degrade RDX in plants, and a bacterial nitroreductase gene *nfsI*	Phytodegradation	Contaminates explosives, hexahydro-1,3,5-triazine (RDX), and 2,4,6-trinitrotoluene (TNT)	U.S.	Zhang *et al.* (2018)

to As treatment, the *PvACR3* transgenic fern showed greatly enhanced tolerance to As. The transgenic fern seed was reported as tolerant and could germinate and grow in the presence of 80μM As(III) or 1200μM arsenate [As(V)] treatment, which was lethal for wild ferns (Chen *et al.*, 2013). The transgenic polar plant has been produced by introducing a yeast *YCF1* gene (ABC transporter) which transports toxic metals into the vacuole for phytoextraction/phytostabilization of Cd and Pb. When the transgenic plant was subjected to agar media containing 0.1 and 0.3mM of $CdCl_2$ the transgenic polar plant grew much better than the wild-type plant. Later on, the field study was carried out and field survival was proven, but the accumulation nature of the transgenes was underway (Shim *et al.*, 2016).

Arabidopsis thaliana was modified by the human *P450CYP1A2* gene and examined for phytoremediation of herbicides, insecticides, and industrial chemical remediation. Transgenic *A. thaliana* plants expressing *CYP1A2* gene showed significant resistance to herbicide simazine (SIM) supplemented either in a plant growth medium or sprayed on foliar parts. The results showed that SIM produces harmful effects on both the rosette diameter and the primary root length of the wild-type plants (Azab *et al.*, 2016). Transgenic alfalfa plants were produced by introducing the *rhlA* gene for the production of rhamnolipids, which helps to reduce the surface tension of the hydrocarbon oil and it's desorption from soil particles, thereby facilitating its recycling by the microorganism. The oil utilization by the wild plant was 4%, while transgenic alfalfa alone resulted in 20% of oil utilization. When transgenic alfalfa was employed with *Candida maltose* for remediation of an oil-polluted environment, it achieved 86% oil degradation by the synergism (Stepanova *et al.*, 2016).

A double genetically modified symbiotic system (*Medicago trucatula* and undulating bacteria) was employed for Cu phytostabilization. In the first genetic modification, the metallothinein gene *mt4a* was transferred from *A. thaliana* to *M. truncatula* plant to improve plant Cu tolerance. On the other hand, a genetically modified *Ensifer medicae* strain, expressed copper resistance genes *copAB* from *Pseudomonas fluorescens*. Double modification enhances Cu phytostabilization (Pérez-Palacios *et al.*, 2017). Transgenic tobacco was used for remediation of As by transferring the *AtACR2* gene (arsenic reductase 2) from *A. thaliana*. Transgenic tobacco proved more tolerant of arsenic (As) than the wild type. The transgenic tobacco could grow on media containing 200μM arsenate. When transgenic tobacco was grown in100μM arsenate for 35 days, the shoot accumulates 28μg/gd wt than the shoot of wild type (40μg/gd wt). However, the arsenic content in the transgenic tobacco roots was found to be higher (2400μg/gd wt) than the wild type (2100μg/gd wt) roots (Nahar *et al.*, 2017). A study was performed to evaluate the effectiveness of transgenic poplars for a controlled field study phytoremediation. Three hydraulically contained test beds were planted with 12 transgenic poplars and 12 wild polar plants and the third bed left unplanted and dosed with an equivalent amount of TCE. The removal of TCE was enhanced in the transgenic tree bed. Total chlorinated ethane removal was 87% in the *CYP2E1* bed, 85% in the wild-type bed, and 34% in unplanted bed. Evapotranspiration of TCE from transgenic leaves was reduced by 80% and the diffusion of TCE from the transgenic stem was reduced to 90% compared to wild type. Cis-dichloroethene and vinyl chloride levels were reduced in the transgenic tree bed. Chloride ion accumulated in the planted bed corresponding to the TCE loss (Legault *et al.*, 2017).

Transgenic Western wheatgrass (*Pascopyrum smithii*) production using the bacterial genes, *xplA* and *xplB*, confer the ability to degrade hexahydro-1, 3, 5-triazine (RDX) in plants, and a bacterial nitroreductase gene *nfsI* enhances the capacity of plants to withstand and detoxify trinitrotoluene (TNT). Transformed plants with *xplA*, *xplB*, and *nfsI* removed significantly more RDX from hydroponic solutions and retained much lower, undetectable, levels of RDX in their leaf tissues when compared to wild types. Moreover, these plants were also more resistant to TNT toxicity and detoxified more TNT than wild-type plants. This was the first study to engineer a field applicable grass species capable of both RDX degradation and TNT detoxification (Zhang *et al.*, 2018).

8.2.5 Plant Genome Editing for Phytoremediation

Gene editing is a remarkable technology to manipulate genomes of a plant and acquire their bioremediation nature to remove pollutants from the environment. Gene editing is an *in vivo* method of site specific mutation made by engineered nucleases. The nucleases, by inserting or deleting, can bring the desired mutation in a specific region of plant genome thereby altering expression or gene output. In this regard, gene editing targeted for the production of plants/crops could tolerate pollutants and can remediate pollutants from an area. The very recent nucleases, frequently used for gene editing, are clustered as regularly interspaced palindromic repeats (CRISPR/Cas9 system) followed by transcription activation-like effector nucleases (TALEN). Recent research from China reported knockout of *OsNramps5* gene using the CRISPR/Cas9 system that produced low Cd content in rice that was safe for human health. The result showed the concentration of Cd (0.05mg/kg) in mutated rice while the wild rice accumulated from 0.33mg/kg to 2.90mg/kg of Cu. The plant yield was not affected significantly during the gene transformation (Tang *et al.*, 2017). This technology is very recent and is highly focused on crop improvement for food sustainability, but little has been done on the phytoremediation issue.

8.3 CHALLENGES OF PHYTOREMEDIATION AND PROSPECTS OF PHYTOREMEDIATION

Phytoremediation is believed to be an environmentally friendly, cost-effective, solar-driven, sustainable, and socially accepted technology (Hussain *et al.*, 2018). However, due to the knowledge gap and the hard field experiments, the technology field applicability is limited (Odoh *et al.*, 2019). Due to the slow growth rate of plants, biomass couldn't increase very fast. The unavailability of optional studied plant genotype and the longer life cycle of the identified plant is another limitation of this technology (Saxena *et al.*, 2020). Furthermore, the big challenge for this technology could be the final disposal of phytoextractors and phytoaccumulators. Moreover, sometimes transgenic plants move from laboratory to field application; the field performance of genetically modified plants to multiple contaminants may not be promising. The environmental concerns of the transgenic plants are still under consideration (Maestri and Marmiroli, 2011; Vazquez-Nunez *et al.*, 2018). A longer time is required for maximum removal of contaminants from the site, so it is a time-consuming process; even then, the complete removal of the contaminants is not guaranteed (Patra *et al.*, 2020). Collecting the disposal might take the maximum time in months to accomplish the task, whereas phytodegradation or phytoextraction might take several years (Bauddh *et al.*, 2017).

The phytoremediation process is dependent on edaphic factors and soil chemistry; whereas, the soil pH, conductivity, porosity, nutrient levels, and presence of soil microbes are instrumental in deciding the uptake mechanisms of the plants. Climate is also a factor to determine the remediation either positively or negatively. Stressed climate reduces the biomass of the plant and prolongs the remediation time. Another factor that might hamper the phytoremediation potential is the age of the plant. Younger plants could remediate better than older plants. Older plants might hold more toxic pollutants. Agronomic practice and soil amendment may negatively influence the mobility of contaminates (Mahar *et al.*, 2016). Transgenic plants may be an environmental concern as well as human or animal health concerns if the horizontal gene pollution happened. In the case of transgenic plants, the possible risk and the proper management techniques during transformation should be considered.

8.4 CONCLUSION

Soil and water pollution with toxic metals and organic pollutants is one of the most burning global concerns due to its risk to human health and the environment (Patra *et al.*, 2020). The phytoremediation is an attractive option for the alleviation of heavy metals and organic pollutants from the contaminated environment (Odoh *et al.*, 2019). Phytoremediation is a technological asset using the natural potential of particular plants and/or their associated microbes or transgenic plants to immobilize, detoxify, transform, extract, sequester, or evaporate various toxic metals and organic pollutants in the environment, especially soil, surface waters, and groundwater (Hussain *et al.*, 2018). The phytoremediation of different pollutants has been frequently performed using techniques like phytostabilization, phytoextraction, and rhizofiltration. The BF should exceed one while the TF should be less than one for both phytostabilization and rhizofiltration processes. Both the BF and the TF should be greater than one for the phytoextraction process. Plant transformation through genetic transformation/gene editing is a recent advancement for environmental protection. Once phytoremediation is performed using any kind of plant, care should be taken for the disposal of pollutant-accumulated plants because they pose risk to human and environmental health.

REFERENCES

Adimalla, N. (2019). Heavy metals contamination in urban surface soils of Medak province, India, and its risk assessment and spatial distribution. *Environmental Geochemistry and Health*, *42*(1), 59–75.

Adu, D. T., and Denkyirah, E. K. (2017). Economic growth and environmental pollution in West Africa: Testing the environmental Kuznets curve hypothesis. *Kasetsart Journal of Social Sciences*. 1–8. DOI: 10.1016/j.kjss.2017.12.008.

Ahemad, M. (2019). Remediation of metalliferous soils through the heavy metal resistant plant growth promoting bacteria: Paradigms and prospects. *Arabian Journal of Chemistry, 12*(7), 1365–1377.

Alemu, A. (2019). *Potentials of Local Rock Substrate and Plant Species for an Integrated Treatment of Chromium (Cr) Containing Tannery Wastewater Using Constructed Wetland Systems (CWSs).*Addis Ababa University, Bahir Dar, Ethiopia.

Alharbi, O. M., Khattab, R. A., and Ali, I. (2018). Health and environmental effects of persistent organic pollutants.*Journal of Molecular Liquids, 263*, 442–453.

Amare, E., Kebede, F., Berihu, T., and Mulat, W. (2018a). Field-based investigation on phytoremediation potentials of *Lemna minor* and *Azolla filiculoides* in tropical, semi-arid regions: Case of Ethiopia. *International Journal of Phytoremediation, 20*(10), 965–972.

Amare, E., Kebede, F., and Mulat, W. (2017). Analysis of heavy metals, physicochemical parameters and effect of blending on treatability of wastewaters in Northern Ethiopia. *International Journal of Environmental Science and Technology, 14*(8), 1679–1688.

Amare, E., Kebede, F., and Mulat, W. (2018b). Wastewater treatment by *Lemna minor* and *Azolla filiculoides* in tropical semi-arid regions of Ethiopia. *Ecological Engineering, 120*, 464–473.

Amberber, M., and Kifle, F. (2016). Assessments of heavy metal accumulation capacity of selected plant species for phytoremediation: A case study in Little Akaki river. *Journal of Bioremediation and Biodegradation, 7*(369), 2.

Andreazza, R., Bortolon, L., Pieniz, S., Bento, F. M., and Camargo, F. A. d. O. (2015). Evaluation of two Brazilian indigenous plants for phytostabilization and phytoremediation of copper-contaminated soils. *Brazilian Journal of Biology 75* (4), 868–877.

Ashraf, S., Ali, Q., Zahir, Z. A., Ashraf, S., and Asghar, H. N. (2019). Phytoremediation: Environmentally sustainable way for reclamation of heavy metal polluted soils. *Ecotoxicology and Environmental Safety, 174*, 714–727.

Awa, S. H., and Hadibarata, T. (2020). Removal of heavy metals in contaminated soil by phytoremediation mechanism: A review. *Water, Air, & Soil Pollution, 231*(2), 47.

Azab, E., Hegazy, A. K., El-Sharnouby, M. E., and AbdElsalam, H. E. (2016). Phytoremediation of the organic xenobiotic simazine by p450-1a2 transgenic *Arabidopsis thaliana* plants. *International Journal of Phytoremediation, 18*(7), 738–746.

Bader, N., Alsharif, E., Nassib, M., Alshelmani, N., and Alalem, A. (2018). Phytoremediation potential of *Suaeda vera* for some heavy metals in roadside soil in Benghazi, Libya. *Asian Journal of Green Chemistry, 3*(1), 82–90.

Bader, N., Faraj, M., Mohamed, A., Alshelmani, N., Elkailany, R., and Bobtana, F. (2020). Evaluation of the phytoremediation performance of *Hammada scoparia* and *Halocnemum Strobilaceum* for Cu, Fe, Zn and Cr accumulation from the industrial area in Benghazi, Libya. *Journal of Medicinal and Chemical Sciences, 3*(2), 138–144.

Basharat, Z., Novo, L. A., and Yasmin, A. (2018). Genome editing weds CRISPR: What is in it for phytoremediation? *Plants, 7*(3), 51.

Basu, S., Rabara, R. C., Negi, S., and Shukla, P. (2018). Engineering PGPMOs through gene editing and systems biology: A solution for phytoremediation? *Trends in Biotechnology, 36*(5), 499–510.

Bauddh, K., Singh, B., and Korstad, J. (2017). *Phytoremediation Potential of Bioenergy Plants.* Springer Nature, Cham, Switzerland.

Bauddh, K., Singh, K., Singh, B., and Singh, R. P. (2015). *Ricinus communis*: A robust plant for bio-energy and phytoremediation of toxic metals from contaminated soil. *Ecological Engineering, 84*, 640–652.

Bekele, M. (2018). *College of Biological and Chemical Engineering Department of Environmental Engineering.* Addis Ababa Science and Technology University, Addis Ababa, Ethiopia.

Benavides, L. C. L., Pinilla, L. A. C., Serrezuela, R. R., and Serrezuela, W. F. R. (2018). Extraction in laboratory of heavy metals through rhizofiltration using the plant *Zea mays* (maize). *International Journal of Applied Environmental Sciences, 13*(1), 9–26.

BenIsrael, M., Wanner, P., Fernandes, J., Burken, J. G., Aravena, R., Parker, B. L., … Dunfield, K. E. (2020). Quantification of toluene phytoextraction rates and microbial biodegradation functional profiles at a fractured bedrock phytoremediation site. *Science of the Total Environment, 707*, 135890.

Cai, L.-M., Wang, Q.-S., Luo, J., Chen, L.-G., Zhu, R.-L., Wang, S., and Tang, C.-H. (2019). Heavy metal contamination and health risk assessment for children near a large Cu-smelter in central China. *Science of the Total Environment, 650*, 725–733.

Capozzi, F., Sorrentino, M. C., Caporale, A. G., Fiorentino, N., Giordano, S., and Spagnuolo, V. (2020). Exploring the phytoremediation potential of *Cynara cardunculus*: A trial on an industrial soil highly contaminated by heavy metals. *Environmental Science and Pollution Research, 27*, 9075–9084

Chang, G., Yue, B., Gao, T., Yan, W., and Pan, G. (2020). Phytoremediation of phenol by *Hydrilla verticillata* (Lf) Royle and associated effects on physiological parameters. *Journal of Hazardous Materials, 388*, 121569.

Chen, Y., Xu, W., Shen, H., Yan, H., Xu, W., He, Z., and Ma, M. (2013). Engineering arsenic tolerance and hyperaccumulation in plants for phytoremediation by a *PvACR3* transgenic approach. *Environmental Science &Technology, 47*(16), 9355–9362.

Chirakkara, R. A., Cameselle, C., and Reddy, K. R. (2016). Assessing the applicability of phytoremediation of soils with mixed organic and heavy metal contaminants. *Reviews in Environmental Science and Bio/Technology, 15*(2), 299–326.

Christian, E., and Beniah, I. (2019). Phytoremediation of polluted waterbodies with aquatic plants: Recent progress on heavy metal and organic pollutants. *Analytical Chemistry.* DOI: 10.20944/preprints201909.0020.v1.

Clay, L., and Pichtel, J. (2019). Treatment of simulated oil and gas produced water via pilot-scale rhizofiltration and constructed wetlands. *International Journal of Environmental Research, 13*(1), 185–198.

Cui, H., Li, H., Zhang, S., Yi, Q., Zhou, J., Fang, G., and Zhou, J. (2020). Bioavailability and mobility of copper and cadmium in polluted soil after phytostabilization using different plants aided by limestone. *Chemosphere, 242*, 125252.

Darajeh, N., Truong, P., Rezania, S., Alizadeh, H., and Leung, D. (2019). Effectiveness of Vetiver grass versus other plants for phytoremediation of contaminated water. *Journal of Environmental Treatment Techniques, 7*, 485–500.

Dimitrova, T., Repmann, F., and Freese, D. (2018). Detoxification of ferrocyanide in asoil–Plant system. *Journal of Environmental Sciences, 77*, 54–64.

Doabi, S. A., Karami, M., Afyuni, M., and Yeganeh, M. (2018). Pollution and health risk assessment of heavy metals in agricultural soil, atmospheric dust and major food crops in Kermanshah province, Iran. *Ecotoxicology and Environmental Safety, 163*, 153–164.

Doty, S. L. (2008). Enhancing phytoremediation through the use of transgenics and endophytes. *New Phytologist, 179*(2), 318–333.

Eid, E. M., Galal, T. M., Sewelam, N. A., Talha, N. I., and Abdallah, S. M. (2020). Phytoremediation of heavy metals by four aquatic macrophytes and their potential use as contamination indicators: A comparative assessment. *Environmental Science and Pollution Research, 27*, 12138–12151.

El-Aassar, M., Fakhry, H., Elzain, A. A., Farouk, H., and Hafez, E. E. (2018). Rhizofiltration system consists of chitosan and natural *Arundo donax* L. for removal of basic red dye. *International Journal of Biological Macromolecules, 120*, 1508–1514.

Farahat, E. A., and Galal, T. M. (2017). Trace metal accumulation by *Ranunculus sceleratus*: Implications for phytostabilization. *Environmental Science and Pollution Research, 25*(5), 4214–4222.

Farraji, H., Zaman, N. Q., Tajuddin, R., and Faraji, H. (2016). Advantages and disadvantages of phytoremediation: A concise review. *International Journal of Environmental Technical Science*, *2*, 69–75.

Fatima, K., Imran, A., Amin, I., Khan, Q. M., and Afzal, M. (2018). Successful phytoremediation of crude-oil contaminated soil at an oil exploration and production company by plants-bacterial synergism. *International Journal of Phytoremediation*, *20*(7), 675–681.

Franchi, E., Cosmina, P., Pedron, F., Rosellini, I., Barbafieri, M., Petruzzelli, G., and Vocciante, M. (2019). Improved arsenic phytoextraction by combined use of mobilizing chemicals and autochthonous soil bacteria. *Science of the Total Environment*, *655*, 328–336.

Francis, E. (2017). Phytoremediation potentials of sunflower (*Helianthus annuus* L.) Asteraceae on contaminated soils of abandoned dumpsites. *International Journal of Scientific Engineering Research*, *8*(1), 1751–1757.

Galal, T. M., Al-Sodany, Y. M., and Al-Yasi, H. M. (2019). Phytostabilization as a phytoremediation strategy for mitigating water pollutants by the floating macrophyte *Ludwigia stolonifera* (Guill.& Perr.) PH Raven. *International Journal of Phytoremediation*, *22*(4), 373–382.

Galal, T. M., Eid, E. M., Dakhil, M. A., and Hassan, L. M. (2017a). Bioaccumulation and rhizofiltration potential of *Pistia stratiotes* L. for mitigating water pollution in the Egyptian wetlands. *International Journal of Phytoremediation*, *20*(5), 440–447.

Galal, T. M., Gharib, F. A., Ghazi, S. M., and Mansour, K. H. (2017b). Phytostabilization of heavy metals by the emergent macrophyte *Vossia cuspidata* (Roxb.)Griff.: a phytoremediation approach. *International Journal of Phytoremediation*, *19*(11), 992–999.

Ghosh, M., Paul, J., Jana, A., De, A., and Mukherjee, A. (2014). Use of the grass, *Vetiveria zizanioides* (L.) Nash for detoxification and phytoremediation of soils contaminated with fly ash from thermal power plants. *Ecological Engineering*, *74*, 258–265.

Girmay, M. (2019). Phytoremediation of heavy metals released from mining waste drainage using selected plant species, in Ethiopia. *International Journal of Mining Science (IJMS)*, *5*(3), 1–10.

Guarino, F., Miranda, A., Cicatelli, A., and Castiglione, S. (2020). Arsenic phytovolatilization and epigenetic modifications in *Arundo donax* L. assisted by a PGPR consortium. *Chemosphere*, *251*, 126310.

Hauptvogl, M., Kotrla, M., Prčík, M., Pauková, Ž., Kováčik, M., and Lošák, T. (2019). Phytoremediation potential of fast-growing energy plants: Challenges and perspectives–A review. *Polish Journal of Environmental Studies*, *29*(1), 505–516.

Haynes, R. (2015). Use of industrial wastes as media in constructed wetlands and filter beds— Prospects for removal of phosphate and metals from wastewater streams. *Critical Reviews in Environmental Science and Technology*, *45*(10), 1041–1103.

Hou, X., Teng, W., Hu, Y., Yang, Z., Li, C., Scullion, J., … Zheng, R. (2020). Potential phytoremediation of soil cadmium and zinc by diverse ornamental and energy grasses. *BioResources*, *15*(1), 616–640.

Hussain, I., Aleti, G., Naidu, R., Puschenreiter, M., Mahmood, Q., Rahman, M. M., … Reichenauer, T. G. (2018). Microbe and plant assisted-remediation of organic xenobiotics and its enhancement by genetically modified organisms and recombinant technology: A review. *Science of the Total Environment*, *628*, 1582–1599.

Ijaz, A., Imran, A., ul Haq, M. A., Khan, Q. M., and Afzal, M. (2016). Phytoremediation: Recent advances in plant-endophytic synergistic interactions. *Plant and Soil*, *405*(1–2), 179–195.

Khan, A. R., Ullah, I., Waqas, M., Park, G.-S., Khan, A. L., Hong, S.-J., … Ur-Rehman, S. (2017). Host plant growth promotion and cadmium detoxification in *Solanum nigrum*, mediated by endophytic fungi. *Ecotoxicology and Environmental Safety*, *136*, 180–188.

Khan, M. I., Cheema, S. A., Anum, S., Niazi, N. K., Azam, M., Bashir, S., ... Qadri, R. (2020). Phytoremediation of agricultural pollutants. *Phytoremediation* (pp. 27–81): Springer Nature, Cham, Switzerland.

Kong, Z., and Glick, B. R. (2017). The role of plant growth-promoting bacteria in metal phytoremediation. *Advances in Microbial Physiology* (Vol. 71, pp. 97–132): Elsevier, Amsterdam, Netherlands.

Kotrba, P., Najmanova, J., Macek, T., Ruml, T., and Mackova, M. (2009). Genetically modified plants in phytoremediation of heavy metal and metalloid soil and sediment pollution. *Biotechnology Advances, 27*(6), 799–810.

Koźmińska, A., Wiszniewska, A., Hanus-Fajerska, E., and Muszyńska, E. (2018). Recent strategies of increasing metal tolerance and phytoremediation potential using genetic transformation of plants. *Plant Biotechnology Reports, 12*(1), 1–14.

Lajayer, B. A., Moghadam, N. K., Maghsoodi, M. R., Ghorbanpour, M., and Kariman, K. (2019). Phytoextraction of heavy metals from contaminated soil, water and atmosphere using ornamental plants: Mechanisms and efficiency improvement strategies. *Environmental Science and Pollution Research, 26*(9), 8468–8484.

Legault, E. K., James, C. A., Stewart, K., Muiznieks, I., Doty, S. L., and Strand, S. E. (2017). A field trial of TCE phytoremediation by genetically modified poplars expressing cytochrome P450 2E1.*Environmental Science &Technology, 51*(11), 6090–6099.

Li, X., Wang, X., Chen, Y., Yang, X., and Cui, Z. (2019). Optimization of combined phytoremediation for heavy metal contaminated mine tailings by a field-scale orthogonal experiment. *Ecotoxicology and Environmental Safety, 168*, 1–8.

Limmer, M., and Burken, J. (2016). Phytovolatilization of organic contaminants. *Environmental Science &Technology, 50*(13), 6632–6643.

Liu, J., Xin, X., and Zhou, Q. (2018). Phytoremediation of contaminated soils using ornamental plants. *Environmental Reviews, 26*(1), 43–54.

Luo, J., Cao, M., Zhang, C., Wu, J., and Gu, X. S. (2020). The influence of light combination on the physicochemical characteristics and enzymatic activity of soil with multi-metal pollution in phytoremediation. *Journal of Hazardous Materials, 393*, 122406.

Luo, Z.-B., He, J., Polle, A., and Rennenberg, H. (2016). Heavy metal accumulation and signal transduction in herbaceous and woody plants: Paving the way for enhancing phytoremediation efficiency. *Biotechnology Advances, 34*(6), 1131–1148.

Ma, Y., Rajkumar, M., Zhang, C., and Freitas, H. (2016). Beneficial role of bacterial endophytes in heavy metal phytoremediation. *Journal of Environmental Management, 174*, 14–25.

Maestri, E., and Marmiroli, N. (2011). Transgenic plants for phytoremediation. *International Journal of Phytoremediation, 13*(Suppl. 1), 264–279.

Mahar, A., Wang, P., Ali, A., Awasthi, M. K., Lahori, A. H., Wang, Q., ... Zhang, Z. (2016). Challenges and opportunities in the phytoremediation of heavy metals contaminated soils: A review. *Ecotoxicology and Environmental Safety, 126*, 111–121.

Misra, S., and Misra, K. G. (2019). Phytoremediation: an alternative tool towards clean and green environment. *Sustainable Green Technologies for Environmental Management* (pp. 87–109): Springer, Singapore.

Morkunas, I., Woźniak, A., Mai, V. C., Rucińska-Sobkowiak, R., and Jeandet, P. (2018). The role of heavy metals in plant response to biotic stress. *Molecules, 23*(9), 2320.

Nahar, N., Rahman, A., Nawani, N. N., Ghosh, S., and Mandal, A. (2017). Phytoremediation of arsenic from the contaminated soil using transgenic tobacco plants expressing ACR2 gene of *Arabidopsis thaliana. Journal of Plant Physiology, 218*, 121–126.

Ng, C. C., Boyce, A. N., Abas, M. R., Mahmood, N. Z., and Han, F. (2020). Evaluation of vetiver grass uptake efficiency in single and mixed heavy metal contaminated soil. *Environmental Processes, 7*(1), 207–226.

Odoh, C. K., Zabbey, N., Sam, K., and Eze, C. N. (2019). Status, progress and challenges of phytoremediation– An African scenario. *Journal of Environmental Management*, *237*, 365–378.

Oluoch, J. O. (2018). *Phytoremediation Potential of Cyperus Alternifolius, Cyperus Dives and Canna Indica in Flamingo Farm Constructed Wetland, Naivasha Sub-County, Kenya*. School of Environmental Studies, Kenyatta University, Nairobi, Kenya.

Pan, P., Lei, M., Qiao, P., Zhou, G., Wan, X., and Chen, T. (2019). Potential of indigenous plant species for phytoremediation of metal (loid)-contaminated soil in the Baoshan mining area, China. *Environmental Science and Pollution Research*,*26*(23), 23583–23592.

Patra, D. K., Pradhan, C., and Patra, H. K. (2018). An in situ study of growth of Lemongrass Cymbopogon flexuosus (Nees ex Steud.) W. Watson on varying concentration of Chromium (Cr^{+6}) on soil and its bioaccumulation: Perspectives on phytoremediation potential and phytostabilisation of chromium toxicity. *Chemosphere*, *193*, 793–799.

Patra, D. K., Pradhan, C., and Patra, H. K. (2020). Toxic metal decontamination by phytoremediation approach: Concept, challenges, opportunities and future perspectives. *Environmental Technology & Innovation*, *18*, 100672.

Paul, M. S., Varun, M., D'Souza, R., Favas, P. J., and Pratas, J. (2014). Metal contamination of soils and prospects of phytoremediation in and around river Yamuna: A case study from North-Central India. *Environmental Risk Assessment of Soil Contamination*. IntechOpen.

Paz-Alberto, A. M., and Sigua, G. C. (2013). Phytoremediation: A green technology to remove environmental pollutants. *American Journal of Climate Change*, *2*, 71–86.

Pérez-Palacios, P., Romero-Aguilar, A., Delgadillo, J., Doukkali, B., Caviedes, M. A., Rodríguez-Llorente, I. D., and Pajuelo, E. (2017). Double genetically modified symbiotic system for improved Cu phytostabilization in legume roots. *Environmental Science and Pollution Research*, *24*(17), 14910–14923.

Qiao, D., Wang, G., Li, X., Wang, S., and Zhao, Y. (2020). Pollution, sources and environmental risk assessment of heavy metals in the surface AMD water, sediments and surface soils around unexploited Rona Cu deposit, Tibet, China. *Chemosphere*, *248*, 125988.

Rahman, M. A., Reichman, S. M., De Filippis, L., Sany, S. B. T., and Hasegawa, H. (2016). Phytoremediation of toxic metals in soils and wetlands: Concepts and applications. *Environmental Remediation Technologies for Metal-Contaminated Soils* (pp. 161–195): Springer, Japan.

Rahman, M. M., Azirun, S. M., and Boyce, A. N. (2013). Enhanced accumulation of copper and lead in amaranth (*Amaranthus paniculatus*), Indian mustard (*Brassica juncea*) and sunflower (*Helianthus annuus*). *PLoS ONE*, *8*(5), e62941.

Rai, P. K., Kim, K.-H., Lee, S. S., and Lee, J.-H.(2020). Molecular mechanisms in phytoremediation of environmental contaminants and prospects of engineered transgenic plants/ microbes. *Science of the Total Environment*, *705*, 135858.

Rastogi, M., and Nandal, M. (2020). Toxic metals in industrial wastewaters and phytoremediation using aquatic macrophytes for environmental pollution control: An eco-remedial approach. *Bioremediation of Industrial Waste for Environmental Safety* (pp. 257–282): Springer, Singapore.

Reddy, P. C. O., Raju, K. S., Sravani, K., Sekhar, A. C., and Reddy, M. K. (2019). Transgenic Plants for Remediation of Radionuclides. *Transgenic Plant Technology for Remediation of Toxic Metals and Metalloids* (pp. 187–237): Elsevier, Amsterdam, Netherlands.

Rissato, S. R., Galhiane, M. S., Fernandes, J. R., Gerenutti, M., Gomes, H. M., Ribeiro, R., and Almeida, M. V. d. (2015). Evaluation of *Ricinus communis* L. for the phytoremediation of polluted soil with organochlorine pesticides. *BioMed Research International*, *2015*, Article ID 549863.

Sakakibara, M., Watanabe, A., Inoue, M., Sano, S., and Kaise, T. (2010). *Phytoextraction and phytovolatilization of arsenic from As-contaminated soils by Pteris vittata*.

Paper presented at the Proceedings of the Annual International Conference on Soils, Sediments, Water and Energy.

Saxena, G., Purchase, D., Mulla, S. I., Saratale, G. D., and Bharagava, R. N. (2020). Phytoremediation of heavy metal-contaminated sites: Eco-environmental concerns, field studies, sustainability issues, and future prospects. *Reviews of Environmental Contamination and Toxicology, 249,* 71–131.

Shim, D., Choi, Y.-I., Park, J., Kim, S., Hwang, J.-U., Noh, E.-W., and Lee, Y. (2016). Engineering poplar plants for phytoremediation. http://www.naro.affrc.go.jp/archive/niaes/marco/marco2009/english/program/W1-05_LeeYoungsook.pdf

Shukla, D., Kesari, R., Mishra, S., Dwivedi, S., Tripathi, R. D., Nath, P., and Trivedi, P. K. (2012). Expression of phytochelatin synthase from aquatic macrophyte *Ceratophyllum demersum* L. enhances cadmium and arsenic accumulation in tobacco. *Plant Cell Reports, 31*(9), 1687–1699.

Siyar, R., Ardejani, F. D., Farahbakhsh, M., Norouzi, P., Yavarzadeh, M., and Maghsoudy, S. (2020). Potential of vetiver grass for the phytoremediation of a real multi-contaminated soil, assisted by electrokinetic. *Chemosphere, 246,* 125802.

Stepanova, A. Y., Orlova, E., Teteshonok, D., and Dolgikh, Y. I. (2016). Obtaining transgenic alfalfa plants for improved phytoremediation of petroleum-contaminated soils. *Russian Journal of Genetics: Applied Research, 6*(6), 705–711.

Subhashini, V., Swamy, A., and Krishna, R. H. (2013). Phytoremediation: Emerging and green technology for the uptake of cadmium from the contaminated soil by plant species. *International Journal of Environmental Sciences, 4*(2), 193–204.

Talano, M. A., Busso, D. C., Paisio, C. E., González, P. S., Purro, S. A., Medina, M. I., and Agostini, E. (2012). Phytoremediation of 2, 4-dichlorophenol using wild type and transgenic tobacco plants. *Environmental Science and Pollution Research, 19*(6), 2202–2211.

Tang, L., Mao, B., Li, Y., Lv, Q., Zhang, L., Chen, C., …Shao, Y. (2017). Knockout of OsNramp5 using the CRISPR/Cas9 system produces low Cd-accumulating indica rice without compromising yield. *Scientific Reports, 7*(1), 1–12.

Thakur, S., Singh, L., Ab Wahid, Z., Siddiqui, M. F., Atnaw, S. M., and Din, M. F. M. (2016). Plant-driven removal of heavy metals from soil: Uptake, translocation, tolerance mechanism, challenges, and future perspectives. *Environmental Monitoring and Assessment, 188*(4), 206.

Tiwari, J., Kumar, S., Korstad, J., and Bauddh, K. (2019). Ecorestoration of polluted aquatic ecosystems through rhizofiltration. *Phytomanagement of Polluted Sites* (pp. 179–201): Elsevier, Amsterdam, Netherlands.

Truu, J., Truu, M., Espenberg, M., Nõlvak, H., and Juhanson, J. (2015). Phytoremediation and plant-assisted bioremediation in soil and treatment wetlands: A review. *The Open Biotechnology Journal, 9*(1), 85–92.

Usman, K., Al-Ghouti, M. A., and Abu-Dieyeh, M. H. (2018). Phytoremediation: halophytes as promising heavy metal hyperaccumulators. *Heavy Metals.* (pp. 201–210): IntechOpen, London, UK.

Vazquez-Nunez, E., Pena-Castro, J. M., Fernandez-Luqueno, F., Cejudo, E., Maria, G., and Garcia-Castaneda, M. C. (2018). A review on genetically modified plants designed to phytoremediate polluted soils: Biochemical responses and international regulation. *Pedosphere, 28*(5), 697–712.

Viktorova, J., Jandova, Z., Madlenakova, M., Prouzova, P., Bartunek, V., Vrchotova, B., … Macek, T. (2016). Native phytoremediation potential of *Urtica dioica* for removal of PCBs and heavy metals can be improved by genetic manipulations using constitutive CaMV 35S promoter. *PLoS ONE, 11*(12), e0167927.

Weyens, N., van der Lelie, D., Taghavi, S., and Vangronsveld, J. (2009). Phytoremediation: Plant–endophyte partnerships take the challenge. *Current Opinion in Biotechnology, 20*(2), 248–254.

Worku, A., Tefera, N., Kloos, H., and Benor, S. (2018). Bioremediation of brewery wastewater using hydroponics planted with vetiver grass in Addis Ababa, Ethiopia. *Bioresources and Bioprocessing*, 5(1), 39.

Zare, K., Sheykhi, V., and Zare, M. (2020). Investigating the heavy metals' removal capacity of some native plant species from the wetland groundwater of Maharlu Lake in Fars province, Iran. *International Journal of Phytoremediation*, 22(7), 781–788.

Zarrin, B., and Azra, Y. (2018). Genome editing weds CRISPR: what is in it for phytoremediation? *Plants*, 7(3), 51–57.

Zhang, J., Martinoia, E., and Lee, Y. (2018). Vacuolar transporters for cadmium and arsenic in plants and their applications in phytoremediation and crop development. *Plant and Cell Physiology*, 59(7), 1317–1325.

Zhang, L., Rylott, E. L., Bruce, N. C., and Strand, S. E. (2017). Phytodetoxification of TNT by transplastomic tobacco (*Nicotiana tabacum*) expressing a bacterial nitroreductase. *Plant Molecular Biology*, 95(1–2), 99–109.

Zhang, L., Rylott, E. L., Bruce, N. C., and Strand, S. E. (2018). Genetic modification of western wheatgrass (*Pascopyrum smithii*) for the phytoremediation of RDX and TNT. *Planta*, 249(4), 1007–1015.

Zhong, T., Xue, D., Zhao, L., and Zhang, X. (2017). Concentration of heavy metals in vegetables and potential health risk assessment in China. *Environmental Geochemistry and Health*, 40(1), 313–322.

9 Utilization of Nanoparticle-Loaded Adsorbable Materials for Leachate Treatment

Kulbir Singh
Guru Jambheshwar University of Science and Technology

CONTENTS

9.1 INTRODUCTION

The excessive generation of municipal solid waste (MSW) has been identified due to rapid economic development in recent years. Around the world, 2.01 billion tons of solid wastes were generated in 2016, and expected to increase by 70% to 3.40 billion tons in 2050 (World Bank, 2019). Generally, MSW is disposed of through local landfills due to cost-effective processes, and this process is used in 70% of the world's countries, especially in developing countries. But due to the generation of huge quantities of solid waste, more existing landfills have touched their maximum capacity faster than their proposed life. So it has become very difficult to construction new landfill sites due to lack of land availability for waste disposal (Moh and Manaf, 2014). MSW landfills encompass more biodegradable and non-biodegradable compounds that gives the ability for a multifarious sequence of microbiological and chemical reactions. However, an inadequacy of this practice is the alterations in microbial communities that require functioning at various decomposition phases (Del Moro et al., 2016). The MSW is prejudiced by their materialization depending on the kind of waste buried, decomposition through microorganism, occurrence of a bio-cover, etc. Afterward, MSW is degraded through microbes and physico-chemical and biological processes that occur. These processes generate contaminated wastewater called landfill leachate, which creates serious health, safety, and environmental consequences. It is estimated that one ton of landfilled MSW produces 0.2 m^3 of leachate as waste (Kurniawan and Lo, 2009). One report showed that waterborne diseases are still a major issue in the developing world where the availability of the safe drinking water is limited (Sadegh et al., 2017).

When moisture content is high in solid waste, more leachate is generated. The leachate may comprise large quantities of hardly biodegradable organic matter. The leachate composition is influenced by various factors that include waste composition, local climatic conditions, landfill practices, and landfill age. Mature landfill leachate has a low (Biological Oxygen Demand)$_5$/(Chemical Oxygen Demand) (BOD$_5$/COD) ratio (<0.1); therefore, conventional biological treatment efficiency is very low, releasing in more concentrations of recalcitrant organic molecules and ammonia nitrogen, which stipulate the requirement of further treatment using chemical and physical methods.

Different types of treatments have been explored such as membrane filtration, chemical precipitation, coagulation/flocculation, ultrasound, advanced oxidation processes (AOPs), and adsorption. In the adsorption process, a substance bonds on a solid material surface from a liquid phase. Numerous types of adsorbent materials such as clay, zeolites, activated carbon, and some bio-sorbents like rice husk, corncob, bagasse, etc. are used for leachate treatment. Based on previous studies, leachate characteristics, the mechanisms of modified adsorbents and their application for leachate treatment, are discussed in this study.

9.2 LEACHATE: CHARACTERIZATION AND AGE

Landfill leachate is usually categorized as high strength wastewater, containing high COD, BOD, grease and oil, as well as total suspended solid (TSS) (both organic and inorganic) as compared to residential wastewater (Shehzad et al., 2016). The general diagram of landfill leachate preparation is shown in Figure 9.1. The leachate quantity depends on water percolation through wastes, the inherent water content of wastes, biochemical processes in waste's cells, and its compaction level on the landfill site. The leachate production, ordinarily larger during the waste, is less compacted since compaction diminishes the filtration rate (Lema et al., 1988). The BOD or COD values of leachate quantified the concentration of organic matters, while degradation of organic substances indicate deamination of amino acids, which is represented by the ammoniacal nitrogen (NH$_3$-N) concentration. In landfill leachate the large quantity

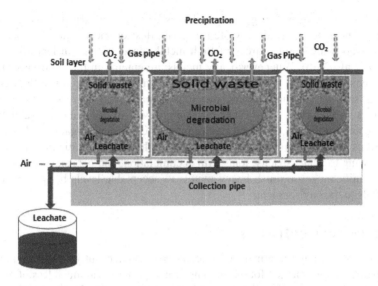

FIGURE 9.1 The mechanism of landfill leachate preparation.

of the ammonia-nitrogen group are in ammonium ion form, usually due to the pH values less than 8.0 (Moreira et al., 2015). This dissolved unionized ammonia is more harmful for anaerobic decomposition as compared to ammonium ions in leachate. Indeed, the existence of humic acid in leachate has extended other consideration (Hilles et al., 2016). Humic acid, having the non-biodegradable segments along with structural complexity, has a greater impact on the action of several pollutants in the environment.

The particular composition of leachate (aerobic, acetogenic, methanogenic, and stabilization) has three types of leachates according to landfill age (young, medium, and old) (Silva et al., 2004, Vazquez et al., 2004). The values of COD 70,900 mgL^{-1}, 5350 mgL^{-1}, and 100 mgL^{-1} was determined for the young, medium, and old age landfill leachate (Tabet et al., 2002; Tatsi et al., 2003).The landfill works as a massive reactor for bacteriological degradation of MSW, and the acidogenic phase occurs for up to five years (Oulego et al., 2016). The leachate formed at this period is described with low pH (due to acid molecules), high COD value (above 10,000 mg/L), and low NH_3-N. The landfill proceeds toward the methanogenic phase with the progression of time. However, the leachate involves a number of changes, including decline of COD as well as an increase in pH and NH_3-N values (Youcai et al., 2000). In addition to that, research also reveals that the COD value can be lower than 3000 mg/L for 10-year-old leachate in landfills. These types of leachate are known as stablished leachate.

9.2.1 IMPACTS OF LEACHATE ON THE ENVIRONMENT

The untreated leachate discharged from landfill sites causes groundwater and surface water pollution that affects water quality in the nearby areas (Sizirici and Yildiz,

2017). Moreover, in waste, growing volumes of scrap, electrical tools, electronic waste, spent batteries, sludge, waste oil, paint, dust of spent catalysts, chemically treated woods are main sources of high metal concentrations in landfill leachate (Mojiri et al., 2016). Whenever heavy metals are entered in the environment, they may accumulate in living systems and originate more adversarial health effects on humans such as anemia, cancers, pancreas damage, and distress to the stomach intestines (Ajima et al., 2015). The escalating levels of pollutants in water bodies have become a serious universal environmental challenge. Organic and inorganic impurities and heavy metals comprising most pollutants are initiating greater concerns owing to acute toxicity, long-term accumulation, and persistence. Hence, it is more significant to diminish heavy metals to acceptable levels by treating earlier disposals to the environment.

9.2.2 TREATMENT OF LEACHATE

Initially, an ordinary treatment of leachate was carried out in combination with domestic sewage, selected for an uncomplicated process and much less of an operating cost for a conventional sewage plant (Ahn et al., 2002). Due to the occurrence of organic inhibitory compounds, along with less biodegradability and heavy metals, this option has been increasingly objectionable. These contents may relegate treatment effectiveness and raise the concentrations of effluent (Cecen and Aktas, 2004).

Another technique based upon the recycling of landfill leachate has been used for its treatment and is considered a low-cost method. Through the recirculation process, the moisture content increases in the reactor system, and there is a substantial lowering in methane production and COD contents at a recirculated leachate volume of 30% of the original waste bed volume (Bae et al., 1998; Chugh et al., 1998). But adverse effects on anaerobic degradation of solid waste has been observed due to high recirculation rates, and, also, problems such as ponding, saturation, and acidic wal., 2002; Renou et al., 2008; San and Onay, 2001). This reliable, simple, and highly cost-effective, biological process (suspended/attached growth) is frequently applied for leachate treatment having high concentrations of BOD. In this process, organic compounds degrade to carbon dioxide and sludge by microorganisms under aerobic conditions and to biogas (CO_2 and CH_4) under anaerobic conditions (Lema et al., 1988). This process is more operative in eliminating organic and nitrogenous contents from young leachates through higher BOD_5/COD ratio (>0.5) and tends to limit process's effectiveness, with time, of the main presence of refractory compounds (largely humic and fulvic acids).

The major portion of the biologically treated young or old leachate having bulky contents of recalcitrant organic molecules is characterized by low BOD, high COD, fairly high alkalinity and NH_3-N, low level of BOD/COD, dark brown or yellow color, and a high oxidation–reduction potential. The various treatment techniques available to treat the leachate waste include chemical oxidation (Forgie, 1988), electrochemical oxidation (Chiang et al., 2001), chemical precipitation and coagulation (Forgie, 1988), reverse osmosis (Chianese et al., 1999; Peters, 1998), nanofiltration (Trebouet et al., 1999, 2001), and adsorption (Halim et al., 2011;

Sizirici and Yildiz, 2017). However, out of these methods the adsorption process is preferred due to its low cost of waste production, cleanness, flexibility in design, and higher efficiency.

9.3 ADSORPTION: BASICS AND APPLICATION

A technique in which a material is moved from liquid part to the surface of a solid and becomes bound through physical and/or chemical interactions is known as adsorption (Kurniawan et al., 2006a). It is implemented extensively due to wide applications, including easy processing, lower operating cost compared to other techniques, and higher removal efficiency (Van Tran et al., 2017; Wendimu et al., 2017; Wong et al., 2017). Adsorption has been commonly applied to eliminate inorganic and organic substances from wastewater at low costs. Besides these, commercial adsorbents have some limitations due to the high treatment cost. The types of adsorbents used for the adsorption should be effective, economical, easy to grow or produce, demanding little handling, and are abundant in the environment like eco-materials (gravel, sand, and zeolite) or waste/by-product of industries. These characteristics make adsorbents competitive with commercially available ones.

9.3.1 ACTIVATED CARBON PRECURSORS

Activated carbons are identified as highly proficient adsorbents due to their appropriate surface area, well developed porosity, and a variable and high level of surface reactivity (Bansal et al., 1988; Singh and Waziri, 2019). These absolute characteristics make activated carbon a more flexible material, which has been considered not only as an adsorbent but also as catalyst and catalyst assistance. It can be used for different applications like the elimination of pollutants from liquid or gaseous phases and the refining or recovery of chemicals (Derbyshire et al., 2001). Activated carbon can be distinguished from elemental carbon using the oxidation process for the outer and inner surfaces carbon atom have (Al-Qodah and Shawabkah, 2009). It is basically a tasteless (Sugumaran et al.2012), amorphous (Campbell et al., 2012), microcrystalline (Jassim et al., 2012), non-graphite (Olafadehan et al., 2012) type of carbon and a black solid material that looks like powder or granular charcoal (WanNik et al., 2006). The activated carbon is a non-graphite state that cannot be transformed into crystalline graphite even at temperatures beyond 3000°C (Harris et al., 2008). In order to reduce the material cost, researchers are continuously trying to find substitutes of commercially activated carbons by producing adsorbents through waste materials. The various adsorbents have been identified from waste materials, including waste carpet (Hassan and Elhadidy, 2017), incineration fly ash (Xue et al., 2014) and tires (Lin and Wang, 2017). Through these processes, waste materials might be renovated to value-added products and also supportive in achieving the waste-to-wealth concept (Egun, 2012). In addition, it also has been recognized that biomass (waste material derived of organic origin) has a massive potential for the creation of adsorbents and more proficiency in decomposing under anaerobic or aerobic conditions (Wong et al., 2018).

9.3.1.1 Agricultural Waste as Activated Carbon Precursors

In commercial activated carbon, the main uses are petroleum residues, wood (Altenor et al., 2009), peat, coal, and lignite, which are more expensive and non-renewable (Ahmedna et al., 2000). Therefore, recently, more attention has been paid to activated carbon preparation through agro waste and lignocelluloses constituents that are cheap and effective (Singh et al., 2018), such as corncob (Tsai et al., 2001), hazel nutshell (Orkun et al., 2012), pruning mulberry shoot (Wang et al., 2010), olivestone (Ubago-Perez et al., 2006), jojoba seed, coconut shell (Li et al., 2008), wood (Yusufu et al., 2012), hazelnut bagasse (Demiral et al., 2008), kenaf fiber (Chowdhury et al., 2011), bamboo (Ademiluyi et al., 2011), rice husk (Yalcin and Sevinc, 2000), petai (Foo and Lee, 2010), ground nutshell (Malik et al., 2006), paper mill sludge (Khalili et al., 2000), Jatropha husk (Ramakrishnan and Namasivayam, 2009), tamarind wood (Sahu et al., 2010), pistachionut (Lua and Yang, 2005), sugarcane bagasse (Chen et al., 2012), jack fruit peel (Prahas et al., 2008), and many more.

9.3.2 PREPARATION OF ACTIVATED CARBON

The activated carbons are prepared applying two main practices, i.e. physical and chemical treatment (Al-Swaidan and Ahmad, 2011; Jun et al., 2010; Carrott and Carrott, 2007). These techniques are capable of changing the size and shape of materials (Imran, 2010). In the physical process, initially raw materials are carbonized followed by activation through carbon dioxide or steam. In comparison, an activating chemical is used in chemical treatment to impregnate the materials; after that, the material is heated in an inert atmosphere (Choy et al., 2005). The chemical treatment has presented more opportunities than physical treatment. The chemical treatment works at less temperatures, provides large surface area coverage (Zhu et al., 2008), more yield (Tsai et al., 1998), constructs well-developed micro porosities (Bello and Ahmad, 2011), and decreases more inorganic contents compared to the physical treatment (Lillo-Rodenas et al., 2003). However, there are some limitations in chemical treatment, like washing, which is required to eliminate the impurities that come from the activating agents and also the corrosiveness properties of the agents (Lozano-Castello et al., 2001).

9.3.2.1 Physical Activation

Generally, carbonization and activation are a two-step process, followed in a physical or "thermal" method (Ketcha et al., 2012). Dry oxidation is basically used in the thermal process that involves the reaction among raw material and gas (CO_2 and air)/steam at temperatures reaching more than 700°C (Al-Qodah and Shawabkah, 2009). A better equality of pore can be attained through CO_2 activation compared to steam (Khezami et al., 2007). The temperature range for carbonization is between 400 and 850°C and around 600–900°C for activation temperature (Ioannidou and Zabaniotou, 2007). Cuhadaroglu and Uygun (2008) stated that carbonization is a technique through which charcoal is attained from the raw material. This product has a low surface area and is not very active. McDougall (1991) discussed that the carbonization process eliminates the volatile contents of the raw materials to renovate the resulting char with more contents of fixed carbon for activation purposes. In the carbonization period, carbon atoms re-arrange into graphitic-like structures and also build primary porosity in the char through processing (Daud et al., 2000). The porosity development in physical activation takes place with the selective elimination of the more reactive carbons of the structure (Martinez et al., 2003). After that, gasification will construct the activated carbon with high porosity (Baseri et al., 2012). The development of pores in the activation process has been divided into three stages; opening of earlier isolated pores, new pore formation by selective activation, and widening of the available pores (Li et al., 2008).

9.3.2.2 Chemical Activation

Chemical activation is also identified as wet oxidation. It includes activating a chemical to be impregnated into the precursor and then washed to develop the activated carbon (Vargas et al., 2011). The chemical activation is mainly processed at a low temperature range from 300 to 700°C (Giraldo and Moreno-Pirajan, 2012) or 400 to 700°C (Girgis et al., 2002), and generally depends upon an inorganic substance used to dehydrate and degrade the cellulosic contents present in the precursor (Bello and Ahmad, 2011). These also could constrain the development of the ash or tar, hence, rising the carbon yield (Joseph et al., 2006). The activating chemicals that are commonly used are K_2CO_3 (Adinata et al., 2007), NaOH, KOH (Zhengrong and Xiaomin, 2013), $ZnCl_2$, H_3PO_4 (Donald et al., 2011), H_2SO_4, K_2S, KCNS (Demiral et al., 2008), HNO_3, H_2O_2, $KMnO_4$, and $(NH4)_2S_2O_8$ (Al-Qodah and Shawabkah, 2009). In the final step, chemically activated carbon materials are washed to remove the chemical components in the activated carbon. The activated materials are mainly washed with acid or alkali, depending on the activating agent applied for preparation, and afterward washed with deionized water. The activating chemicals basically occupied the porosity in the carbon structure of the prepared active carbon. Therefore, the washing process is required in chemical activation to construct porosity in activated carbon (Ahmadpour and Do, 1997). In the case of lingocellulosic materials, $ZnCl_2$ and H_3PO_4, chemical-activating agents are commonly used for activation purposes (Williams and Reed, 2004). Al-Qodah and Shawabkah (2009) reported that $ZnCl_2$ could produce a higher surface area in comparison to H_3PO_4. Donald et al. (2011) stated that $ZnCl_2$ is more capable of generating microporous structures in the activated carbon and generating greater surface area. So H_3PO_4 is efficiently generating the mesopores and creating the higher pore volumes and diameters. H_3PO_4 is used more frequently because $ZnCl_2$ could show unfavorable environmental impacts (Asadullah et al., 2007); the activated carbon that is prepared through $ZnCl_2$ cannot be implemented in food and pharmaceutical industries (Al-Qodah and Shawabkah, 2009). The carbon product recovery is also easier through H_3PO_4 treatment, which only requires washing with water (Wang et al., 2010). Table 9.1 shows precise activation conditions of several agricultural residues and their use for leachate treatment.

TABLE 9.1

Adsorption of Organic and Inorganic Pollutants from Leachate on Agro-Waste Adsorbents

Precursor	Pollutant Removal from Leachate	Removal Efficiency (%)	References
Banana pseudo stem ($ZnCl_2/76°C$)	COD, NH_3-N, and color	88.9% (color), 85.8% (COD)	Ghani et al. (2017)
Sea mango(700°C, N_2/KOH/ microwave, N_2)	COD, NH_3-N, and color	67.28% (color) 68.98% (COD) 79.77% (NH_3-N)	Shehzad et al. (2016)
Sugarcane bagasse(700°C, N_2/ KOH/687°C, N_2)	COD, NH_3-N, and color	87.3% (color) 77.8% (COD) 41.05% (NH_3-N)	Azmi et al. (2015)
Sugarcane bagasse(700°C, N_2/ KOH/microwave, N_2)	COD, NH_3-N, and color	97.83% (color) 88.14% (COD) 47.25% (NH_3-N)	Azmi et al. (2016b)
Oil palm empty fruit bunch derived activated carbon temperatures (600–900°C)	COD, NH_3-N, and color	81% (COD), 32% (NH_3-N), 74% (color)	Ismail et al. (2018)
Sugarcane bagasse (700°C / KOH/600°C)	COD, NH_3-N, and color	94.74% (color) 83.61% (COD) 46.65% (NH_3-N)	Azmi et al. (2016a) Ferraz and Yuan (2020)
Spent coffee grounds activated carbon (500°C/ H_3PO_4)	COD and color	94% (COD) 100% (color)	

9.3.3 OTHER USED ADSORBENTS

9.3.3.1 Clay

Clay materials have received increased attention mainly because they are cheap, available in large quantities in many regions of the world, and easy to prepare. On the other hand, clay materials have a high specific area and their surface contains a variety of functional groups which can represent the binding sites in adsorption processes (Nguyen et al., 2013). Geologically, clay materials are defined as particles with smaller dimensions up to 2 μm, massive, or stratified rocks, characterized by high porosity (Nafees and Waseem, 2014). The chemical structure of clay materials involves hydrous aluminosilicates that encompass a variety of minerals and metal oxides (Al_2O_3, MgO, SiO_2, Fe_2O_3, CaO). The basic units of the clay are networks of silicon tetrahedra and aluminum octahedra that interlock forming layers (Nafees et al., 2013). The length of these layers is several microns and thicknesses are in small nanometers and arranged into stacks, and the gaps filled with exchangeable metal cations (Kyzas et al., 2012). Due to these characteristics, the clay materials have excellent adsorption properties and can be successfully be used for the retention of various heavy metal ions from aqueous media (Azanfire et al., 2020; Kumar et al., 2012).

9.3.3.2 Zeolite

Zeolites are a cluster of aluminosilicate, microporous minerals that have good adsorbing properties. They encompass various alkali and alkaline-earth ions like Mg^{2+}, Ca^{2+}, Na^+, K^+, Al^{3+}, and some zeolites also, including natrolite, stilbite, and phillipsite (Mariner and Surdam, 1970). The zeolites are commonly developed at places where volcanic ash reacts with water after contact. They have a large number of pores with different pore diameters, so it easily adsorbs substances based on the molecular size; these are known as molecular sieves (Rollmann et al., 2007). This effective adsorbing ability can be a cheap, promising, and environmentally friendly approach of fixing the catalytic agent on the substrate (Huang et al., 2008; Mohamed and Mohamed, 2008).

9.3.3.3 NANOPARTICLE

The advancement of nanotechnology provides a good substitute to improve the treatment efficiency. The nanomaterials (usual size: 1–100 nm) have distinct physical and chemical characteristics like a large surface area, surface actions, and specific equality. Hence, nanomaterials have received substantial attention for finding and removing heavy metals and organics and other inorganic pollutants from water or wastewater along with environmental sustainability (Lu and Astruc, 2018). More review studies have recently been published especially on carbon nanotube, metal oxide nanomaterials, and graphene/graphene oxide. (Liu et al., 2011; Raya and Shipley, 2015). Two types of nanomaterials were discussed, namely inorganic that includes transition metal/oxide/sulfide, and organic carbon/silicon-based nanomaterials that include organic polymer and organic polymer-supported nanocomposites(Lu and Astruc, 2018). The transition metal or metal oxides nanoparticles have unique features; they are largely applied to remove heavy metal ions from wastewater, such as magnetic ferric oxide (Feng et al., 2012), iron, silver (Fabrega et al., 2011), manganese oxide (Gupta et al., 2011), titanium oxide (Luo et al., 2010), magnesium oxide (Gao et al., 2008), copper oxide (Goswamia et al., 2012), gold (Daniel and Astruc, 2004), cerium oxide (Cao et al., 2010), and zinc oxide (Mahajan et al., 2017; Sharma and Nain, 2017).

9.3.4 SURFACE-MODIFIED ADSORBENT

Adsorption through surface-modified, easily available, cheap adsorbents has been used to eliminate anions and cations in aqueous solutions. Dispersion of nanomaterials onto porous supporting materials, i.e. sand, biochar, brick, cement, sponge silica, zeolite, etc., shown in Figure 9.2 are being used to improve the surface area, reduce the aggregation of metals, as well as enhance the treatment capacity (Liu et al., 2020; Qu et al., 2013). The nanomaterials embedded with activated carbon tend to be involved by many researchers (Nethaji et al., 2013). Firstly, nanoparticle-loaded activated carbon can disperse and stabilize the loaded metal nanoparticles that is indicated by reducing aggregation, leaching, and surface passivation of metal nanoparticles (Ho et al., 2017). In this condition, activated carbon performs as the supporter to the dispersion and anchoring of metal nanoparticles with diminished aggregation. Secondly, nanoparticle-loaded activated carbon may also change the characteristics and escalation of both the oxygen-containing functional groups and active sites, which

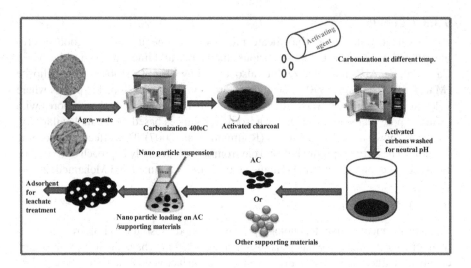

FIGURE 9.2 Synthesis of activated carbon from agro waste and loading of nanoparticles on supporting materials.

formulate superior functionality of nanoparticle-loaded activated carbon (Rodriguez-Narvaez et al., 2019). Thirdly, nanoparticle-loaded activated carbon performs as catalytic/redox that assists the reaction rate for the contaminants. In addition, nanoparticle metals incorporated in biochar can increase the thermal stability and biochar yield (Liu et al., 2020). The coating of iron oxide to surface modification has been applied in many studies, because iron oxide improves the surface area, surface site reactivities and surface defects on adsorbents (Ahmedzeki, 2013). It was also found that iron oxide-coated adsorbents were capable of removing bacteria, metals ions like mercury, zinc, cadmium, lead, nickel, iron, copper, chromium, and manganese, and oxyanionic heavy metals like arsenite and chromate in aqueous solutions (Norris et al., 2013).

9.4 LEACHATE TREATMENT: USING NANOPARTICLE-LOADED ADSORBENT

Landfill leachates have mostly organic matter as well as ammonia-nitrogen (NH_3-N), heavy metals, chlorinated organic, and inorganic salts (Renou et al., 2008). Ma et al. (2019) investigated an innovative process to degrade the refractory compounds contained in biologically pretreated leachate using cow dung ash composites loaded with nano-Fe_3O_4 (nano-Fe_3O_4@CDA) as a catalyst. The optimal conditions applied were nano-Fe_3O_4@CDA dosage of 0.8 g/L, ozone input of 3.0 g/L, and time of reaction at 120 min. This environment yielded the COD and color number removal by 53% and 89%, respectively, and the BOD_5/COD increased from 0.05 to 0.32. The phenolic compounds values diminished sharply from 28.08% to 8.56%, mainly due to the opening of the ring as well as to the formation of organic intermediates with a low molecular weight. The increasing quantity of pharmaceuticals, electronic waste, and xenobiotics in solid waste produces landfill leachate, as high-strength wastewater that has xenobiotics and heavy metals with high chemical oxygen demand. This kinds of leachate can pollute nearby ground/surface water resources and soil if it is not treated appropriately (Sizirici and Tansel, 2015). In addition, the accumulation of heavy metals in living tissues/organs could cause cancer and brain damage. Therefore, suitable treatment processes must be considered to prevent the release of toxic pollutants. Sizirici and Yildiz (2017) determined the adsorption ability of iron oxide-coated gravel for metals like as Cd(II), Cu(II), Fe(II), Ni(II), and Zn(II) in high-strength leachate samples. The level of metal adsorption on iron oxide-coated gravel at pH 7 was in the order Cu(II) > Cd(II) > Fe(II) > Zn(II) > Ni(II). The Freundlich model fit soundly with the investigational data, representing multilayer adsorption practices and the heterogeneity of the surface (R^2 ranging 0.57–0.94). The Temkin model also fit well to the investigation (R^2 ranging 0.67–0.98), showing that the adsorption is an exothermic process. Adam et al. (2019) determined the potential of nano-ionic copper-doped oil palm frond activated carbon (*n*-OPFAC) with the highest adsorption percentage of COD and color and it showed 82% at 45 min and 57% at 60 min, respectively, in semi-aerobic leachate. Table 9.2 recapitulates nanoparticle-loaded adsorbents that have been used for leachate treatment.

TABLE 9.2

Adsorption of Organic and Inorganic Pollutants from Leachate on Nanoparticle-Loaded Adsorbent

Precursor	Pollutant Removal from Leachate	Optimum Conditions	Removal Efficiency (%) or Adsorption Capacity (mg/g)	Reference
W-Doped TiO$_2$nanoparticles	COD	pH (6.63), tungsten content of (2.64%), temperature (472°C), time (34 h)	46%	Azadi et al. (2017)
Immobilized *Phanerochaete chrysosporium* loaded with N-TiO$_2$	NH$_3$-N, TOC	COD (200 ppm), pH (6 for TOC, 7 for NH$_3$-N), temperature (37°C), time (72h)	74% (TOC), 75% (NH$_3$-N)	Hu et al. (2016)
Zeolite and activated carbon composite	COD, NH$_3$-N	pH (7), shaking speed (200), contact time (105 min)	22.99 mg/g (COD) 32.89 mg/g (NH$_3$-N)	Halim et al. (2010a)
Zeolite and activated carbon composite fixed bed column adsorption performance	COD, NH$_3$-N	Diameter of column (6.4 cm), flow rate of 8 ml/min, length (36 cm), contact time (75 min)	86.4% and 3.23 mg/g (COD) 92.6% and 4.46 mg/g (NH$_3$-N)	Halim et al. (2010b)
Carbon–minerals composite adsorbent (AC, zeolite, RHC, and limestone)	COD, NH$_3$-N and color	pH (7), shaking speed (200), contact time (5 h)	98% (color) 65% (COD) 70% (NH$_3$-N)	Halim et al. (2012)

N: Nitrogen, RHC: Rice husk carbon, W: Tungsten

9.5 CONCLUSION

Landfill leachate is the liquid produced from the site; it has high refractory organic molecules, inorganic salts, and toxicities that are not easily removed through biological treatment. The requirement of further treatment using chemical and physical methods depends upon the age and the leachate composition. The adsorption process has been applied to treat landfill leachate and its effectiveness depends on the recalcitrant pollutants removal rate, cost-effectiveness, and eco-friendliness. The use of commercial adsorbents for leachate treatment is limited due to high cost. These adsorbents have been replaced by eco-materials (zeolite, sand, and gravel) or by-product/waste from other industries or agriculture and has been reported by many researchers. The nanoparticles also provide an efficient alternative to improve the treatment efficiency. To reduce the aggregation, leaching, and surface passivation of nanoparticles, dispersion of nanoparticles onto porous supporting materials like as biochar, brick, zeolite, sand, cement, and sponge silica, are being used by many researchers. Above all, it can be concluded that nanoparticle-loaded activated carbon/or other supporting materials are an efficient technology to diminish the values of various leachate parameters such as COD, color, ammonia nitrogen, and heavy metals.

REFERENCES

Adam, N.H., Yusoff, M.S., Bakar, S.N.A., Aziz, H.A. and Halim, H.2019. Potential of nano-ionic copper doped activated carbon as adsorbent in leachate treatment. *Materials Today: Proceedings* 17:1169–75.

Ademiluyi, F.T., Amadi, S.A. and Amakama, J.N.2011. Adsorption and treatment of organic contaminants using activated carbon from waste Nigerian bamboo. *Journal of Applied Science Environment Management* 3:39–47.

Adinata, D., Daud, W.M.A.W. and Aroua, M.K.2007. Preparation and characterization of activated carbon from palm shell by chemical activation with K_2CO_3. *Bioresource Technology* 98:145–9.

Ahmadpour, A. and Do, D.D.1997. The preparation of activated carbon from Macadamia nutshell by chemical activation. *Carbon* 35:1723–32.

Ahmedna, M., Marshall, W.E. and Rao, R.M.2000. Production of granular activated carbons from select agricultural by-products and evaluation of their physical, chemical and adsorption properties. *Bioresource Technology* 71:113–23.

Ahmedzeki, N.S.2013. Adsorption filtration technology using iron coated sand for the removal of lead and cadmium ions from aquatic solutions. *Desaline Water Treatment* 51:5559–65.

Ahn, W.Y., Kang, M.S., Yim, S.K. and Choi, K.H., 2002. Advanced landfill leachate treatment using integrated membrane process. *Desalination* 149:109–14.

Ajima, M.N.O., Nnodi, P.C., Ogo, O.A., Adaka, G.S., Osuigwe, D.I. and Njoku, D.C.2015. Bioaccumulation of heavy metals in Mbaa River and the impact on aquatic ecosystem. *Environment Monitoring Assessment* 187(12):768.

Al-Qodah, Z. and Shawabkah, R.2009. Production and characterization of granular activated carbon from activated sludge. *Brazilian Journal of Chemical Engineering* 26:127–36.

Al-Swaidan, H.A. and Ahmad, A.2011. Synthesis and characterization of activated carbon from Saudi Arabian dates tree's fronds wastes. *Proceedings of the 3rd International Conference on Chemical, Biological and Environmental Engineering* 20:25–31.

Altenor, S., Carene, B., Emmanuel, E., Lambert, J., Ehrhardt, J.J. and Gaspard, S.2009. Adsorption studies of methylene blue and phenol onto vetiver roots activated carbon prepared by chemical activation. *Journal of Hazardous Materials* 165:1029–39.

Asadullah, M., Rahman, M.A., Motin, M.A. and Sultan, M.B.2007. Adsorption studies on activated carbon derived from steam activation of jute stick char. *Journal of Surface Science and Technology* 23:73–80.

Azadi, S., Jashni, A.K. and Javadpour, S.2017. Photocatalytic treatment of landfill leachate using W-doped TiO_2 nanoparticles. *Journal of Environmental Engineering* 143(9):04017049.

Azanfire, B., Bulgariu, D., Nemeş, L. and Bulgariu, L.2020. Optimization of experimental parameters for retention of Pb(II) ions from aqueous solution on clay adsorbent. *Technium* 2:38–47.

Azmi, N.B., Bashir, M.J., Sethupathi, S. and Ng, C.A., 2016a. Anaerobic stabilized landfill leachate treatment using chemically activated sugarcane bagasse activated carbon: kinetic and equilibrium study. *Desalination and Water Treatment* 57 (9):3916–27.

Azmi, N.B., Bashir, M.J.K., Sethupathi, S., Aun, N.C. and Lam, G.C.2016b. Optimization of preparation conditions of sugarcane bagasse activated carbon via microwave induced KOH activation for stabilized landfill leachate remediation. *Environmental Earth Science* 75(10):902.

Azmi, N.B., Bashir, M.J.K., Sethupathi, S., Wei, L.J. and Aun, N.C.2015. Stabilized landfill leachate treatment by sugarcane bagasse derived activated carbon for removal of color, COD and NH_3-N –optimization of preparation conditions by RSM. *Journal of Environment Chemical Engineering* 3(2):1287–94.

Bae, J.H., Cho, K.W., Bum, B.S., Lee, S.J. and Yoon, B.H.1998. Effects of leachate recycle and anaerobic digester sludge recycle on the methane production from solid waste. *Water Science Technology* 38:159–68.

Bansal, R.C., Donnet, J.B. and Stoeckli, F.1988. A review of: "Active Carbon.". Marcel Dekker, New York. *Journal of Dispersion Science and Technology* 11:482.

Baseri, J.R., Palanisamy, P.N. and Sivakumar, P. 2012. Preparation and characterization of activated carbon from *Thevetiaperuviana* for the removal of dyes from textile waste water. *Advances in Applied Science Research* 3:377–83.

Bello, O.S. and Ahmad, M.A.2011. Adsorption of dyes from aqueous solution using chemically activated mango peels. *Proceedings of the 2nd International Conference on Environment Science and Technology* 6:103–6.

Campbell, Q.P., Bunt, J.R., Kasaini, H. and Kruger, D.J.2012. The preparation of activated carbon from South African coal. *Journal of the Southern African Institute of Mining and Metallurgy* 112:37–44.

Cao, C.Y., Cui, Z.M., Chen, C.Q., Song, W.G. and Cai, W.2010. Ceria hollow nanospheres produced by a template-free microwave-assisted hydrothermal method for heavy metal ion removal and catalysis. *The Journal of Physical Chemistry C* 114:9865–70.

Carrott, P.J.M. and Carrott, M.M.R.2007. Lignin – from natural adsorbent to activated carbon: A review. *Bioresource Technology* 98(12):2301–12.

Cecen, F. and Aktas, O.2004. Aerobic co-treatment of landfill leachate with domestic wastewater. *Environmental Engineering Science* 21:303–12.

Chan, G.Y.S., Chu, L.M. and Wong, M.H.2002. Effects of leachate recirculationon biogas production from landfill co-disposal of municipal solid waste, sewage sludge and marine sediment. *Environmental Pollution* 118:393–99.

Chen, C.X., Huang, B., Li, T. and Wu, G.F.2012. Preparation of phosphoric acid activated carbon from sugarcane bagasse by mechano-chemical processing. *Bio Resources* 7:5109–16.

Chianese, A., Ranauro, R. and Verdone, N.1999. Treatment of landfill leachate by reverse osmosis. *Water Resources* 33:647–52.

Chiang, L.C., Chang, J.E. and Chung, C.T.2001. Electrochemical oxidation combined with physical-chemical pretreatment processes for the treatment of refractory landfill leachate. *Environmental Engineering Science* 18:369–79.

Chowdhury, Z.Z., Zain, S.M., Khan, R.A. and Ashraf, M.A.2011. Preparation, characterization and adsorption performance of the KOH-activated carbons derived from kenaf fiber for lead (II) removal from wastewater. *Scientific Research and Essays* 6:6185–96.

Choy, K.K.H., Barford, J.P. and McKay, G.2005. Production of activated carbon from bamboo scaffolding waste-process design, evaluation and sensitivity analysis. *Chemical Engineering Journal* 109:147–65.

Chugh, S., Clarke, W., Pullammanappallil, P. and Rudolph, V.1998. Effect of recirculated leachate volume on MSW degradation. *Waste Management and Resource* 16:564–73.

Cuhadaroglu, D. and Uygun, O.A.2008. Production and characterization of activated carbon from a bituminous coal by chemical activation. *African Journal of Biotechnology* 7:3703–10.

Daniel, M.C. and Astruc, D., 2004. Gold nanoparticles: assembly, supramolecular chemistry, quantum-size-related properties, and applications toward biology, catalysis, and nanotechnology. *Chemical Reviews* 104:293–346.

Daud, W.M.A.W., Ali, W.S.W. and Sulaiman, M.Z.2000. The effects of carbonization temperature on pore development in palm-shell-based activated carbon. *Carbon* 38:1925–32.

Del Moro, G., Prieto-Rodríguez, L., De Sanctis, M., Di Iaconi, C., Malato, S. and Mascolo, G.2016. Landfill leachate treatment: Comparison of standalone electrochemical degradation and combined with a novel biofilter. *Chemical Engineering Journal* 288:87–98.

Demiral, H., Demiral, I., Tumsek, F. and Karabacakoglu, B.2008. Pore structure of activated carbon prepared from hazelnut bagasse by chemical activation. *Surface Interface Anal* 40:616–9.

Derbyshire, F., Jagtoyen, M., Andrews, R., Rao, A., Martın-Gullon, I. and Grulke, E.2001. Carbon Materials in Environmental Applications. In: Radovic L.R. (ed) *Chemistry and Physics of Carbon*. Marcel Decker, New York 27: 1–66.

Donald, J., Ohtsuka, Y. and Xu, C.C.2011. Effects of activation agents and intrinsic minerals on pore development in activated carbons derived from a Canadian peat. *Materials Letters* 65:744–7.

Egun, N.K.2012. The waste to wealth concept: Waste market operation in Delta State, Nigeria. *Greener Journal of Social Science* 2(6):206–12.

Fabrega, J., Luoma, S.N., Tyler, C.R., Galloway, T.S.2011. Lead, silver nanoparticles: Behaviour and effects in the aquatic environment. *Environment International* 37:517–31.

Feng, L., Cao, M., Ma, X.Y. and Hu, C.2012. Superparamagnetic high-surface- area Fe_3O_4 nanoparticles as adsorbents for arsenic removal. *Journal of Hazardous Matter* 217–218:439–46.

Ferraz, F.M. and Yuan, Q.2020. Organic matter removal from landfill leachate by adsorption using spent coffee grounds activated carbon. *Sustainable Materials and Technologies* 23:00141.

Foo, P.Y.L. and Lee, L.Y. 2010. Preparation of activated carbon from Parkiaspeciosa pod by chemical activation. Proceedings of the *World Congress in Engineering and Computational Science* 2:1–3.

Forgie, D.J.L.1988. Selection of the most appropriate leachate treatment methods. Part 2: A review of recirculation, irrigation and potential physical-chemical treatment methods. *Water Pollution and Resource* 23(2):329–40.

Gao, C., Zhang, W., Li, H., Lang, L. and Xu, Z.2008. Controllable fabrication of mesoporousMgO with various morphologies and their absorption performance for toxic pollutants in water. *Crystal Growth & Design* 8:3785–90.

Ghani, Z.A., Yusoff, M.S., Zaman, N.Q., Zamri, M. and Andas, J.2017. Optimization of preparation conditions for activated carbon from banana pseudo-stem using response surface methodology on removal of color and COD from landfill leachate. *Waste Management* 62:177–87.

Giraldo, L. and Moreno-Pirajan, J.C.2012. Synthesis of activated carbon mesoporous from coffee waste and its application in adsorption zinc and mercury ions from aqueous solution. *European Journal of Chemistry* 9(2):938–48.

Girgis, B.S., Yunis, S.S. and Soliman, A.M.2002. Characteristics of activated carbon from peanut hulls in relation to conditions of preparation. *Materials Letters* 57:164–72.

Goswamia, A., Raul, P.K. and Purkait, M.K.2012. Arsenic adsorption using copper (II) oxide nanoparticles. *Chemical Engineering Research and Design* 90:1387–96.

Gupta, K., Bhattacharya, S., Chattopadhyay, D., Mukhopadhyay, A., Biswas, H., Dutta, J., Ray, N.R. and Ghosh, U.C.2011. Ceria associated manganese oxide nanoparticles: Synthesis, characterization and arsenic(V) sorption behavior. *Chemical Engineering* 172:219–29.

Halim, A.A., Abidin, N.N.Z., Awang, N., Ithnin, A., Othman, M.S. and Wahab, M.I.2011. Ammonia and COD removal from synthetic leachate using rice husk composite adsorbent. *Journal of Urban Environmental Engineering* 5:24–31.

Halim, A.A., Aziz, H.A., Johari, M.A.M. and Ariffin, K.S. 2010a. Comparison study of ammonia and COD adsorption on zeolite, activated carbon and composite materials in landfill leachate treatment. *Desalination* 262:31–35.

Halim, A.A., Aziz, H.A., Johari, M.A.M., Ariffin, K.S. and Adlan, M.N. 2010b. Ammoniacal nitrogen and COD removal from semi-aerobic landfill leachate using a composite adsorbent: Fixed bed column adsorption performance. *Journal of Hazardous Materials* 175:960–64.

Halim, A.A., Aziz, H.A., Johari, M.A.M., Ariffin, K.S. and Bashir, M.J.2012. Semi-aerobic landfill leachate treatment using carbon–minerals composite adsorbent. *Environmental Engineering Science* 29:306–12.

Harris, P.J.F., Lie, Z. and Suenaga, K. 2008. Imaging the atomic structure of activated carbon. *Journal of Physics: Condensed Matter* 20:1–5.

Hassan, A.F. and Elhadidy, H. 2017. Production of activated carbons from waste carpets and its application in methylene blue adsorption: Kinetic and thermodynamic studies. *Journal of Environmental Chemistry Engineering* 5:955–63.

Hilles, A.H., Amr, S.S.A., Hussein, R.A., El-Sebaie, O.D. and Arafa, A.I.2016. Performance of combined sodium persulfate/H_2O_2 based advanced oxidation process in stabilized landfill leachate treatment. *Journal of Environmental Management* 166:493–98.

Ho, S.H., Zhu, S. and Chang, J.S.2017. Recent advances in nanoscale-metal assisted biochar derived from waste biomass used for heavy metals removal. *Bioresource Technology* 246(2017):123–34.

Hu, L., Zeng, G., Chen, G., Dong, H., Liu, Y., Wan, J., Chen, A., Guo, Z., Yan, M., Wu, H. and Yu, Z. 2016. Treatment of landfill leachate using immobilized *Phanerochaete chrysosporium* loaded with nitrogen-doped TiO_2 nanoparticles. *Journal of Hazardous Materials* 301:106–18.

Huang, M., Xu, C., Wu, Z., Huang, Y., Lin, J. and Wu, J., 2008. Photocatalytic discolorization of methyl orange solution by Pt modified TiO_2 loaded on natural zeolite. *Dyes and Pigments* 77: 327–334. http://dx.doi.org/10.1016/j.dyepig.2007.01.026.

Imran, A.2010. The quest for active carbon adsorbent substitutes: Inexpensive adsorbents for toxic metal ions removal from wastewater. *Separation & Purification Reviews* 39:95–171.

Ioannidou, O. and Zabaniotou, A.2007. Agricultural residues as precursors for activated carbon production – a review. *Renewable & Sustainable Energy Reviews* 11:1966–2005.

Ismail, Y.M.N.S., Ngadi, N., Hassim, M.H., Kamaruddin, M.J.A., Johari, A. and Aziz, M.A.A.2018. Preparation of activated carbon from oil palm empty fruit bunch by physical activation for treatment of landfill leachate. *IOP Conference Series: Materials Science and Engineering* 458:012036.

Jassim, A.N., Amlah, L.K., Ali, D.F. and Aljabar, A.T.A.2012. Preparation and characterization of activated carbon from Iraqi apricot stones. *Chemical Engineering Technology* 3:60–5.

Joseph, C.G., Zain, H.F.M. and Dek, S.F.2006. Treatment of landfill leachate in Kayu Madang, Sabah: Textural and physical characterization (Part I). *Malaysian Journal of Analytical Sciences* 10(1):1–6.

Jun, T.Y., Arumugam, S.D., Latip, N.H.A., Abdullah, A.M. and Latif, P.A.2010. Effect of activation temperature and heating duration on physical characteristics of activated carbon prepared from agriculture waste. *Environment Asia* 3:143–8.

Ketcha, J.M., Dina, D.J.D., Ngomo, H.M. and Ndi, N.J.2012. Preparation and characterization of activated carbons obtained from maize cobs by zinc chloride activation. *American Chemical Science Journal* 2(4):13660.

Khalili, N.R., Campbell, M., Sandi, G. and Golas, J. 2000. Production of micro-and meso-porous activated carbon from paper mill sludge I. Effect of zinc chloride activation. *Carbon* 38:1905–15.

Khezami, L., Ould-Dris, A. and Capart, R.2007. Activated carbon from thermo-compressed wood and other lignocellulosic precursors. *Bioresources* 2(2):193–209.

Kumar, P.S., Ramalingam, S., Sathyaselvabala, V., Kirupha, S.D., Murugesan, A. and Sivanesan, S. 2012. Removal of Cd (II) from aqueous solution by agricultural waste cashew nut shell. *Korean Journal of Chemical Engineering* 29:756–68.

Kurniawan, T.A., Chan, G.Y.S., Lo, W.H. and Babel, S.2006a. Comparison of low-cost adsorbents for treating wastewater laden with heavy metals. *Science of the Total Environment* 366(2–3):407–24.

Kurniawan, T.A., and Lo, W.2009. Removal of refractory compounds from stabilized landfill leachate using an integrated H_2O_2 oxidation and granular activated carbon (GAC) adsorption treatment. *Water Research* 43:4079–91.

Kyzas, G.Z., Lazaridis, N.K. and Mitropoulos, A.C.2012. Removal of dyes from aqueous solutions with untreated coffee residues as potential low-cost adsorbents: Equilibrium, reuse and thermodynamic approach. *Chemical Engineering Journal* 189–190:148–59.

Lema, J.M., Mendez, R. and Blazquez, R.1988. Characteristics of landfill leachates and alternatives for their treatment: A review. *Water Air and Soil Pollution* 40:223–50.

Li, W., Yang, K., Peng, J., Zhang, L., Guo, S. and Xia, H.2008. Effects of carbonization temperatures on characteristics of porosity in coconut shell chars and activated carbons derived from carbonized coconut shell chars. *Industrial Crops and Products* 28:190–8.

Lillo-Rodenas, M.A., Cazorla-Amoros, D. and Linares-Solano, A.2003. Understanding chemical reactions between carbons and NaOH and KOH: An insight into the chemical activation mechanism. *Carbon* 41:267–75.

Lin, J.H. and Wang, S.B.2017. An effective route to transform scrap tire carbons into highly-pure activated carbons with a high adsorption capacity of ethylene blue through thermal and chemical treatments. *Environment Technology and Innovation* 8:17–27.

Liu, J., Jiang, J., Meng, Y., Aihemaiti, A., Xu, Y., Xiang, H. and Chen, X.2020. Preparation, environmental application and prospect of biochar supported metal nanoparticles: A review. *Journal of Hazardous Materials* 388:122026.

Liu, Y., Su, G., Zhang, B., Jiang, G. and Yan, B.2011. Nanoparticle-based strategies for detection and remediation of environmental pollutants. *Analyst* 136:872–77.

Lozano-Castello, D., Lillo-Rodenas, M.A., Cazorla-Amoros, D. and Linares-Solano, A. 2001. Preparation of activated carbons from Spanish anthracite I. Activation by KOH. *Carbon* 39:741–9.

Lu, F. and Astruc, D.2018. Nanomaterials for removal of toxic elements from water precipitaion adsorption heavy metals removal ion exchange filtration coagulation biosorprtion. *Coordination Chemistry Reviews*. Elsevier B.V. 356:147–64

Lua, A.C. and Yang, T.2005. Characteristics of activated carbon prepared from pistachio-nut shell by zinc chloride activation under nitrogen and vacuum conditions. *Journal of Colloid Interface Science* 290:505–13.

Luo, T., Cui, J., Hu, S., Huang, Y. and Jing, C.2010. Arsenic removal and recovery from copper smelting wastewater using TiO_2. *Environment Science and Technology* 44:9094–98.

Ma, C., Yuan, P., Jia, S., Liu, Y., Zhang, X., Hou, S., Zhang, H. and He, Z.2019.Catalytic micro-ozonation by Fe_3O_4 nanoparticles @ cow-dung ash for advanced treatment of biologically pre-treated leachate. *Waste Management* 83:23–32.

Mahajan, A., Nain, K.S. and Lohchab, R.K.2017. Photocatalytic degradation of pulp & paper mill wastewater using ZnO photocatalyst. *International Journal of Emerging Technology and Advanced Engineering* 9:623–28.

Malik, R., Ramteke, D.S. and Wate, S.R.2006. Physico-chemical and surface characterization of adsorbent prepared from groundnut shell by $ZnCl_2$ activation and its ability to adsorb color. *Indian Journal of Chemical Technology* 13:319–28.

Mariner, R.H. and Surdam, R.C.1970. Alkalinity and formation of zeolites in saline alkaline lakes.*Science* 170:977–80. http://dx.doi.org/10.1126/science.170.3961.977.

Martinez, M.L., Moiraghi, L., Agnese, M. and Guzman, C.2003. Making and some properties of activated carbon produced from agricultural industrial residues from Argentina. *Journal of the Argentine Chemical Society* 91(4–6):103–8.

McDougall, G.J.1991. The physical nature and manufacture of activated carbon. *Journal of the South African Institute of Mining and Metallurgy* 91(4):109–20.

Moh, Y.C. and Manaf, L.A.2014. Overview of household solid waste recycling policy status and challenges in Malaysia. *Resource Conservation Recycle* 82:50–61.

Mohamed, R.M. and Mohamed, M.M., 2008. Copper (II) phthalocyanines immobilized on alumina and encapsulated inside zeolite-X and their applications in photocatalytic degradation of cyanide: A comparative study. *Applied Catalysis A: General* 340:16–24.

Mojiri, A., Aziz, H.A., Zaman, N.Q., Aziz, S.Q. and Zahed, M.A.2016. Metals removal from municipal landfill leachate and wastewater using adsorbents combined with biological method. *Desalination and Water Treatment* 57:2819–33.

Moreira, F.C., Soler, J., Fonseca, A., Saraiva, I., Boaventura, R.A. and Brillas, E.2015. Incorporation of electrochemical advanced oxidation processes in a multistage treatment system for sanitary landfill leachate. *Water Research* 81:375–87.

Nafees, M. and Waseem, A.2014. Organoclays as sorbent material for phenolic compounds: A review. *Clean – Soil Air Water* 42:1500–08.

Nafees, M., Waseem, A. and Khan, A.R.2013. Comparative study of laterite and bentonite based organoclays: Implications of hydrophobic compounds remediation from aqueous solutions. *The Scientific World Journal* 2013: ID: 681769.

Nethaji, S., Sivasamy, A. and Mandal, A.B.2013. Preparation and characterization of corn cob activated carbon coated with nano-sized magnetite particles for the removal of Cr(VI). *Bioresource Technology* 134:94–100.

Nguyen, T.A.H., Ngo, H.H., Guo, W.S., Zhang, J., Liang, S., Yue, Q.Y., Li, Q. and Nguyen, T.V.2013. Applicability of agricultural waste and byproducts for adsorptive removal of heavy metals from wastewater. *Bioresource Technology* 148:574–85.

Norris, M.J., Pulford, I.D., Haynes, H., Dorea, C.C. and Phoenix, V.R.2013. Treatment of heavy metals by iron oxide coated and natural gravel media in sustainable urban drainage systems. *Water Science Technology* 68:674–80.

Olafadehan, O.A., Jinadu, O.W., Salami, L. and Popoola, O.T.2012. Treatment of brewery waste water effluent using activated carbon prepared from coconut shell. *International Journal of Applied Science and Technology* 2(1):65–78.

Orkun, Y., Karatepe, N. and Yavuz, R.2012. Influence of temperature and impregnation ratio of H_3PO_4 on the production of activated carbon from hazel nutshell. *Acta Physica Polonica A* 121:277–80.

Oulego, P., Collado, S., Laca, A. and Diaz, M.2016. Impact of leachate composition on the advanced oxidation treatment. *Water Resource* 88:389–402.

Peters, T.A.1998. Purification of landfill leachate with reverse osmosis and nanofiltration. *Desalination* 119:289–93.

Prahas, D., Kartika, Y., Indraswati, N. and Ismadji, S.2008. Activated carbon from jackfruit peel waste by H_3PO_4 chemical activation: Pore structure and surface chemistry characterization. *Chemical Engineering Journal* 140:32–42.

Qu, X., Alvarez, P.J.J. and Li, Q.2013. Applications of nanotechnology in water and wastewater treatment. *Water Research* 47:3931–46.

Ramakrishnan, K. and Namasivayam, C.2009. Development and characteristics of activated carbons from Jatropha Husk, an agro industrial solid waste, by chemical activation methods. *Journal of Environmental Engineering and Management* 19(3):173–8.

Raya, P.Z. and Shipley, H.J.2015. Inorganic nano-adsorbents for the removal of heavy metals and arsenic: A review. *RSC Advances* 5:29885–907.

Renou, S., Givaudan, J.G., Poulain, S., Dirassouyan, F. and Moulin, P.2008. Landfill leachate treatment: Review and opportunity. *Journal of Hazardous Materials* 150:468–93.

Rodriguez-Narvaez, O.M., Peralta-Hernandez, J.M., Goonetilleke, A. and Bandala, E.R. 2019. Biochar-supported nanomaterials for environmental applications. *Journal of Industrial and Engineering Chemistry* 78:21–33.

Rollmann, L.D., Valyocsik, E.W. and Shannon, R.D., 2007. Zeolite molecular sieves. In: *Inorganic syntheses*. Wiley, New York, pp 61–8.

Sahu, J.N., Acharya, J. and Meikap, B.C.2010. Optimization of production conditions for activated carbons from tamarind wood by zinc chloride using response surface methodology. *Bioresource Technology* 101:1974–82.

Sadegh, H., Ali, G.A.M., Gupta, V.K., Makhlouf, H.A.S., Ghoshekandi, R.S., Nadagouda, N.M., Sillanpaa, M. and Megiel, E.2017. The role of nanomaterials as effective adsorbents and their applications in wastewater treatment. *Journal of Nanostructure in Chemistry* 7:1–14.

San, I. and Onay, T.T.2001. Impact of various leachate recirculation regimes on municipal solid waste degradation. *Journal of Hazardous Materials* 87:259–71.

Sharma, S. and Nain, K.S.2017. Photodegradation of real textile wastewater by using ZnOcatalyst. *International Journal of Emerging Technology and Advanced Engineering* 7:202–07.

Shehzad, A., Bashir Mohammed, J.K., Sethupathi, S. and Lim, J.W.2016. Simultaneous removal of organic and inorganic pollutants from landfill leachate using sea mango derived activated carbon via microwave induced activation. *International Journal of Chemical Reactor Engineering* 14(5):991–1001.

Silva, A.C., Dezotti, M., and Sant Anna, G.L.2004.Treatment and detoxification of a sanitary landfill leachate. *Chemosphere* 55:207–14.

Singh, K. and Waziri, S.A.2019. Activated carbons precursor to corncob and coconut shell in the remediation of heavy metals from oil refinery wastewater. *Journal of Materials and Environmental Sciences* 10(7):657–67.

Singh, K., Waziri, S.A. and Ram,C.2018. Removal of heavy metals by adsorption using agricultural based residue: A review. *Research Journal of Chemistry and Environment* 22(5):65–74.

Sizirici, B. and Tansel, B.2015. Parametric fate and transport profiling for selective groundwater monitoring at closed landfills: A case study. *Waste Management* 38:263–70.

Sizirici, B. and Yildiz, I.2017. Adsorption capacity of iron oxide-coated gravel for landfill leachate: Simultaneous study. *International Journal of Environment Science and Technology* 14:1027–36.

Sugumaran, P., Susan, V.P., Ravichandran, P. and Seshadri, S.2012. Production and characterization of activated carbon from banana empty fruit bunch and Delonixregia fruit pod. *Journal of Sustainable Energy & Environment* 3:125–32.

Carrott, S.P.J.M. and Carrott, M.M.L.R. 2007. Lignin – From natural adsorbent to activated carbon: A review. *Bioresource Technology* 98(12):2301–12.

Tabet, K., Moulin, P., Vilomet, J.D., Amberto, A. and Charbit, F.2002. Purification of landfill leachate with membrane processes: Preliminary studies for an industrial plant. *Separation Science and Technology* 37:1041–63.

Tatsi, A.A., Zouboulis, I., Matis, K.A. and Samaras, P.2003. Coagulation–flocculation pretreatment of sanitary landfill leachates. *Chemosphere* 53:737–44.

Trebouet, D., Schlumpf, J.P., Jaouen, P. and Quemeneur, F.2001. Stabilized landfill leachate treatment by combined physicochemical nano filtration processes. *Water Resource* 35(12):2935–42.

Trebouet, D., Schlumpf, J.P., Jaouen, P.F., Maleriat, J.P. and Quemeneur, F.1999. Effect of operating conditions on the nanofiltration of landfill leachates: Pilot-scale studies. *Environmental Technology* 20(6):587–96.

Tsai, W.T., Chang, C.Y. and Lee, S.L.1998. A low cost adsorbent from agricultural waste corn cob by zinc chloride activation. *Bioresource Technology* 64:211–7.

Tsai, W.T., Chang, C.Y., Wang, S.Y., Chang, C.F., Chien, S.F. and Sun, H.F.2001. Cleaner production of carbon adsorbents by utilizing agricultural waste corn cob. *Resources, Conservation& Recycling* 32:43–53.

Ubago-Perez, R., Carrasco-Marin, F., Fairen-Jimenez, D. and Moreno-Castilla, C.2006. Granular and monolithic activated carbons from KOH-activation of olive stones. *Microporous and Mesoporous Materials* 92:64–70.

VanTran, T., Bui, Q.T.P., Nguyen, T.D., Le, N.T.H. and Bach, L.G.2017. A comparative study on the removal efficiency of metal ions (Cu^{2+}, Ni^{2+}, and Pb^{2+}) using sugarcane bagasse-derived $ZnCl_2$ activated carbon by the response surface methodology. *Adsorption Science and Technology* 35:72–85.

Vargas, A.M.M., Cazetta, A.L., Garcia, C.A., Moraes, J.C.G., Nogami, E.M and Lenzi, E.2011. Preparation and characterization of activated carbon from a new raw lignocellulosic material: Flamboyant (*Delonixregia*) pods. *Journal of Environment and Management* 92:178–84.

Vazquez, H.A., Jefferson, B., and Judd, S.J.2004. Membrane bioreactors vs conventional biological treatment of landfill leachate: A brief review. *Journal of Chemical Technology & Biotechnology* 79:1043–49.

Wang, J., Wu, F.A., Wang, M., Qiu, N., Liang, Y. and Fang, S.Q.2010. Preparation of activated carbon from a renewable agricultural residue of pruning mulberry shoot. *African Journal of Biotechnology* 9(19):2762–7.

WanNik, W.B., Rahman, M.M., Yusof, A.M., Ani, F.N. and Adnan, C.M.C.2006. Production of activated carbon from palm oil shell waste and its adsorption characteristics. In: *Proceedings of the 1st international conference on natural research and engineering technology*, Putrajaya, Malaysia, 646–54.

Wendimu, G., Zewge, F. and Mulugeta, E.2017. Aluminium-iron-amended activated bamboo charcoal (AIAABC) for fluoride removal from aqueous solutions. *Journal of Water Process Engineering* 16:123–31.

Williams, P.T. and Reed, A.R.2004. High grade activated carbon matting derived from the chemical activation and pyrolysis of natural fibre textile waste. *Journal of Analytical and Applied Pyrolysis* 71:971–86.

Wong, S., Lee, Y., Ngadi, N., Inuwa, I.M. and Mohamed, N.B., 2017. Synthesis of acti-
    vated carbon from spent tea leaves for aspirin removal. *Chinese Journal of Chemical
    Engineering* 26(5):1003–11.
Wong, S., Ngadi, N., Inuwa, I.M. and Hassan, O.2018. Recent advances in applications of
    activated carbon from biowaste for wastewater treatment: A short review. *Journal of
    Cleaner Production* 175:361–75.
World Bank. 2019 WHAT aWASTE: A Global Review of Solid Waste Management. Urban
    Development Series, September 2019. http://hdl.handle.net/10986/17388.
Xue, Q., Li, J., Wang, P., Liu, L. and Li, Z.Z.2014. Removal of heavy metals from landfill
    leachate using municipal solid waste incineration fly ash as adsorbent. *Clean: Soil, Air,
    Water* 42(11):1626–31.
Yalcin, N. and Sevinc, V.2000. Studies of the surface area and porosity of activated carbons
    prepared from rice husks. *Carbon* 38:1943–5.
Youcai, Z., Jianggying, L., Renhua, H. and Guowei, G.2000. Long-term monitoring and pre-
    diction for leachate concentrations in Shanghai refuse landfill. *Water Air Soil Pollution*
    122 (3):281–97.
Yusufu, M.I., Ariahu, C.C. and Igbabul, B.D.2012. Production and characterization of acti-
    vated carbon from selected local raw materials. *African Journal of Pure and Applied
    Chemistry* 6(9):123–31.
Zhengrong, G. and Xiaomin, W.2013. Carbon materials from high ash bio-char: A nano-
    structure similar to activated grapheme. *American Journal of Engineering and Applied
    Sciences* 2(1):15–34.
Zhu, Z., Li, A., Xia, M., Wan, J. and Zhang, Q.2008. Preparation and characterization of
    polymer based spherical activated carbons. *Chinese Journal of Polymer Science*
    26(5):645–51.

10 Application of Fe-TiO$_2$ Nanoparticle Composite Encompassing the Dual Effect

Steffi Talwar and Anoop Kumar Verma
Thapar Institute of Engineering and Technology

Vikas Kumar Sangal
Malaviya National Institute of Technology

CONTENTS

10.1 INTRODUCTION

Being the emerging producer of pharmaceutical drugs, India's economy is creating resources for the middle class to afford western medicines. With the huge development in this sector come other problems related to health care, as most of the population is ignorant about the ill effects of the excessive use of these compounds (Nag and Ghosh, 2013; Singh et al., 2014). This leads to the detection of various pharmaceutical compounds in the various fresh water sources, such as lakes and wells in the villages. The major root of these pollutants is the incapability of the traditional treatment plants to completely manage the waste generated. Due to these inabilities, researchers have focused upon advanced treatment technologies. The non-selective •OH radicals produced during various processes like photocatalysis, photo-Fenton, ozonation, sonolysis, and many more, have proven its credentials to treat various kinds of pollutants (Diao et al., 2017; Sarkar, Chakraborty, and Bhattacharjee, 2015). But the extortionate nature of these processes in terms of increased chemical consumption, high cost, and increased energy consumption are hindering their scale-up in the industries (Bansal, Verma, and Talwar, 2018; Talwar, Sangal, and Verma, 2018).

In this context, taking into concern the pitfalls of both advanced oxidation processes (AOPs) (Yang et al., 2018), the technology involving the combination of two processes is showing wide popularity among researchers. Among these AOPs, photocatalysis (Buriak, Kamat, and Schanze, 2014; Lin et al., 2016; Luo et al., 2017) has proven its eminence in degrading pharmaceutical active ingredients and drugs mainly in slurry mode (Maroga Mboula et al., 2012; MiarAlipour et al., 2018). But there are certain problems in scaling up process like cumbersome separation of TiO_2 (Daneshvar et al., 2005), a larger quantity of catalyst being utilized (Rastegar et al., 2012; Su et al., 2018; Zhang et al., 2017; Zhao et al., 2018), and process drawbacks like electron-hole recombination (Elmolla and Chaudhuri, 2010). Hence, the application of photocatalysis is mainly preferred in fixed bed mode where the catalyst is supported on the inert support material (Bansal et al., 2016). But again, some problems pose a crunch for large scale applications as well, like an increase in the treatment time, mass transfer limitations, and inertness of the support (Kovacic et al., 2016; Vaiano et al., 2015). In the case of homogeneous photocatalysis, i.e. photo-Fenton (Bokare and Choi, 2014), although the time taken for the reaction is less as compared to the other AOPs, there are certain shortcomings like a larger quantity of catalyst being utilized, iron sludge production, and high doses of H_2O_2 that are restricting field-scale applications. Hence, the researchers are focusing on the new innovations, taking into consideration the disadvantages for the efficacious treatment of pharmaceutical compounds.

Therefore, for an energy efficient, as well as cost effective solution, the recent studies have shifted the spotlight upon the immobilization of the nanoparticles over inert surfaces, helping to overcome the constraint of the cumbersome catalyst separation (Gadiyar et al., 2013; Mozia et al., 2012; Mukherjee, Barghi, and Ray, 2013). Henceforth, taking into consideration the advantages and disadvantages of the processes of photo-Fenton and photocatalysis, the integration of these processes using in-situ have proven to be a new concept. The dual effect helps in curbing most of the problems faced by both the processes of photo-Fenton and photocatalysis. Thus, for the present research work, a novel composite support made up of clay, foundry sand (FS), and fly ash (FA) has been employed. The prepared composite beads are coated with TiO_2 for the photocatalysis process to take place. The continuous leaching of iron from the composite beads leads to the photo-Fenton process; thus, leading to the in-situ dual effect (photocatalysis and photo-Fenton) to take place. This then leads to intensification in the reaction rate along with a decrease in the degradation time of pollutants. Hence, this study shows the novel in-situ process of inducing the dual effect of photo-Fenton and photocatalysis for the degradation of ofloxacin, a model pharmaceutical compound.

The basic mechanism for the dual effect of photocatalysis and photo-Fenton involves the photocatalysis taking place from the surface-active TiO_2 layer, thus leading to the generation of hydroxyl radicals. Besides this, there would be subsequent leaching of iron from these composite beads. The composite beads prepared from the iron composite have both Fe(III) and Fe(II) (Sharma et al., 2020; Talwar, Sangal, and Verma, 2019). Thus, to execute photo-Fenton studies, these Fe(III) ions need to be converted into Fe(II). The electron required for this reduction comes from the conduction band of TiO$_2$ (Kim et al., 2012). This step helps in covering the shortcoming of the photocatalysis process of the electron-hole recombination, as an electron gets involved in reduction. This will eventually lead to an increased number of hydroxyl radicals, thus increasing the reaction rate. This also helps in the oxidation of water molecules to form hydroxyl radicals. Hence, proving the dual effect of photocatalysis and photo-Fenton process to take place simultaneously.

The effect of numerous parameters like the covered surface area by the beads, treatment time, the dose of H$_2$O$_2$, and area/volume ratio of the reactor, UV intensity on the degradation of ofloxacin is difficult to evaluate due to the complexity of the reactions involved. This may hinder the scaling up operations of the process reactors. Being multivariate in nature and forming the non-linear correlations, these reactions are difficult to solve. Hence, in this case, artificial intelligence (AI) is required to solve the problems. In the recent past, applications of AI have attracted researchers from various fields, as the computers are trained to behave more intelligently using various techniques (Abiodun et al., 2018). Artificial neural networks (ANNs), being one of the model parts of machine learning, are being used in place of traditional regression and statistical analysis (Shojaeimehr et al., 2014). ANN has been widely used for classifications, recognition of patterns, and for the prediction of various models. The promising capability of the ANN, including high-speed processing provided in massive implementation, has increased the research for the same. The application of ANN has grown widely in the field of treatment technologies (Talwar et al., 2018). The system of treatment technologies are complex in nature and include challenges that cannot be solved by the computational ability of traditional procedures and conventional mathematics (Homem and Santos, 2011). But ANN is not capable of solving all the problems related to the treatment of wastewaters, as it does not provide the optimal solutions to the problem. In traditional methods of optimization, the effect of only one variable is studied at a time, and it does not consider interactive affects between the variables, so that it could not exhibit an overall effect of variables on the particular response. Response surface methodology (RSM) is one of the statistical tools, used to rectify this issue. RSM is commonly used for the experimental design and optimization of chemical and mechanical operations in industrial processes and also applied for wastewater treatment methods (Kaur, Sangal, and Kushwaha, 2015). Modeling helps in the predicting results as well as reducing the number of experiments required. In the present work, research based on the preparation and application of a novel nanocomposite of Fe-TiO$_2$ for the degradation of ofloxacin was carried out. ANN modeling, along with the optimization using BBD, was performed to maximize the degradation by minimizing the error and optimizing various parameters required for the treatment of pollutants.

10.2 MATERIALS AND METHODS

The pharmaceutical drug ofloxacin was kindly supplied by the pharmaceutical manufacturing unit in Baddi (Himachal Pradesh, India). The nano-photocatalyst P25 – TiO_2 was obtained from the Degussa (Frankfurt, Germany). Sodium acetate (> 98.5%) and acetic acid (> 99.5%) were procured from Ranchem Chemicals (India) and were used for the preparation of the acetate buffer. Hydrogen peroxide, i.e. H_2O_2 (30% w/v), used as an oxidant, was obtained from Ranbaxy (India). The waste foundry sand (FS) and fly ash (FA) were obtained from the nearby local manufacturing industries in Patiala, Punjab, India. The waste products, along with the clay, were utilized for the preparation of inert fixed-bed materials. Double distilled water was used for all the preparation of all reagents and for conducting all reactions.

10.2.1 ANN MODELING

Deep learning is an aspect of AI; the term "deep learning" refers to ANN with complex multilayers. The best advantage of ANN is that it makes the model easier and more accurate than other non-traditional methods. The model behaves like the human brain with neurons intelligently performing various operations. A neural network consists of different layers, i.e. input, hidden, and output. These layers are independent of each other. The hidden layer consists of numerous neurons which varies depending on minimizing the mean square error. The information from the input layer gets transferred to hidden layers where the processing of data takes place, and in the end it is received by the output layer using the transfer functions.

The most commonly used types of ANNs include multilayered feed-forward neural networks and Kohonen self-organizing mapping that are trained by the back-propagation algorithm. The majority of ANN architectures are feed-forward networks. Error back-propagation algorithm is mostly used to train feed-forward networks. ANN, which uses the back-propagation algorithm for learning the appropriate weights, is one of the most common models used in ANN. The most widely used transfer function for the input and hidden layers are the sigmoid transfer function.

All the calculations for ANN were performed with MATLAB R2017b mathematical software with ANN toolbox. The various input parameters were covered surface areas by the beads, treatment time, the dose of H_2O_2, area/volume ratio of the reactor, and UV intensity with % degradation as an output parameter.

10.2.2 OPTIMIZATION USING RSM

Conventional manual optimization takes a lot of time as well as resources for the calculations of the model. Also, no interactions between the process parameters are studied. RSM incorporating Box-Behnken designs (BBD) was utilized to optimize the process parameters (Sangal, Kumar, and Mishra, 2013). Process parameters for different processes have been optimized using BBD (Talwar, Sangal, and Verma,

2018). The statistical Design-Expert software version 6.06 (STAT-EASE Inc., Minneapolis, MN, U.S.) was also utilized for this experiment.

There are three levels of BBD under RSM with a range of operational parameters as H$_2$O$_2$ dose (450–1350 mg L^{-1}), time t (5–120 min), the number of beads (50–150), and output as % degradation; where 50, 100, and 150 beads correspond to 50%, 100%, and 150% of area covered by the beads. The response, i.e. output Z$_i$ which is the function of input factors T$_1$, T$_2$ ....T$_i$....T$_f$, is obtained from the following correlation in Equation 10.1:

$$Z_i = \phi\,(T_1, T_2, ... T_i ... T_f) \tag{10.1}$$

The above relations between input and output variables are considered to be quadratic in nature as shown in Equation 10.2:

$$Z = v_0 + \sum_{i=1}^{k} v_i T_i + \sum_{i}^{k} v_{ii} T_i^2 + \sum \sum_{i<j} v_{ij} T_i T_j + \kappa_t \tag{10.2}$$

where Z is the response; v_i is the linear effect of T$_i$, v_0 is the constant, v_{ij} is the linear by linear interaction effect among input factors T$_i$ and T$_j$, and v_{ii} is the quadratic effect of input factors T$_i$. The simulated results from ANN used for predicting the optimization and using inputs experiments were designed in RSM (Table 10.1).

TABLE 10.1

Set of Experiments Suggested by Box Behnken Design

S. No.	Time (min)	H$_2$O$_2$ Dose	Number of Beads	% Degradation
1	120.00	75.00	100.00	78
2	5.00	750.00	100.00	20
3	62.50	75.00	50.00	65.7
4	62.50	75.00	150.00	75
5	120.00	412.50	50.00	80.2
6	62.50	750.00	150.00	64.56
7	62.50	750.00	50.00	60.54
8	62.50	412.50	100.00	79
9	62.50	412.50	100.00	89
10	62.50	412.50	100.00	87
11	5.00	412.50	50.00	36.5
12	62.50	412.50	100.00	81
13	120.00	750.00	100.00	69.3
14	120.00	412.50	150.00	85.3
15	62.50	412.50	100.00	85.6
16	5.00	412.50	150.00	37.67
17	5.00	75.00	100.00	43.5

10.2.3 Statistical Analysis

The Design-Expert software was employed. All the experimental analysis was performed three times. Analysis of variance (ANOVA) was used to analyze the data. $P < 0.05$ indicated the value had a significant effect. The analysis of the optimum conditions was evaluated using the response surface analysis of dependent and independent variables.

10.2.4 Experimental Procedure

For the dual effect process to proceed, the composite beads made up of clay, FA, and FS were used. FS and FA were used as the natural iron source and help in proving the strength to the composite beads. The beads spherical in shape were prepared by using clay, FA, and FS in the ratio (2:1:1, respectively). The composite beads were baked in a muffle furnace at 800°C for 3 h for providing the strength to the beads. Further, the beads were soaked in the water for 36 h for providing even more strength to the beads. These composite beads were used for the photo-Fenton reactions to proceed in the acidic pH. The prepared composite beads were coated with TiO_2 for the photocatalysis reaction to take place. The standard dip-coating method (Verma, Prakash, and Toor, 2014) was utilized for coating the composite beads, and beads were coated two times for the appropriate coating of nanoparticles.

For the dual effect experiments to take place, the prepared composite beads were dipped in the 200 mL of ofloxacin solution ($C_0 = 25$ mg L^{-1}) providing the flow rate of air as 3 L min^{-1}, 300 mg L^{-1} of H_2O_2 addition, and maintain the pH of 3–3.5 utilizing acetate buffer. The concentration of ferrous and ferric ions was determined during the reaction for confirming the photo-Fenton reactions to take place.

10.3 RESULTS AND DISCUSSION

For confirming the effectiveness of dual effect, various experiments were conducted for the ofloxacin degradation. Various preliminary studies including adsorption, photolysis, photocatalysis, photo-Fenton, and dual effect were performed. Adsorption was studied by keeping the solution in darkness and only 5% of the degradation was observed at 1 h. Photolysis utilizing UV light leads to 16% removal of the ofloxacin in 1 h. Further photo-Fenton reactions were conducted utilizing the uncoated beads along with the addition of H_2O_2 in acidic conditions leading to 45% degradation, which is due to hydroxyl radical production due to the iron content leaching. Further, the photocatalysis leads to 25% degradation of ofloxacin in 1 h. By combing the two processes of photo-Fenton and photocatalysis leading to the in-situ dual effect showed 80% degradation of ofloxacin in 1 h. This confirmed the synergistic effect of the in-situ dual effect of individual processes.

10.3.1 ANN Modeling

In the present study modeling of the in-situ dual effect of photocatalysis and photo-Fenton using ANN has been performed. MATLAB (Mathworks 9.3.0) was used to generate the network from the data. The ratio of the input data for training validation

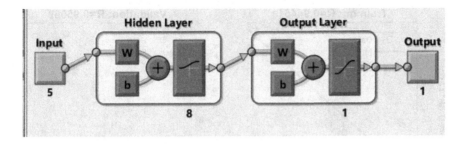

FIGURE 10.1 Layers of the neural network.

and testing was kept as 75%:15%:15%, respectively (Talwar, Verma, and Sangal, 2019). The number of layers is an important part of the topology which gives information about the transfer function as well as the number of neurons used. The network consisted of one input layer, one hidden layer, and one output layer (Figure 10.1). The Levenberge Marquardt algorithm was used to train the neurons, and the output performance was evaluated using root mean square error. The algorithm used was back-propagation.

One hidden layer with three-layer feed-forward back-propagation in ANN was used for the modeling of the data (Agatonovic-Kustrin and Beresford, 2000). The training function used was "Traingdx." Figure 10.1 shows the network with one hidden layer. Inputs required for the model were the surface area covered, treatment time, A/V ratio, intensity, and H$_2$O$_2$ dose.

Neurons were varied from 4 to 10 for the optimization of the number of neurons required. The model was repeated thrice for minimizing the mean square error (MSE). MSE was minimum for the eight neurons. With increasing the number of neurons, fitting was deteriorated and the error was increasing. Hence, a network with eight neurons was used for modeling the various parameters for obtaining the best results. After that, the training of the network was performed. A number of training sessions were performed for the best framework and finally, after 900 iterations, the optimal network was achieved. The values of the regression coefficient were 0.94451, 0.95069, 0.94057, and 0.94421 for training, validation, testing (Figure 10.2), and all. This gave confirmation that the model was best suited for the simulation of the output.

10.3.2 OPTIMIZATION USING THE RESPONSE SURFACE METHODOLOGY

The advantage of data modeling is the potential to find the optimized conditions for maximum degradation of ofloxacin. Hence, after the modeling, the final goal of the study was to optimize the various operating conditions used for the modeling. For this purpose, RSM was used after ANN.

The simulated data obtained from ANN was used for optimization using the Design Expert software. BBD was used for the optimization analysis. Seventeen experiments were suggested and analyzed using simulated data for % degradation as output (Table 10.1).

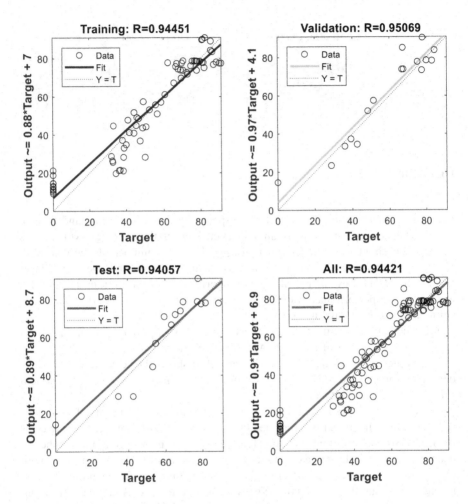

FIGURE 10.2 Regression analysis for ANN modeling.

Various models were evaluated using the sequential sum of square analysis to evaluate the relevant model terms. To find out about the adequacy of models, statistics were used and p-values were evaluated. The p-value came out to be less than 0.01 suggesting the quadratic model to be best. The predicted as well as actual responses were in the appropriate range and under residual proximity.

The value of R^2, as well as predicted R^2, were 98.29% and 87.29%, suggesting the quadratic model was acceptable for the predicted model. The value of p was <0.001 for the model, hence, leading it to be significant in nature. The analysis of variance showed that with the H_2O_2 dose, time taken for degradation were highly significant terms (Table 10.2).

For evaluating the effect of various parameters like time, H_2O_2 dose, and the surface area covered by the beads, the degradation was evaluated using 3-D response graphs.

TABLE 10.2
ANOVA Table for the Percentage (%) Degradation Response using BBD

	Sum of Square	DF	Mean Square	F Value	Prob > F	
Model	6752.64	9	750.29	44.59	<0.0001	Significant
Time	3833.81	1	3833.81	227.83	<0.0001	
H$_2$O$_2$ dose	285.60	1	285.60	16.97	0.0045	
No. of beads	47.97	1	47.97	2.85	0.1352	
Time ^{2}	1532.22	1	1532.22	91.05	<0.0001	
H$_2$O$_2$ dose ^{2}	662.51	1	662.51	39.37	0.0004	
No. of beads ^{2}	119.45	1	119.45	7.10	0.0323	
Time × H$_2$O$_2$ dose	54.76	1	54.76	3.25	0.1142	
Time × No. of beads	3.86	1	3.86	0.23	0.6465	
H$_2$O$_2$ dose × No. of beads	6.97	1	6.97	0.41	0.5404	
Residual	117.79	7	16.83			
Lack of fit	47.75	3	15.92	0.91	0.5118	Not significant
Pure error	70.05	4	17.51			
Cor total	6870.43	16				

With an increase in the time from 5 min to 90 min, the degradation rate increases. From Figure 10.3a, it was observed that up to 90 min the degradation rate becomes constant, and after that, there was not much change in the rate of degradation. Time is the very important parameter and was also confirmed from ANOVA analysis. From Figure 10.3a, the dose of oxidant was studied with respect to the change in time. The oxidant dose plays an important role in both processes, i.e. in photo-Fenton

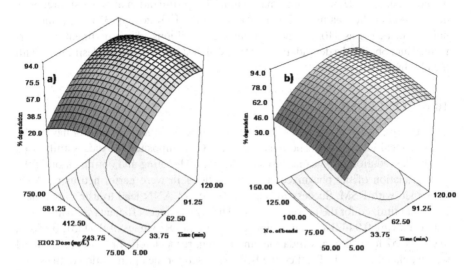

FIGURE 10.3 Surface plots showing interaction between various parameters for % degradation as response.

as well as in photocatalysis. The dose of H_2O_2 varied from 75 mg L^{-1} to 750 mg L^{-1}. It was observed that there was approximately a 27% increment in the rate constant in the degradation of ofloxacin when the concentration of the oxidant was increased up to 300 mg L^{-1}. But on further increase in the dose of H_2O_2 the degradation starts to decrease. The decrement might be due to the effect of scavenging of the excess oxidant.

The covered surface area is an important parameter for the application of the proposed technology and has also been studied in large-scale in the present research work. In fact, the rate of degradation of the ofloxacin varies with the surface area covered by the catalyst coated beads in the batch reactor. Hence, for this purpose, the beads were varied from 50 to 150 representing the covered surface area to be 50% to 150%, respectively. From Figure 10.3b, it was observed that, as the number of beads increased from 50 to 100, the degradation rate increases. Approximately 71% of the degradation was observed by covering 50% of the area that further increased to 92% as the surface area covered by the beads increased to 100% with almost 100 beads. This might be due to the presence of a layer of TiO_2 along with the iron leaching from the beads, producing Fe(II) and Fe(III) ions. The electron produced from the photocatalysis reacts with Fe(III) to produce Fe(II) helping in photo-Fenton reactions. This leads to increased production of hydroxyl radicals. With a further increase in the number of beads to 150, i.e. overlapping the beads, the degradation rate slowed down by 24%. This might be due to the formation of dead zones and blockage of active sites.

10.3.3 OPTIMIZATION ANALYSIS

For optimizing these parameters, BBD was used. At the optimum conditions, the % degradation response ought to be the maximum. Therefore, the response was optimized keeping % degradation as maximum. The optimization analysis showed 91% degradation at the treatment time of 94 min with H_2O_2 dose of 354 mg/L and the number of beads as 110. At these optimized conditions, the experiment was performed and obtained a degradation of 89%, which showed complete desirability with the obtained analysis.

10.4 CONCLUSION

The effective usage of the nano-Fe-TiO_2 composite was the main goal of having the technological advantages. The fixed bed Fe-TiO_2 composite was successfully used for the degradation of pharmaceutical pollutants. Modeling and optimization of the % degradation of the pharmaceutical drug ofloxacin were performed using ANN combined with RSM. Successful implementation of ANN lead to using eight neurons as optimized for the present process. The regression coefficients were also evaluated showing 0.94421 for the complete process and R^2 values of 0.94451, 0.95069, and 0.94057 for training, validation, and testing, respectively. Indicating the model was evaluated correctly. Further, the BBD was used for the optimization of the simulated data. The quadratic model with R^2 of 98.29% also suggested the significance of the model. Adequacy of simulated data of ANN with RSM was also confirmed. The optimization analysis showed 91% degradation at the treatment time of 94 min with H$_2$O$_2$ dose of 354 mg/L and the number of beads as 110. The rate of reaction was significantly increased using a dual effect as compared to an individual of photo-Fenton and photocatalysis. Hence, it was concluded that the dual effect technique was successful in the treatment of pharmaceutical wastewater and has the potential for large-scale applications.

REFERENCES

Agatonovic-Kustrin, S., and R. Beresford. 2000. "Basic Concepts of Artificial Neural Network (ANN) Modeling and Its Application in Pharmaceutical Research." *Journal of Pharmaceutical and Biomedical Analysis*. 22(5): 717–27. https://doi.org/10.1016/S0731-7085(99)00272-1.

Bansal, Palak, Anoop Verma, Kashish Aggarwal, Amanjit Singh, and Saurabh Gupta. 2016. "Investigations on the Degradation of an Antibiotic Cephalexin Using Suspended and Supported TiO$_2$: Mineralization and Durability Studies." *Canadian Journal of Chemical Engineering*. 94. https://doi.org/10.1002/cjce.22512.

Bansal, Palak, Anoop Verma, and Steffi Talwar. 2018. "Detoxification of Real Pharmaceutical Wastewater by Integrating Photocatalysis and Photo-Fenton in Fixed-Mode." *Chemical Engineering Journal*. 349: 838–848. https://doi.org/10.1016/j.cej.2018.05.140.

Bokare, Alok D., and Wonyong Choi. 2014. "Review of Iron-Free Fenton-like Systems for Activating H$_2$O$_2$ in Advanced Oxidation Processes." *Journal of Hazardous Materials*. 275: 121–35. https://doi.org/10.1016/j.jhazmat.2014.04.054.

Buriak, Jillian M., Prashant V. Kamat, and Kirk S. Schanze. 2014. "Best Practices for Reporting on Heterogeneous Photocatalysis." *ACS Applied Materials and Interfaces*. 2014. https://doi.org/10.1021/am504389z.

Daneshvar, N, D. Salari, A. Niaei, M. H. Rasoulifard, and a. R. Khataee. 2005. "Immobilization of TiO2 Nanopowder on Glass Beads for the Photocatalytic Decolorization of an Azo Dye C.I. Direct Red 23." *Journal of Environmental Science and Health. Part A, Toxic/ hazardous Substances & Environmental Engineering*. 40(8): 1605–17. https://doi.org/10.1081/ESE-200060664.

Diao, Zeng-hui, Xiang-rong Xu, Dan Jiang, Jin-jun Liu, Ling-jun Kong, Gang Li, Lin-zi Zuo, and Qi-hang Wu. 2017. "Simultaneous Photocatalytic Cr (VI) Reduction and Ciprofloxacin Oxidation over TiO$_2$/Fe^0 Composite under Aerobic Conditions: Performance, Durability, Pathway and Mechanism." *Chemical Engineering Journal*. 315: 167–76. https://doi.org/10.1016/j.cej.2017.01.006.

Elmolla, Emad S., and Malay Chaudhuri. 2010. "Photocatalytic Degradation of Amoxicillin, Ampicillin and Cloxacillin Antibiotics in Aqueous Solution Using UV/TiO$_2$ and UV/H$_2$O$_2$/TiO$_2$ Photocatalysis." *Desalination*. 252(1–3): 46–52. https://doi.org/10.1016/j.desal.2009.11.003.

Gadiyar, Chethana, Bhanupriya Boruah, Calvina Mascarenhas, and Vidya Shetty K. 2013. "Immobilized Nano TiO$_2$ for Photocatalysis of Acid Yellow-17 Dye in Fluidized Bed Reactor." *International Journal of Current Engineering and Technology*. 4: 84–87.

Homem, Vera, and Lúcia Santos. 2011. "Degradation and Removal Methods of Antibiotics from Aqueous Matrices – A Review." *Journal of Environmental Management*. 92: 2304–47. https://doi.org/10.1016/j.jenvman.2011.05.023.

Kaur, Parminder, Vikas Kumar Sangal, and Jai Prakash Kushwaha. 2015. "Modeling and Evaluation of Electro-Oxidation of Dye Wastewater Using Artificial Neural Networks RSC Advances Modeling and Evaluation of Electro-Oxidation." *RSC Advances*. 5: 34663–71. https://doi.org/10.1039/c4ra14160a.

Kim, Hyung Eun, Jaesang Lee, Hongshin Lee, and Changha Lee. 2012. "Synergistic Effects of TiO₂ Photocatalysis in Combination with Fenton-like Reactions on Oxidation of Organic Compounds at Circumneutral pH." *Applied Catalysis B: Environmental.* 115–116: 219–24. https://doi.org/10.1016/j.apcatb.2011.12.027.

Kovacic, Marin, Subhan Salaeh, Hrvoje Kusic, Andraz Suligoj, Marko Kete, Mattia Fanetti, Urska Lavrencic Stangar, Dionysios D. Dionysiou, and Ana Loncaric Bozic. 2016. "Solar-Driven Photocatalytic Treatment of Diclofenac Using Immobilized TiO₂-Based Zeolite Composites." *Environmental Science and Pollution Research.* 23(18): 17982–94. https://doi.org/10.1007/s11356-016-6985-6.

Lin, Wei Hao, Yi Hsuan Chiu, Pao Wen Shao, and Yung Jung Hsu. 2016. "Metal-Particle-Decorated ZnO Nanocrystals: Photocatalysis and Charge Dynamics." *ACS Applied Materials and Interfaces.* 8: 32754–63. https://doi.org/10.1021/acsami.6b08132.

Luo, Chengzhi, Xiaohui Ren, Zhigao Dai, Yupeng Zhang, Xiang Qi, and Chunxu Pan. 2017. "Present Perspectives of Advanced Characterization Techniques in TiO₂-Based Photocatalysts." *ACS Applied Materials and Interfaces.*9: 23265–86. https://doi.org/10.1021/acsami.7b00496.

Maroga Mboula, V., V. Héquet, Y. Gru, R. Colin, and Y. Andrès. 2012. "Assessment of the Efficiency of Photocatalysis on Tetracycline Biodegradation." *Journal of Hazardous Materials.* 209–210: 355–64. https://doi.org/10.1016/j.jhazmat.2012.01.032.

MiarAlipour, Shayan, Donia Friedmann, Jason Scott, and Rose Amal. 2018. "TiO2/porous Adsorbents: Recent Advances and Novel Applications." *Journal of Hazardous Materials.* 341: 404–23. https://doi.org/10.1016/j.jhazmat.2017.07.070.

Mozia, Sylwia, Piotr Brozek, Jacek Przepiórski, Beata Tryba, and Antoni W. Morawski. 2012. "Immobilized TiO₂ for Phenol Degradation in a Pilot-Scale Photocatalytic Reactor." *Journal of Nanomaterials.* 2012. https://doi.org/10.1155/2012/949764.

Mukherjee, Debjani, Shahzad Barghi, and Ajay Ray. 2013. "Preparation and Characterization of the TiO₂ Immobilized Polymeric Photocatalyst for Degradation of Aspirin under UV and Solar Light." *Processes.* 2(1): 12–23. https://doi.org/10.3390/pr2010012.

Nag, Tanmay, and Arnab Ghosh. 2013. "Cardiovascular Disease Risk Factors in Asian Indian Population: A Systematic Review." *Journal of Cardiovascular Disease Research.* 4(4): 222–28. https://doi.org/10.1016/j.jcdr.2014.01.004.

Rastegar, M., K. Rahmati Shadbad, a. R. Khataee, and R. Pourrajab. 2012. "Optimization of Photocatalytic Degradation of Sulphonated Diazo Dye C.I. Reactive Green 19 Using Ceramic-Coated TiO₂ Nanoparticles." *Environmental Technology.* 33(7–9): 995–1003. https://doi.org/10.1080/09593330.2011.604859.

Sangal, Vikas, Vineet Kumar, and Mani Mishra. 2013. "Optimization of a Divided Wall Column for the Separation of C4–C6 Normal Paraffin Mixture Using Box-Behnken Design." *Chemical Industry and Chemical Engineering Quarterly.* 19(1): 107–19. https://doi.org/10.2298/CICEQ121019047S.

Sarkar, Santanu, Sudip Chakraborty, and Chiranjib Bhattacharjee. 2015. "Photocatalytic Degradation of Pharmaceutical Wastes by Alginate Supported TiO₂ Nanoparticles in Packed Bed Photo Reactor (PBPR)." *Ecotoxicology and Environmental Safety.* 121: 263–70. https://doi.org/10.1016/j.ecoenv.2015.02.035.

Sharma, Kritika, Steffi Talwar, Anoop Kumar Verma, Diptiman Choudhury, and Borhan Mansouri. 2020. "Innovative Approach of in-Situ Fixed Mode Dual Effect (Photo-Fenton and Photocatalysis) for Ofloxacin Degradation." *Korean Journal of Chemical Engineering.* 37: 350–7. https://doi.org/10.1007/s11814-019-0427-3.

Shojaeimehr, Tahereh, Farshad Rahimpour, Mohammad Ali Khadivi, and Marzieh Sadeghi. 2014. "A Modeling Study by Response Surface Methodology (RSM) and Artificial Neural Network (ANN) on Cu²⁺ Adsorption Optimization Using Light Expended Clay Aggregate (LECA)." *Journal of Industrial and Engineering Chemistry.* 20: 870–80. https://doi.org/10.1016/j.jiec.2013.06.017.

Singh, Kunwar P., Premanjali Rai, Arun K. Singh, Priyanka Verma, and Shikha Gupta. 2014. "Occurrence of Pharmaceuticals in Urban Wastewater of North Indian Cities and Risk Assessment." *Environmental Monitoring and Assessment*. 186(10): 6663–82. https:// doi.org/10.1007/s10661-014-3881-8.

Su, Tongming, Qian Shao, Zuzeng Qin, Zhanhu Guo, and Zili Wu. 2018. "Role of Interfaces in Two-Dimensional Photocatalyst for Water Splitting." *ACS Catalysis*. 8: 2253–76. https://doi.org/10.1021/acscatal.7b03437.

Talwar, S., V. K. Sangal, and A. Verma. 2018. "Feasibility of Using Combined TiO$_2$ photocatalysis and RBC Process for the Treatment of Real Pharmaceutical Wastewater." *Journal of Photochemistry and Photobiology A: Chemistry*. 353. https://doi.org/10.1016/j.jphotochem.2017.11.013.

Talwar, Steffi, Vikas Kumar Sangal, Anoop Verma, Parminder Kaur, and Alok Garg. 2018. "Modeling, Optimization and Kinetic Study for Photocatalytic Treatment of Ornidazole Using Slurry and Fixed-Bed Approach." *Arabian Journal for Science and Engineering*. 43: 6191–202. https://doi.org/10.1007/s13369-018-3388-7.

Talwar, Steffi, Anoop Kumar Verma, and Vikas Kumar Sangal. 2019. "Modeling and Optimization of Fixed Mode Dual Effect (Photocatalysis and Photo-Fenton) Assisted Metronidazole Degradation Using ANN Coupled with Genetic Algorithm." *Journal of Environmental Management*. 250: 109428. https://doi.org/10.1016/j.jenvman.2019.109428.

Talwar, Steffi, Vikas Kumar Sangal, and Anoop Kumar Verma. 2019. "In-Situ Dual Effect of Novel Fe-TiO$_2$ Composite for the Degradation of Phenazone." *Separation and Purification Technology*. 211: 391–400. https://doi.org/10.1016/j.seppur.2018.10.007.

Vaiano, V., O. Sacco, D. Sannino, and P. Ciambelli. 2015. "Nanostructured N-Doped TiO2 Coated on Glass Spheres for the Photocatalytic Removal of Organic Dyes under UV or Visible Light Irradiation." *Applied Catalysis B: Environmental*. 170–171: 153–61. https://doi.org/10.1016/j.apcatb.2015.01.039.

Verma, A, N. T. Prakash, and A. P. Toor. 2014. "An Efficient TiO$_2$ Coated Immobilized System for the Degradation Studies of Herbicide Isoproturon: Durability Studies" *Chemosphere*. 109: 7–13.

Yang, Longkai, Xin Wang, Xianmin Mai, Tan Wang, Chao Wang, and Xin Li. 2018. "Constructing Efficient Mixed-Ion Perovskite Solar Cells Based on TiO$_2$ Nanorod Array." *Journal of Colloid and Interface Science*. 534: 459–68 https://doi.org/10.1016/j.jcis.2018.09.045.

Zhang, Li, Mingke Qin, Wei Yu, Qinghong Zhang, Hongyong Xie, and Zhiguo Sun. 2017. "Heterostructured TiO$_2$/WO$_3$ Nanocomposites for Photocatalytic Degradation of Toluene under Visible Light." *Journal of Electrochemical Society*. 164(14): 1086–90. https://doi.org/10.1149/2.0881714jes.

Zhao, J., Shengsong Ge, Duo Pan, Qian Shao, Jing Lin, Zhikang Wang, Zhen Hu, Tingting Wu, Zhanhu Guo. 2018. "Solvothermal Synthesis, Characterization and Photocatalytic Property of Zirconium Dioxide Doped Titanium Dioxide Spinous Hollow Microspheres with Sunflower Pollen as Bio-Templates." *Journal of Colloid and Interface Science*. 529: 111–21 https://doi.org/10.1016/j.jcis.2018.05.091.

Index

Printed in the United States
By Bookmasters